USA in Space

USA in Space

Edited by
Frank N. Magill
and
Russell R. Tobias

VOLUME 3

Spa - Z
Indexes

617 - 890

Salem Press, Inc.

Pasadena, CA   Englewood Cliffs, NJ

The paper used in these volumes conforms to the American
National Standard for Permanence of Paper for Printed
Library Materials, Z39.48-1984.

Library of Congress Cataloging-in-Publication Data
USA in space/edited by Frank N. Magill and
Russell R. Tobias.
p. cm.
Includes bibliographical references and index
1. Astronautics–United States.   I. Magill, Frank
Northen, 1907-.   II. Tobias, Russell R.
TL789.8.U5U83 1996
387.8'0973–dc20   96-42044
CIP
ISBN 0-98356-924-0 (set)
ISBN 0-89356-927-5 (vol 3)

First Printing

PRINTED IN THE UNITED STATES OF AMERICA

CONTENTS

USA in Space

SPACE SHUTTLE MISSION
STS 41-B

Date: February 3 to February 11, 1984
Type of mission: Manned Earth-orbiting spaceflight

STS 41-B, the tenth mission of the space shuttle, displayed the operational potential of the Space Transportation System. Although it was not the first operational flight, STS 41-B could be considered one of the most important. The crew not only returned from space to land for the first time where they were launched, at Kennedy Space Center, Florida, but also demonstrated the ability of astronauts to maneuver in space.

PRINCIPAL PERSONAGES

VANCE D. BRAND, Commander
ROBERT L. "HOOT" GIBSON, Pilot
RONALD E. MCNAIR,
BRUCE MCCANDLESS II, and
ROBERT L. STEWART, Mission Specialists

Summary of the Mission

The National Aeronautics and Space Administration (NASA) launched *Challenger* on Space Transportation System (STS) mission 41-B at 8:00 A.M. February 3, 1984, from Launchpad 39A at Kennedy Space Center, Florida. The space shuttle carried five astronauts, two communications satellites, and a variety of scientific payloads into a circular orbit 305 kilometers above Earth. Vance Brand was the commander of STS 41-B. Brand had also commanded STS-5, the fifth flight and first operational flight of the space shuttle, in November, 1982. Robert L. "Hoot" Gibson, making his first spaceflight, was the pilot. Bruce McCandless, Ronald McNair, and Robert Stewart, also making their first flights, were mission specialists. STS 41-B was the fourth flight of the orbiter *Challenger.*

Approximately eight hours into the mission, at 3:59 P.M. February 3, a Western Union communications satellite, Westar 6, was deployed from the payload bay while the space shuttle was passing over the far western Pacific Ocean on its sixth orbit. As planned, the satellite and its attached Payload Assist Module, Delta class (PAM-D), rocket motor were set spinning on a turntable in the payload bay and released by springs to coast away from the orbiter. Also as planned, the PAM-D's perigee kick motor fired 45 minutes later to boost the satellite toward its intended circular orbit 35,900 kilometers above Earth. The solid-fueled rocket, however, burned for only 16 seconds of the normal 85-second burn period, and although

the satellite was working properly, it ended in an orbit that was both too low and too elliptical.

The failure prompted NASA to delay the deployment of a nearly identical satellite owned by the Republic of Indonesia, Palapa B2, from the second day until the fourth day of the mission, February 6. Palapa B2 was also deployed normally, but a PAM-D failure robbed it of the thrust necessary to reach the proper orbit. Both satellites were retrieved and returned to Earth by space shuttle astronauts on STS 51-A in November, 1984.

Palapa B2, the Indonesian national telecommunications satellite, was the second Palapa to be deployed by *Challenger.* The first was deployed on STS 7. Palapa B2, built by Hughes Communications International of Los Angeles, California, was to provide a communications network to Indonesia and the Association of Southeast Nations, which includes the Philippines, Thailand, Malaysia, Singapore, and Papua, New Guinea. Palapa B2 and its PAM-D booster weighed 4,366 kilograms. In orbit, the satellite weighed 630 kilograms. With its antennae deployed, it was 6.8 meters in height and 2.16 meters in diameter.

Westar 6, built by Hughes Aircraft Company of El Segundo, California, was the third advanced Western Union communications satellite to be placed in space and was the first Westar to be deployed from the space shuttle. The Westar system provided continuous video, facsimile, data, and voice communications service throughout the United States, Puerto Rico, and the Virgin Islands. Westar 6 was similar in size and weight to Palapa B2.

Much of STS 41-B was devoted to practicing for *Challenger's* next flight, a mission to retrieve and repair the Solar Maximum Mission (SMM) satellite (sometimes called "Solar Max"). The satellite's attitude control system, the computerized system that maintains a spacecraft's posture in orbit, failed shortly after SMM was launched to study the Sun in 1980. The STS 41-C rescue mission would require *Challenger* to fly within 90 meters of the spinning satellite while an astronaut wearing a Manned Maneuvering Unit (MMU) docked with and held the satellite so that other crew members could grasp it with the space shuttle's 15-meter remote manipulator arm.

On the second day of the STS 41-B mission, February 4, the crew members tried to inflate the Integrated Rendezvous

View of Astronaut Bruce McCandless during EVA. (NASA)

Target (IRT), a Mylar balloon 2 meters in diameter that they would use to practice maneuvering *Challenger.* The orbiter was to move about 14.5 kilometers from the balloon, close back to about 9.7 kilometers, move out again, then drift to a distance of slightly more than 225 kilometers before closing back to only 244 meters. The IRT burst, however, and the crew used radar to track pieces of the thin, aluminum-like material as it floated into space.

Also on the second day, Bruce McCandless and Ronald

McNair (who would die aboard *Challenger* on STS 51–L in January, 1986) used two 35-millimeter motion-picture cameras with "fish-eye" lenses to begin filming portions of their mission for a documentary being developed in a 360-degree format especially for planetarium viewing. They used the cameras on the sixth day. Cinema 360, a consortium of planetariums, later produced the film *The Space Shuttle: An American Adventure* in cooperation with NASA.

In addition to the two satellites that it deployed — Palapa

B2 and Westar 6 — *Challenger* carried the West German Shuttle Pallet Satellite (SPAS-01A) and a variety of minor payloads in its bay and crew cabin. SPAS-01A carried eight scientific experiments and served as a substitute for the Solar Maximum Mission satellite when the STS 41-B crew members practiced retrieval and repair procedures on the seventh day. The experiments operated automatically, and most, with the exception of those on SPAS-01A, were activated by the crew on the second day. Five Get-Away Special (GAS) canisters in the payload bay contained small physics, biology, technology, and materials science experiments designed to investigate or take advantage of the low-gravity environment in space. Three materials processing experiments and a life-sciences experiment with rats were carried in the pressurized cabin.

On the fifth day, at 5:25 A.M. February 7, McCandless and Stewart became the first people to take an untethered spacewalk. They wore MMUs, 136-kilogram jet-propelled backpacks that allow free, detached flight. Of their six hours outside the orbiter, McCandless and Stewart spent more than two and one-half hours testing the MMUs. Spacewalkers normally breathe oxygen for an extended period to remove nitrogen from the bloodstream before beginning their Extravehicular Activity (EVA). By lowering the orbiter's cabin pressure, McCandless and Stewart achieved the same goal. They abbreviated the "pre-breathe" period and gave themselves more time to test the MMUs.

To test his MMU maneuvering capabilities, McCandless flew about 15 meters from the orbiter, returned, flew 96 meters away, and returned while being tracked by *Challenger*'s radar. McCandless also practiced docking himself with another spacecraft. To perform this feat, he latched himself to an equipment box in the payload bay with a portable tool called a trunnion pin acquisition device. Meanwhile, Stewart practiced using a manipulator foot restraint. The restraint is a set of straps at the end of the armlike Remote Manipulator System (RMS) that creates for astronauts a stable working platform similar to the "cherry pickers" used by telephone line technicians. Later, McCandless and Stewart traded positions and repeated the routines.

The crew spent most of the next day working with SPAS-01A and the eight scientific experiments it carried. Built by a West German firm, SPAS-01A was the first satellite ever to be refurbished and flown again. It had been released as a free-flying space platform during STS-7; during STS 41-B, however, the satellite remained attached to *Challenger*. SPAS-01A weighed 1,448 kilograms. It was 4.8 meters long, 3.4 meters high, and 1.5 meters wide.

On the seventh day, February 9, SPAS-01A stood in for Solar Max while Stewart and McCandless conducted the mission's second EVA. SPAS was to have been hoisted outside the payload and set spinning slowly on the remote manipulator arm while McCandless used the trunnion pin acquisition device to dock with it. An electrical problem in the arm's

wrist joint made hoisting impossible, so McCandless practiced docking with SPAS in the payload bay.

During their EVA, which began at 3:40 A.M., McCandless and Stewart practiced more free-flight maneuvers in the MMUs. They practiced stopping precisely, holding a steady position, and controlling speed and direction. Stewart tested a tool being developed for fueling satellites inside the payload bay. The second spacewalk lasted more than six hours and was captured on film by the Cinema-360 cameras.

The eight-day STS 41-B mission ended on schedule, milliseconds past 7:17 A.M., February 11, on *Challenger*'s 128th orbit. The mission had lasted 7 days, 23 hours, 15 minutes, and 54 seconds. *Challenger* had traveled more than 5.3 million kilometers. Brand landed the orbiter, gliding faster than 350 kilometers per hour, less than 6.5 centimeters off the center line of runway 15 at Kennedy Space Center, with 690 meters of pavement to spare.

Knowledge Gained

The STS 41-B mission turned Buck Rogers-type fantasies into reality and set the stage for the space shuttle to mature into a reliable, increasingly economical tool for research and commerce. Despite the fact that equipment problems prevented the completion of several tasks that would have provided valuable information about the space shuttle and its capabilities, STS 41-B demonstrated that untethered EVAs are feasible.

In spite of several equipment failures on hardware launched from the space shuttle during the missions, most of the major test objectives of the flight were accomplished. Among their tasks, the five-member crew deployed two commercial communications satellites and practiced satellite repair procedures.

Westar 6, owned by Western Union, and Palapa B2, owned by the Republic of Indonesia, did not reach proper orbit because their PAM-D rocket motors failed. A failure investigation committee appointed by the manufacturer of the payload assist modules — McDonnell Douglas Astronautics Company of Huntington Beach, California — released its detailed findings in a final report in September, 1984.

Exhibiting their prowess with MMUs in the Solar Maximum Mission satellite repair rehearsal, McCandless and Stewart gave their fellow astronauts and the public a preview of mission tasks that were expected to become typical of future space shuttle flights. Their satellite repair practice led to the successful rescue of SMM on the next mission, STS 41-C in April, 1984.

For the scientists who flew the variety of experiments aboard *Challenger*, STS 41-B revealed valuable information about microgravity and its potential benefits for society, ranging from specialized manufacturing to biomedical therapy. The small, automatic GAS experiments included studies of how proteins crystallize, how liquids move, how light scatters, how cosmic rays affect memory, and how radish seeds sprout in microgravity; other tests had commercial applications, such as that aimed at the development of an energy-efficient arc lamp. These experiments were sponsored by GTE Laboratories, the U.S. Air Force, NASA's Goddard Space Flight Center in Greenbelt, Maryland, and high school and college students from Utah and the University of Aberdeen in Scotland.

The Monodisperse Latex Reactor (MLR) payload was one of three materials processing payloads flown inside the pressurized crew cabin. MLR, carried on four previous space shuttle flights, was designed to develop monodisperse, or identically sized, beadlike rubber particles for use in medical and industrial research. Another materials processing payload, the Acoustic Containerless Experiment System (ACES), was activated on the second day; it tested a materials processing furnace. The third materials processing payload, the Isoelectric Focusing (IEF) experiment, tested electrophoresis, a process for separating protein fluids by their diverse electric charges.

STS 41-B was the second mission to carry live rats for scientific experimentation. On this flight, six rats were the subject of a student experiment to study whether weightlessness relieves the symptoms of arthritis. Dan Weber, who designed the experiment while he was attending Hunter College High School in New York City, flew three healthy rats and three arthritic rats in a life-support cage called an Animal Enclosure Module (AEM). Before the rats were launched, Weber injected Freund's adjuvant, a substance used to induce arthritis, into the animals' hind paws. After the flight, the rats were killed and Weber conducted autopsies, comparing the space shuttle rats with twelve controls — six healthy and six arthritic rats.

The results of the rat experiment were inconclusive, partly because the flight was not long enough to allow arthritis to develop fully in the animals. During postflight examination, Weber found no significant differences in the degree of swelling between the flown rats and the ground controls, and reported that spaceflight did not inhibit the development of arthritis. The flight crew, however, observed that the arthritis spread less extensively in the space shuttle rats than in rats they examined in earlier ground-based tests — an indication of possible beneficial effects of weightlessness.

Six experiments on SPAS-01A were sponsored by West Germany's Ministry of Research and Technology. Materialwissenschaftliche Autonome Experiments Unter Schwerelosigkist (MAUS) 1 and 2 investigated materials processing in microgravity. MAUS 1 mixed bismuth and manganese, two metals that are difficult to combine on Earth, in an attempt to create a new permanent magnetic alloy. MAUS 2 was a basic crystal-growth experiment. Another experiment, with friction loss, studied how pneumatic or air-operated conveyor systems operate in the absence of gravity. An electronic remote-sensing camera, the Modular Optoelectric Multispectral Scanner (MOMS), was used to scan land masses on Earth. The Bonn neutral mass spectrometer measured the intensity and composition of gaseous contaminants in and around *Challenger*'s payload bay. Another experiment studied heat transfer in microgravity.

Two experiments on SPAS-01A were sponsored by the European Space Agency. The first measured the yaw, or drift, of a stabilized spacecraft, with the goal of gathering information to help simplify attitude control systems. The other measured solar power cells in direct sunlight to set calibration standards for systems on Earth.

NASA learned how much more easy it would be to maintain space shuttle flight schedules with routine Florida landings. *Challenger*'s unprecedented touchdown at Kennedy Space Center's 4.8-kilometer-long Shuttle Landing Facility (SLF) eliminated the need for a cross-country ferry from one of NASA's other landing sites and enabled crews to prepare the orbiter for another flight in fifty-five days. The return-to-launch-site landing allowed the shortest turnaround time for a space shuttle orbiter to that date and demonstrated the effectiveness of a design characteristic that had been unproved until then.

Context

With its disappointing equipment failures, STS 41-B was far from the most successful of the first twenty-five space shuttle missions at accomplishing predetermined tasks. Nevertheless, the mission was full of spectacular "firsts" for the United States' space program. Happily for NASA, those historical events — and breathtaking film footage from the motion-picture cameras — were enough to leave a lasting impression of mission success. STS 41-B included the first untethered spacewalks in program history, the first flight of an astronaut wearing an MMU, the first use of foot restraints mounted on the RMS, the first EVA without an extended "pre-breathe" period, and the first space shuttle landing at Kennedy Space Center, Florida.

The tenth space shuttle mission was an important educational and political tool for NASA, which continually found itself in battles with the U.S. Congress over funding. For lawmakers who might be questioning the need for a space shuttle, the mission demonstrated program continuity. For example, during the satellite repair practice, crew members not only rehearsed procedures that would be used on the next flight but also tested equipment and techniques that eventually would be used to build and service a space station. A few days before *Challenger* was launched, President Ronald Reagan had announced plans for a space station in his "State of the Union" address. Reagan had reaffirmed the shuttle's convenience and necessity as a "truck" to carry parts of the space station into orbit. By the time *Challenger* was launched, almost three years had passed since the first space shuttle, *Columbia*, blasted into orbit. News media coverage, often a determinant of public interest in the space shuttle program, had begun to wane. In the Cinema-360 project, NASA had seen an effective means of communicating its goals to the American public. STS 41-B produced approximately 914 meters of film, 80 percent of which was of high quality. The footage was incorporated into a 32-minute documentary and shown at planetariums across the country. The film gives its audience a striking vantage point — unprecedented, undistorted, and unobstructed views of Earth and space. Through *The Space Shuttle: An American Adventure*, NASA was able to document its reusable spaceship as the necessary predecessor to a permanently manned space station.

STS 41-B was the first of sixteen flights to take place under a coded designation. The mission was designated as *4* (for 1984, the fiscal year of the launch), *1* (for Kennedy Space Center, one of two launch sites), and *B* (for the second launch scheduled in the fiscal year). In the past, NASA had numbered missions one through nine consecutively, beginning with STS-1. The new alphanumeric system, instituted as interest in the space shuttle as a commercial transport peaked, supported the belief that the space shuttle truly was a reliable and increasingly economical way to put payloads into orbit. The new system was intended to reduce the public confusion created when numerically designated missions flew out of order because of payload changes or cancellations.

Bibliography

Braun, Wernher von, et al. *Space Travel: A History*. Rev. ed. New York: Harper & Row, Publishers, 1985. This is the fourth, updated edition of *History of Rocketry and Space Travel*. Von Braun, a German rocket pioneer, was a key figure in the development of the U.S. space program. One of his collaborators, Ordway, worked with von Braun at the Army Ballistic Missile Agency, the forerunner of NASA's Marshall Space Flight Center in Huntsville, Alabama. Suitable for general audiences, this illustrated edition takes the reader from the rocketry of the ancient Chinese and Greek civilizations through the space shuttle eras of the United States and the Soviet Union.

Furniss, Tim. *Space Shuttle Log*. New York: Jane's Publishing Co., 1986. This book covers the first twenty-two missions of the space shuttle program. The design concepts for the entire Space Transportation System are reviewed, as well as specific data on each flight. Black-and-white photographs from each mission accompany the summaries.

Kerrod, Robin. *Space Shuttle*. New York: Gallery Books, 1984. Components of the space shuttle vehicles and highlights from the first dozen missions are detailed in this well-illustrated volume. Its full-color photographs are perhaps this book's greatest asset.

National Aeronautics and Space Administration. *Mission 41-B: Alone in Space*. MR-41B. Washington, D.C.: Author, 1984. A summary of the tenth space shuttle mission. With general information about payloads aboard orbiter *Challenger*, the failed deployment of two commercial satellites, and two precedent-setting untethered spacewalks. Suitable for general audiences.

————. *NASA Activities: March, 1984*. Washington, D.C.: Government Printing Office, 1984. Official monthly publication for employees of the National Aeronautics and Space Administration. Contains STS 41-B mission facts and a pictorial chronology from prelaunch preparations through landing.

Beth Dickey

SPACE SHUTTLE MISSION
STS 41-C

Date: April 6 to April 13, 1984
Type of mission: Manned Earth-orbiting spaceflight

Space shuttle mission 11 (STS 41-C) launched into low Earth orbit the Long-Duration Exposure Facility to study materials degradation. The crew then performed a repair on the Solar Maximum Mission satellite, achieving the first in-orbit repair of a serviceable free-flying spacecraft.

PRINCIPAL PERSONAGES
   ROBERT L. CRIPPEN, Commander
   FRANCIS R. SCOBEE, Pilot
   TERRY J. HART,
   JAMES D. VAN HOFTEN, and
   GEORGE D. NELSON, Mission Specialists
   FRANK J. CEPOLLINA, Multimission Modular
      Spacecraft and Flight Support System Project
      Manager, Goddard Space Flight Center
   FRANCIS J. LOGAN, Deputy Project Manager
   PETER E. O'NEILL, Mission Manager
   WILLIAM N. STEWART, Mission Operations Manager
   GERALD P. KENNEY, Payload Integration Manager
   JAY H. GREENE, Flight Director
   JOHN B. BIBATTISTA, Long-Duration Exposure Facility
      Experiment Manager
   WILLIAM H. KINARD, Long-Duration Exposure
      Facility Chief Scientist

Summary of the Mission

To give an account of the Solar Maximum Repair Mission, one must begin with the launch of the Solar Maximum Mission (SMM, or Solar Max) observatory, which employed for the first time a serviceable multimission modular spacecraft. Launched on February 14, 1980, from the Cape Canaveral Air Force Station, the vehicle achieved an altitude of 576 kilometers, with an inclination of 28.5 degrees. The SMM scientific payload consisted of seven instruments designed to make in-orbit measurements of solar phenomena — including flares, the corona, and the total solar output — during a period near the height of maximum solar activity. These operations were successfully completed during the first eight months of Solar Max's lifetime.

Prior to the launch of the SMM observatory, all nations with space programs had designed their low-orbit, free-flying spacecraft as expendable vehicles; when the spacecraft's opera-

tional life was over or when it experienced failure, it was abandoned, left to decay, or burn up in Earth's atmosphere. The multimission modular spacecraft was conceived as a multipurpose, low-cost, reusable spacecraft. It was an assembly of modular subsystems that would be individually replaceable in the laboratory and in space. These modules are the attitude control system, the communications and data handling subsystem, the modular power system, and the propulsion module. The multimission modular spacecraft and its SMM experiment payload were cooperatively designed by the National Aeronautics and Space Administration (NASA), American industry, and American and foreign experimenters for several years of operation.

During the fourth to seventh months of orbital operations, the coronagraph/polarimeter instrument in the payload section experienced successive failures of two four-stage integrated circuits, causing its Main Electronics Box (MEB) to fail. (The main purpose of the MEB was to create television-type pictures of the occulted solar corona for transmission to Earth.) A few months later, the spacecraft suffered a series of three failures because of a long-term fuse derating problem. This problem caused the loss of precise three-axis stabilization of the satellite, rendering selected viewing of regions on the Sun impossible. Engineers managed to devise a system which would allow three of the solar instruments to continue limited measurements, but extensive repairs were definitely necessary.

Under the direction of Frank Cepollina, a plan for the retrieval of SMM by the space shuttle using the so-called flight support system was changed to an orbital repair plan. Engineers estimated that for an expenditure of $45 million, in addition to launch costs, the SMM observatory, whose replacement cost would have been $235 million, could be restored to nominal operation. At launch, the SMM observatory, including the development of the serviceable multimission modular spacecraft, had cost NASA $79 million. In August, 1982, the U.S. Congress gave NASA permission to plan the rendezvous with and repair of the SMM spacecraft. These measures would be taken after more than four years of space operations; the observatory would thus be operational during the period of minimum solar activity in the eleven-year solar cycle.

The Solar Maximum Repair Mission was scheduled to fly

with the Langley Research Center's Long-Duration Exposure Facility. Use would be made of the Space Transportation System (STS), more popularly known as the space shuttle, with the multimission modular spacecraft's spare attitude control system module, a new main electronics box, tools for Extravehicular Activity (EVA), the flight support system, and the Long-Duration Exposure Facility installed in its cargo bay. The flight support system would be used to berth and service the malfunctioning solar observatory. This mission would be the eleventh flight of the space shuttle, scheduled for April 4, 1984. Other experiments would be performed in the crew cabin, including a nineteen-year-old student's separate middeck honeybee experiment to determine bees' adaptability to weightless conditions. Also included in the payload were a 360-degree camera in the shuttle bay and a 70-millimeter camera for making exposures through the portholes in the aft flight deck. Mission planners agreed that the Long-Duration Exposure Facility would function as an excellent secondary payload because it required the same 28.5-degree launch inclination and close to the same altitude that the Solar Maximum Repair Mission required.

Mission planners were confident about the schedule and the flight plan. An available Landsat D attitude control system module was chosen to replace Solar Max's malfunctioning module. This spare would still serve to support the Earth resources Landsat D spacecraft through their launches. After the building and testing of a new coronagraph/polarimeter main electronics box, engineers determined that substituting it for the faulty system aboard Solar Max would restore six of the seven SMM instruments to full operational capability. Numerous training simulations were performed on the ground with various elements of the servicing mission. These simulations indicated the practicality of performing the required servicing activities during two six-hour EVA periods.

On the morning of April 6, following the usual prelaunch preparations at the Kennedy Space Center, the space shuttle *Challenger* was launched for the STS 41-C mission — two days later than scheduled. A direct-ascent burn was employed, and the external tank was separated from the shuttle and left to burn up over the southern tip of Hawaii. Upon arrival at the desired 28.5-degree, 491-kilometer-apogee orbit, the space shuttle cargo bay doors were opened. The remote manipulator system and the flight support system were tested before they were used for deployment of the Long-Duration Exposure Facility and the SMM berthing and repair operations.

On April 7, mission specialist Terry J. Hart released the Long-Duration Exposure Facility using the Canadian-built remote manipulator system. This huge 9,707-kilogram payload, designed for gravity-gradient orientation, stabilized perfectly upon release. With the secondary payload deployed, the space shuttle performed maneuvers to increase its altitude to 502 kilometers to rendezvous with Solar Max.

On April 8, James D. van Hoften and George D. Nelson donned their EVA space-suits in the air lock and went into the cargo bay. Nelson put on the manned maneuvering unit and, untethered, flew his planned ten-minute, sixty-meter flight to approach Solar Max. The plan called for the astronaut to put the maneuvering unit controls into an autopilot mode to stop the spiraling spacecraft. The space shuttle would be maneuvered to approach Solar Max and use its remote manipulator system to grasp a fixture located on the satellite. Although the maneuvering unit flew perfectly and synchronized with the spin rate of the satellite, after three attempts at increasing velocities, the shuttle failed to lock onto a spacecraft trunion located between the two solar arrays. Yet another unsuccessful attempt was made to grasp the uncontrolled, tumbling observatory with the remote manipulator. A last, desperate attempt failed when Nelson held onto a solar array panel and the spacecraft structure and attempted to stabilize the SMM using the manned maneuvering unit. As the maneuvering fuels were running low, Commander Robert Crippen ordered Nelson to return to the shuttle.

Directed by William Stewart, the SMM Payload Operations Control Center at Goddard and worldwide ground stations performed a difficult feat in regaining attitude control of the observatory before the SMM batteries lost too much power. Alternate magnetic control programs were loaded into the spacecraft's computer memory and commanded to develop about one-half the previous spin rate. Engineers on the ground wanted to achieve the rates used in backup computer simulations of the remote manipulator system capturing a spinning SMM.

Over the next sixteen hours on April 9 and 10, space planners at Goddard took measures to improve the power and thermal situation in the observatory. On April 10, after the desired SMM spin rate was established and the latest space shuttle approach maneuver was attempted, Crippen performed a fuel savings maneuver to rendezvous with the observatory. Fortunately, Terry Hart was able to grasp the SMM on his first attempt. The SMM was then berthed to the flight support system for the planned series of repair operations.

On April 11, van Hoften and Nelson again went into the cargo bay and compressed nearly two days of scheduled EVA activities into one. Riding the manipulator foot restraint attached to the remote manipulator, van Hoften, assisted by Nelson, used a battery-powered module service tool to remove the spacecraft's attitude control module and put it onto a temporary holding conical nut on the side of the flight support system. He removed the replacement module from a holding fixture and installed it on the observatory. The faulty module was next stowed onto the holding fixture on the side of the flight support system.

A second repair was performed within four minutes by van Hoften, with a gloved-hand installation of a 0.5-kilogram plasma baffle over a small rectangular propane exhaust port in the X-ray polychromator instrument panel. This repair cor-

View of the Long Duration Exposure Facility in orbit above the Earth (left); Astronaut van Hoften using tool to work on capture Solar Maximum satellite. (NASA)

rected an interference in that instrument's data output.

For the third repair, the failed main electronics box was removed by van Hoften, who used a specially adapted, battery-powered screwdriver and other EVA hand tools. Nelson then removed a new main electronics box from a storage locker and performed the delicate installation. The complicated electronics box, which has 365 connections through eleven electrical connectors, had been built and tested on the ground after the coronagraph/polarimeter instrument failure. The new box had never operated with the rest of that instrument until it was installed in space. A thirty-page EVA sequence had been written by space planners for this complicated exchange.

The cargo bay activities were performed so smoothly that the astronauts were able to complete their work earlier than planned, giving van Hoften an opportunity to don the manned maneuvering unit and perform a familiarization flight down the cargo bay. A planned space shuttle maneuver to reboost Solar Max to a 528-kilometer altitude, in order to extend its orbital lifetime, was canceled when the shuttle's maneuvering fuel became so low that the maneuver would have jeopardized the safe reentry of the shuttle and its crew. Only four short burns were performed to circularize the shuttle's orbit before SMM was released.

Before the conclusion of the Solar Max repairs, its computer memory was loaded and the spacecraft was examined carefully. On April 12, once it had been determined that the attitude control system module and the new electronics box appeared to be working satisfactorily, the remote manipulator was used to engage the space-craft and hold it suspended while the high-gain antenna system was deployed. This system would be used with the new Tracking and Data-Relay Satellite System. The observatory redeployment went smoothly at the space shuttle's altitude of 502 kilometers, and Solar Max was tested yet again by the ground stations while the space shuttle was still in the vicinity. Upon final ground confirmation of the successful observatory repairs, the crew closed the shuttle, and *Challenger* returned to a landing at Edwards Air Force Base on April 13. (The plan had originally called for a landing at the Kennedy Space Center, but weather there was hazardous.)

The retrieved attitude control module was returned to Goddard Space Flight Center for measurement of the degradation of its mechanical, electrical, and thermal components. Technicians planned to refurbish it and possibly fly it on a subsequent scientific mission. The faulty electronics box was also returned to Goddard, where technicians from the High Altitude Observatory in Boulder, Colorado, examined it. Three circuit boards were reworked by installing a new counter chip on each board. The fifteen individual boards, groups of boards, and the entire box were retested using automated computer programs. Engineers determined that this unit could have been reflown. A report was prepared following the completion of a number of degradation studies, which used the retrieved SMM attitude control module main electronics box components, materials, and surfaces. The general condition of the retrieved electrical and mechanical parts led engineers to believe that long-life operation of space components can be expected with use of proper design techniques.

Knowledge Gained

With STS 41-C, the space shuttle for the first time used a direct-ascent burn to orbit, followed by the release of its external tank to burn up at high altitude. This mode of higher-than-usual orbital injection allows for more efficient servicing of free-flying spacecraft above the area of residual atmospheric drag of the shuttle's parking orbit.

At an altitude of 491 kilometers and an orbital inclination of 28.5 degrees, the space shuttle successfully deployed the

Long-Duration Exposure Facility. This satellite was designed for a series of in-orbit studies of materials degradation. The satellite's low-cost, twelve-sided open structure, outfitted with experiment trays, was easily deployed by the shuttle's remote manipulator system; it was designed to operate unattended for years before its recovery.

The space shuttle crew demonstrated the ability to rendezvous with and retrieve Solar Max at an altitude of about 502 kilometers, using the remote manipulator system in a backup capture mode. This maneuver was followed by the successful forty-two-minute exchange of an attitude control subsystem module, proving that repairs could be made on a serviceable spacecraft in orbit. In addition, the mission showed that with considerable planning, replacement hardware and tool development, simulations, and crew training, some EVA repairs can be made on payload instruments not designed for in-orbit servicing. To ensure successful servicing in orbit, however, work areas must be well within astronauts' operational constraints.

One of the most serious problems during the mission arose because of the lack of a secondary, manual means of triggering the trunion pin attachment device during attempts to capture the free-flying observatory. Also, the astronauts were forced to adjust the shuttle's fuel consumption and drift control during various rendezvous modes.

For an in-orbit repair mission to proceed in a timely and economical manner, good photographs and other documentation of the free-flyer must be available prior to launch. Replacement units and spare components must also be available. Proper repair planning and sufficient use of simulations are essential for an in-orbit repair to be successful. Planned alternate backup modes were found necessary during several contingency situations during STS 41-C. The mission specialists' limited EVA time during this mission indicated the need for additional power tools and servicing aids to increase astronauts' productive servicing time in orbit.

In addition to the highly successful in-orbit repairs of Solar Max at less than 20 percent of its replacement cost, the return to Earth of the malfunctioning observatory hardware opened the possibility of ground repair, refurbishment, and subsequent reuse of space resources. Returned materials and components were evaluated and generally found to be in excellent condition, except for degradation of thermal blanket materials and a high micrometeoroid count. The results of these studies, which followed SMM's four years of in-orbit operations, are aiding the design and operation of long-life serviceable spacecraft, scientific instrumentation, and applications payloads.

Context

In late 1969, Joseph Purcell, Project Manager for the Orbiting Astronomical Observatories, developed the concept of a modular spacecraft that could be reconfigured to handle various types and sizes of payloads. With the advent of the space shuttle, Frank Cepollina advocated extending the modular concept to include the capacity for in-orbit replacement of the modules. The idea for modular and serviceable spacecraft was adopted after an intercenter competition for new space exploration concepts. After several years of refinement, the multimission modular spacecraft was adopted as the spacecraft to support the Solar Maximum and Landsat D missions. Following SMM's successful launch and nine months of operation before the attitude control failure, the Solar Maximum Repair Mission demonstrated the space shuttle's capability to rendezvous with, capture, repair, and redeploy a malfunctioning spacecraft.

After winning the approval of Congress and undergoing two years of intense preparations, the STS 41-C Solar Maximum Repair Mission was performed. This pathfinder mission first proved the capability of the space shuttle to deploy the large, passive Long-Duration Exposure Facility and then to perform the repairs on Solar Max, extending the observatory's scientific investigations. The astronauts, employing in-orbit EVA and intravehicular activity servicing techniques, proved the practicality of exchanging an attitude control system module designed for servicing. The difficult exchange of the 12.4-kilogram coronagraph/polarimeter main electronics box was also accomplished, showing the possibility of in-orbit repair of payload instruments not specifically designed for in-orbit servicing.

The successful stable deployment of the Long-Duration Exposure Facility demonstrated that a low-cost, passive spacecraft could be orbited by the space shuttle, permitting the performance of long-term materials degradation experiments in space. The student honeybee experiment indicated that, after an initial difficulty with adaptation, the honeybees were able to make a honeycomb while in a weightless state—proving that, with certain environmental conditions controlled, life-forms can adapt to the space environment.

While scientists are very concerned about the amount of scientific data returned, they are also conscious of the need for timely return of data to the investigators who spend so much time designing and building experiments for space. By the time a new replacement spacecraft and payload can be funded, produced, and qualified for flight (which usually takes from three to seven years), the experiment aboard a malfunctioning spacecraft could become outdated. In this context, in-orbit repair and servicing of payloads is clearly valuable to the experimenters.

The STS 41-C mission demonstrated the technical, scientific, and economic benefits of using the space shuttle to retrieve, repair, and redeploy free-flying spacecraft that were designed for routine servicing. The mission also demonstrated the more limited capability of repairing payload instruments not intended for in-orbit servicing. More nations are expected to use these techniques of designing spacecraft and their payloads for routine servicing and repairs, thereby extending the benefits derived from their nations' initial investments.

Bibliography

Allen, Joseph P., and Russell Martin. *Entering Space: An Astronaut's Odyssey*. New York: Stewart, Tabori and Chang, 1984. A personal account of space travel aboard a U.S. space shuttle, with details of the experience of launch, the work accomplished in and around the spacecraft, and the thrilling return to Earth. Illustrated with color photographs.

Chaikin, Andrew. "Solar Max: Back from the Edge." *Sky and Telescope* 67 (June, 1984): 494- 497. This article provides an overview of the Solar Maximum Repair Mission, recounting its successes and failures for the general reader.

Furniss, Tim. *Space Shuttle Log*. London: Jane's Publishing Co., 1986. An in-depth history of the space shuttle and its flights, from STS-1 through STS 61-A. Features details about the design of the space shuttle vehicle and an overview of the shuttle systems. A concise account of the shuttle flights is provided, along with black-and-white photographs from each mission covered.

McMahan, Tracy, and Valerie Neal. *Repairing Solar Max: The Solar Maximum Repair Mission*. Washington, D.C.: Government Printing Office, 1984. Describes the preparations for the Solar Max repair and the people who worked on STS 41-C. Tells of the successful operations and the problems encountered. This reference also discusses the new era of orbital repairs of spacecraft.

Maran, Stephen P., and Bruce F. Woodgate. "A Second Chance for Solar Max." *Sky and Telescope* 67 (June, 1984): 498-500. This article offers two astronomers' views of the SMM repair and the promise it held for future repair missions.

Robert E. Davis

SPACE SHUTTLE MISSION
STS 51-A

Date: November 8, 1984 to November 16, 1984
Type of mission: Manned Earth-orbiting spaceflight

STS 51-A was the first space retrieval mission. Using techniques devised on the spot, the crew captured and returned to Earth two satellites that otherwise would have been useless forever. The impressive success of this flight was an excellent demonstration of the value of people working in space.

PRINCIPAL PERSONAGES
FREDERICK H. HAUCK, Commander
DAVID M. WALKER, Pilot
JOSEPH P. ALLEN,
ANNA L. FISHER,
DALE A. GARDNER, Mission Specialists

Summary of the Mission

When five astronauts were assigned to STS 51-A in late 1983, they expected their mission to be a routine cargo-hauling trip into space. Yet with the first flight of 1984, STS 41-B, it began to appear as though STS 51-A might become something more unusual. The crew of STS 41-B saw perfect deployments of their payload of two communications satellites, Indonesia's Palapa-B2 (Indonesian for "fruit of the effort") and Western Union's Westar-6. In both cases, however, the rocket motors designed to propel them from the Challenger to their orbital destinations 35,800 kilometers above Earth shut down prematurely. The two satellites were left in useless orbits about one thousand kilometers high. During 1984, the crew of STS 51-A prepared for the retrieval of these two satellites. Meant for orbits out of reach of the space shuttle, they had not been designed for recapture, but, through creative thinking and clever design by engineers and the astronauts, an apparently workable solution was devised. While plans were being formulated and equipment manufactured, ground controllers gradually lowered the orbits of the two satellites to an altitude the space shuttle orbiter could reach. On November 8, 1984, *Discovery* began its second trip to space less than one-tenth of a second after the scheduled time of 7:15 A.M. eastern standard time. This fourteenth mission of the Space Transportation System was commanded by Frederick Hauck. His pilot was David Walker, and the mission specialists were Joseph Allen, Anna Fisher, and Dale Gardner. As soon as the *Discovery* was settled into its orbit 302 kilometers above Earth, they began preparing for their job in space.

The task of rendezvousing with an orbiting spacecraft is very complicated. In addition to getting to the right point in space to meet the target, it is essential to be in the same orbit as the target. Many fine adjustments are required to make the entire orbits match and not simply intersect at one point. Matching orbits ensures that the two spacecraft stay together once they meet. It would require a total of forty-four carefully timed burns of the *Discovery* rocket engines over several days to bring it to its first quarry and still allow it to be in the correct positions at the correct times to deploy its payload of two communications satellites.

The first satellite the astronauts were to deploy was owned by Telesat of Canada. It was the third Canadian communications satellite taken into space by the space shuttle. This one was designated Anik-D2 (Inuit for "brother") or Telesat-H. After confirming that the satellite was in good condition, the crew deployed it on schedule at 4:04 P.M. on November 9 and fired *Discovery*'s engines to move away to a safe distance. This separation burn served the additional purpose of contributing to the intricate orbital dance required for the rendezvous. Forty-five minutes later, a timer on the Anik fired its rocket engine, and, unlike the satellites being pursued on this mission, it went smoothly to its targeted orbit.

The second satellite onboard was to be leased to the United States Navy. Thus the manufacturer (Hughes Aircraft Company) called it Leasat 1, while the Navy referred to it as Syncom IV-1. It was released from the payload bay on November 10 at 7:56 A.M. and successfully attained its intended orbit.

After completing the delivery of the two satellites, the crew could devote full attention to the retrieval plans. While *Discovery* continued to stalk Palapa on November 10, Allen and Gardner tested the spacesuits they would use for their Extravehicular Activities (EVA's, or spacewalks). On November 11, the orbital chase progressed.

As Hauck and Fisher completed the rendezvous with Palapa on November 12, Walker helped Allen and Gardner into their bulky spacesuits. By the time the two spacewalkers had exited the airlock and entered *Discovery*'s payload bay, the orbiter had closed to within 9 meters of Palapa. Allen donned the Manned Maneuvering Unit (MMU). With its small thrusters, this backpack allowed him to "fly" in space. Gardner

Astronauts Gardner and Allen on the RMS after recapture of Westar VI. (NASA)

helped him attach the 2-meter-long "stinger" to the front of his MMU. The stinger was a pole with extendible fingers like the ribs of an umbrella, and it projected in front of Allen like a lance in front of a knight. It would be used to dock with Palapa.

Palapa was a cylinder 2.7 meters high and 2.1 meters in diameter. At one end was the nozzle of the spent rocket engine, and the other end held the fragile antennas that would have been put into service had the satellite reached its intended orbit. The astronauts would have liked to capture the satellite with the orbiter's 15-meter mechanical arm (the Remote Manipulator System, or RMS), but the satellite's

smooth exterior presented nothing for the arm to snag. It was the job of the EVA astronauts to attach something for the RMS to hold and then to stop the satellite from rotating at 1.5 revolutions per minute so the RMS could capture it.

After confirming that all of his equipment was working as intended, Allen waited until an orbital sunrise and used his MMU to travel to the nozzle end of Palapa. He slowly approached the satellite and inserted the stinger rod into the nozzle of the rocket engine. With a control lever, he opened the rod's fingers inside the engine. The force of the extended prongs against the interior of the satellite provided a connection between the astronaut and the spacecraft. He had successfully docked with the errant satellite.

At that point, Allen began rotating with the satellite. By using the gyroscopic stabilization system in his MMU, he was able to stop both himself and Palapa from spinning. Fisher then was able to guide the RMS to a fixture on the side of the stinger. *Discovery* finally had control over Palapa. To secure the satellite in the payload bay, it would be necessary to connect the nozzle end to a cradle in the bay. The antennas made the other end too fragile. Therefore, the next goal was to attach a temporary fixture (the Antenna Bridge Structure, or ABS) over the antennas so the RMS could hold it by that end. Allen and Gardner then would connect mounting brackets to the nozzle end that would allow the satellite to attach to the cradle waiting in the payload bay, and the RMS would lower it into place.

While Allen relaxed in his position at one end of the satellite (still connected to it with the stinger), Fisher positioned the entire assembly so that Gardner could attach the ABS to the other end. To his surprise and frustration, a small protrusion on the spacecraft prevented the ABS from fitting properly. He struggled but was unable to make the critical connection. Without it, the RMS would not be able to place Palapa into its cradle.

After some quick discussion, the crew, with the concurrence of mission controllers, chose a backup procedure that had been practiced briefly before the flight. With Walker orchestrating the operations from inside *Discovery*, Allen detached himself from the stinger and returned the MMU to its station. He positioned himself in a foot restraint attached to the side of the payload bay, and Fisher used the RMS to hand him the 597-kilogram Palapa. He took it by the antenna end and held it while Gardner worked at the other end.

In orbit, objects have essentially no weight. They do, however, still have mass. Allen found that he was able to control the huge mass of the satellite through slow, careful movements. He was in an uncomfortable position, with the satellite above his head, but he endured long enough for his coworker to complete the necessary tasks. For one orbit of Earth, lasting ninety minutes, he held the satellite while Gardner removed the stinger from the nozzle and attached the bracket for the cradle. This chore had been planned for the two crewmen working together, but because Allen was now replacing the function of the ABS, Gardner had to do this task alone. The two men then managed to position the satellite in its holder and lock it into the payload bay.

The astronauts collected their tools and finished the EVA six hours after it had begun. Palapa, despite their difficulties, was secured in the payload bay. Yet there was no time for celebrating. Westar was in orbit more than 1,100 kilometers ahead. The crew would spend the next day chasing it and recharging their spacesuits in preparation for capturing that satellite. Meanwhile, they were faced with the question of whether there might be an interfering protrusion on Westar as there had been on Palapa. If there were, the ABS would not

fit again. With Palapa berthed in the bay, the working space would be more crowded, and it had been fatiguing for Allen to keep the satellite positioned for Gardner. The astronauts and ground controllers devised a new plan.

On November 14, with Westar rotating only 9 meters from Gardner's home in space, he flew his MMU/stinger combination to the nozzle end. He docked with it and stopped its rotation just as Allen had done two days earlier. Instead of guiding the RMS to the grapple target on the stinger, Fisher used it to hold a foot restraint. Allen fixed himself in it as if he were on a cherrypicker, and Fisher positioned him so he could grab the antenna end of Westar. While he held it, Gardner removed and stowed his MMU. With the RMS holding Allen and Allen holding Westar, Gardner attached the mounting bracket. Allen was in a more comfortable position than he had been in when he had held Palapa, and Gardner had the benefit of his experience as he attached the mounting bracket by himself once again. Five hours and forty-three minutes after it had begun, the EVA ended, with both satellites safely in the payload bay awaiting return to Earth.

The next day, the proud crew prepared for the return to Earth, held a press conference, and frequently looked into the payload bay to assure themselves they had indeed captured the satellites. At 5:55 A.M. on November 16, while above the Indian Ocean, the crew fired *Discovery*'s engines for 184 seconds to bring them back to Earth. Thirteen minutes after sunrise, at 6:59 A.M., the orbiter touched down at Kennedy Space Center. *Discovery* had begun its 5.3-million-kilometer journey only 5 kilometers away with two satellites to deploy and a hopeful crew. It ended with two satellites retrieved and a jubilant crew.

Knowledge Gained

The two communications satellites *Discovery* carried into space attained their intended orbital positions. At the orbital altitude of both satellites (35,800 kilometers), it takes twenty-four hours to circle the globe. The orbital speed then matches the rotational speed of Earth, and each satellite appears stationary in the sky. Many communications satellites are placed in such geosynchronous orbits, where they can act as relay antennas in continuous view of widely separated ground stations.

The Anik-D2 was stored in orbit for almost two years before being used. This allowed Telesat to launch it while launch costs were still relatively low, and it was available in space should an unexpectedly early need to use it have arisen. It was brought into service on November 1, 1986, and formed part of a network of communications satellites serving Canada. It is located above the equator at longitude 110.5° west. The Leasat assumed duties immediately after it was checked out in orbit and continued to function as expected, aiding in communications for the Navy. It is stationed in orbit over the Atlantic Ocean at longitude 15° west.

Shortly after the *Discovery* reached orbit, the crew activated the diffusive mixing of organic solutions experiment. It ran throughout the flight with little attention required from the astronauts. The experiment was designed to investigate the growth of crystals from organic compounds in the near weightlessness of orbit. The failure of some valves prevented some of the chemicals from crystallizing, but other samples produced hundreds of crystals. As scientists had hoped, the crystals grown in space, without the distorting effects of gravity, were larger and purer than those produced in Earth-based laboratories. The detailed analysis of their properties is expected to shed light on the complex process of crystal formation and aid in producing better crystals both on Earth, and, when space manufacturing becomes a reality, in that environment.

The knowledge that astronauts could manipulate extremely massive objects in space was an unexpected benefit of the mission. If everything had worked according to plan, this need would never have arisen. After Allen held the satellites and was able to position them wherever Gardner asked, it was realized that this ability could be applied in many other situations.

Context

That ability to handle large, massive objects in orbit did indeed prove very useful for NASA. Relying on the results of STS 51-A, the crew of STS 51-I successfully captured the malfunctioning Leasat 3 deployed on STS 51-D. It was the confidence and experience gained from STS 51-A that made it possible to develop a plan for having the astronauts hold and move this 6,890-kilogram satellite. Again devising procedures during the mission, three spacewalking astronauts on STS-49 manually captured and maneuvered the 4,064-kilogram INTELSAT VI/F-3. Astronauts on STS-61 were able to accurately position nearly 300-kilogram instruments in their EVA work to repair and upgrade the Hubble Space Telescope. The direct handling of massive objects, as demonstrated first on STS 51-A, is expected to permit great flexibility in future space retrieval and construction tasks.

The failure of Palapa and Westar to reach their targeted orbits after their deployment on STS 41-B, although not NASA's fault, did tarnish the agency's image. The recovery of the satellites restored the "can-do" reputation NASA reveled in. It also was another example of the value of having people in space, as the recovery was clearly beyond the realm of remotely controlled machines.

Palapa had been purchased by the Indonesian government to help tie together its many isolated regions. Westar was built for Western Union's use in its commercial work. Both satellites belonged to insurance underwriters after their rocket failures. Insurers had paid $180 million to the original owners, and, by paying a total of $10.5 million for the recovery operation, they hoped to resell the satellites and recoup much of their losses. The failures of these satellites occurred in a year in which spacecraft insurers paid out almost three hundred million dollars in claims, and there was great concern over the enormity of these payments. The retrieval helped alleviate that problem.

The recovery agreements had specified that the two salvaged satellites would be relaunched with the Space Transportation System. After the reevaluation of the United States' space policy in the wake of the STS 51-L accident, it was decided that commercial communications satellites would no longer be taken into space on these human flights. Yet there were not enough expendable launch vehicles in the nation's inventory to loft all of the satellites that were ready, and the unexpected hiatus in the capability of the United States to launch satellites caused long delays in the relaunching of Palapa and Westar. Westar 6 eventually was purchased by the Asia Satellite Telecommunications Co., Ltd. and was refurbished and launched as AsiaSat 1 on April 7, 1990. This launch, using the People's Republic of China's Long March 3 launch vehicle, marked the first time that a spacecraft made in the United States was launched on a Chinese launch vehicle. The refurbished Palapa B2, renamed Palapa B2R, was launched on a Delta 6925 six days later and joined two other Palapa communications satellites that were already in orbit. The commander of this challenging mission, Frederick Hauck, gained a great deal of experience from the complexity of the flight. His next trip into space was as commander of the very successful STS-26, the first flight after the failed STS 51-L. After the *Challenger* was destroyed and the crew killed on STS 51-L, the Space Transportation System underwent many changes, and greater emphasis was placed on missions requiring people in space rather than routine cargo-hauling trips. STS 51-A continues to shine as a stellar example both of the kind of mission that requires people in space and of the ability of NASA to accomplish impressive and valuable tasks in space.

Bibliography

Allen, Joseph P. and Russell Martin. *Entering Space: An Astronaut's Odyssey.* New York: Stewart, Tabori, and Chang, 1984. This exquisite book was written by one of the STS 51-A mission specialists and includes his firsthand account of the flight. Suitable for all audiences, it describes the experiences of spaceflight, both the routine chores of living and working and the excitement and drama of being in space. It presents more than two hundred color photographs displaying the beautiful views available in space as well as the activities performed by astronauts.

Cooper, Henry S. F., Jr. *Before Lift-off: The Making of a Space Shuttle Crew.* Baltimore: The Johns Hopkins University

Press, 1987. The long process of training astronauts for a flight on the space shuttle is described by following the crew of STS 41-G, the flight immediately before STS 51-A. Although astronauts train for a specific mission, the book contains a great deal of interesting insight into the difficulty and importance of the preflight training for any mission. All readers will gain an appreciation of the challenges that precede a space shuttle flight and how hard the astronauts have to work in order to make their mission look so easy and smooth.

Furniss, Tim. *Manned Spaceflight Log.* Rev. ed. London: Jane's Publishing Company, 1986. With a description of every human mission into space through Soyuz T-15 in March, 1986, this book is entertainingly written and should be enjoyed by general audiences. It provides the essential facts from each flight and allows the reader to understand any spaceflight in the context of humankind's efforts to explore and work in space.

Joels, Kerry M., and Gregory P. Kennedy. *The Space Shuttle Operator's Manual.* Rev. ed. New York: Ballantine Books, 1987. This book contains a wealth of information on space shuttle systems and flight procedures. It is written as a manual for imaginary crew members on a generic mission and will be appreciated by anyone interested in how the astronauts fly the orbiter, deploy satellites, conduct spacewalks, and live in space. The book contains many drawings and some photographs of equipment.

Powers, Robert M. *Shuttle: The World's First Spaceship.* Harrisburg, Pa.: Stackpole Books, 1979. Despite having been written before the first flight of the Space Transportation System, this book contains excellent descriptions of the space shuttle systems and the types of missions conducted. There are very good explanations of why and how the environment of space is used for a variety of scientific and technological applications. The book includes many paintings and drawings, a glossary, and an index.

Marc D. Rayman

SPACE SHUTTLE FLIGHTS
JANUARY – JUNE, 1985

Date: January 24 to June 24, 1985
Type of mission: Manned Earth-orbiting spaceflight

STS 51-C was the first shuttle mission entirely dedicated to a classified U.S. Department of Defense payload. Mission 51-D was the first flight to have an unscheduled spacewalk and the first to carry a non-pilot, nonscientist astronaut, U.S. Senator Edwin Garn of Utah. STS 51-B carried the Spacelab 3 science payload into orbit, including rats and monkeys as well as human beings. STS 51-G was one of the most successful of the series. Four satellites were released, one was retrieved, and many important technological and scientific investigations were completed.

PRINCIPAL PERSONAGES

THOMAS K. "KEN" MATTINGLY, STS 51-C
    Commander
LOREN J. SHRIVER, STS 51-C, Pilot
ELLISON S. ONIZUKA, STS 51-C Mission Specialist-1
JAMES F. BUCHLI, STS 51-C Mission Specialist-2
GARY E. PAYTON, STS 51-C Payload Specialist-1
KAROL J. BOBKO, STS 51-D Commander
DONALD E. WILLIAMS, STS 51-D Pilot
M. RHEA SEDDON, STS 51-D Mission Specialist-1
S. DAVID GRIGGS, STS 51-D Mission Specialist-2
JEFFREY A. HOFFMAN, STS 51-D Mission Specialist-3
CHARLES D. WALKER, STS 51-D Payload Specialist-1
E. JAKE GARN, STS 51-D Payload Specialist-2
ROBERT F. OVERMYER, STS 51-B Commander
FREDERICK D. GREGORY, STS 51-B Pilot
DON L. LIND, STS 51-B Mission Specialist-1
NORMAN E. THAGARD, STS 51-B Mission Specialist-2
WILLIAM E. THORTON, STS 51-B Mission Specialist-3
LODEWIJK VAN DEN BERG, STS 51-B Payload
    Specialist-1
TAYLOR WANG, STS 51-B Payload Specialist-2
DANIEL C. BRANDENSTEIN, STS 51-G Commander
JOHN O. CREIGHTON, STS 51-G Pilot
JOHN M. FABIAN, STS 51-G Mission Specialist-1
STEVEN R. NAGEL, STS 51-G Mission Specialist-2
SHANNON W. LUCID, STS 51-G Mission Specialist-3
PATRICK BAUDRY, STS 51-G Payload Specialist-1
SULTAN SALMAN A. AL-SAUD, STS 51-G Payload
    Specialist-2

Summary of the Missions

According to the National Aeronautics and Space Act of 1958, "The aeronautical and space activities of the United States shall be conducted so as to contribute…to the expansion of human knowledge of phenomena in the atmosphere and space. The [National Aeronautics and Space] Administration shall provide for the widest practicable and appropriate dissemination of information concerning its activities and the results thereof. "

Unlike its Soviet counterpart, NASA opted to make public all of its activities relating to manned spaceflight. This would allow the budding agency a chance to showcase its greatest accomplishments, as well as to lay open to criticism its shortcomings.

The chief personnel within NASA understood politics and budgets. They knew that the amount of money they would be allocated by the U.S. Congress would, for the most part, be dictated by the will of the American public. If the citizens believed that a project was worthwhile (for example, landing an American on the Moon before the Russians), NASA would be given support to accomplish the goal. From Mercury-Redstone 3, the first manned spaceflight, in May, 1961, through the spectacular retrieval of two stranded satellites by the STS 51-A crew in November, 1984, every manned journey into space was duly publicized. Information about the crew and mission was presented well in advance of the launch, and every aspect of flight, from lift-off to landing, was broadcast live around the world.

This open policy came to an end, at least temporarily, for the fifteenth flight of the space shuttle, STS 51-C, the first flight of a dedicated Department of Defense (DOD) payload. Since the DOD was in the habit of deploying secret satellites, the flight of 51-C would be shrouded in mystery. At least, that was the plan.

The flight that became STS 51-C was originally manifested as STS-10, to be flown in December, 1983. Its major payload would be a classified satellite, launched into geosynchronous orbit (that is, an orbit in which the satellite's velocity and altitude make it appear to hover over one spot on Earth's surface) by the Inertial Upper Stage (IUS). The IUS is a two-stage solid propellant payload booster that can be taken into low Earth orbit by either an expendable launch vehicle, such as the Titan, or the space shuttle.

The military crew for STS-10, announced in October, 1982, included veteran Ken Mattingly, who had flown previously as command module pilot of Apollo 16 and commander of STS-4. He would be the commander of this flight. His pilot would be Loren Shriver, and the mission specialists would be Ellison Onizuka and James Buchli. The payload specialist, a U.S. Air Force manned spaceflight engineer, would be named at a later date.

Following the failure of the IUS on STS-6 to place the first Tracking and Data-Relay Satellite (TDRS-A) into geosynchronous orbit, the Air Force asked NASA to delay the launch of STS-10 until the problem had been resolved.

At the beginning of 1984, the crew for STS-10 (which now included payload specialist Gary Payton) and its payload were designated STS 41-E under the new numbering system (where the "4" represented Fiscal Year 1984, the "1" Kennedy Space Center and "E" the fifth mission of the fiscal year). It was scheduled for launch on July 7, 1984, aboard *Challenger*.

When *Challenger* returned from its STS 41-G mission, loose tiles were found all over its skin. Nearly four thousand of them had to be replaced before the spacecraft could be flown again. NASA decided to use *Discovery*, just back from STS 51-A, for the now-designated STS 51-C. This would push the launch back to January, 1985.

NASA announced that the launch of STS 51-C would take place between 1:15 and 4:15 P.M., eastern standard time, on January 23, 1985; the administration would not reveal the exact launch time until the countdown resumed after its last scheduled hold at nine minutes before launch. No information about the military payload would be released. On December 18, the *Washington Post* revealed that the payload for STS 51-C was an electronic monitoring satellite, the advanced Signal Intelligence Satellite (SigInt). Although many in the DOD came close to calling the announcement an act of treason, the *Washington Post* had not actually given any details about the satellite.

On January 23, the mission was delayed for twenty-four hours because of what NASA called "extreme weather conditions." This was the first time a manned spaceflight had been postponed because of cold weather. At 11:40 A.M., eastern standard time, on January 24, the crew transfer van carrying the astronauts arrived at Launch Complex 39A. The crew usually arrived two and one-half hours prior to launch, but that mark passed and the public countdown clocks remained blank. Suddenly, at 2:41 P.M., a NASA launch commentator announced, "T minus nine minutes and counting."

Discovery was launched at 2:50 P.M., 1 hour and 35 minutes into the 3-hour launch window. Two problems involving orbiter systems caused the delays, but each was corrected on its own. All systems performed normally during launch, and the orbiter was placed into an orbit of 332 by 341 kilometers. NASA announced only that the shuttle was in orbit and that everything was going well. Status reports were given every eight hours thereafter.

Although never confirmed officially, at about 7:00 A.M. on January 25, the SigInt satellite was deployed along with its IUS. The IUS fired about forty-five minutes later and placed the satellite into its proper orbit. Later, the DOD announced that the IUS had been deployed and had successfully completed its mission.

At 12:23 P.M. on January 27, NASA announced that *Discovery* would be landing at 4:23 P.M. the next day. Prior to the flight, NASA had indicated that it would announce the landing time sixteen hours before it was to take place. The de-orbit burn took place at 3:18 P.M. over the Indian Ocean, and *Discovery* began its long glide back to the Kennedy Space Center. At 4:23:23 P.M., its main landing gear touched down 792.48 meters beyond the threshold on runway 15, bringing the mission to an official end after 3 days, 1 hour, 33 minutes, and 23 seconds. There was no postflight interview of the astronauts, who were quickly whisked away for debriefing.

The mission of STS 51-D, *Discovery*'s fourth flight, got off to a shaky start but clearly demonstrated the overall flexibility of the space shuttle program. Originally, the crew of STS 51-D was supposed to fly STS 51-E, on *Challenger*, in early March, 1985. Only six days before launch, severe problems were discovered with the primary payload: the second Tracking and Data-Relay Satellite (TDRS-B). The flight, already delayed five times previously, was finally canceled outright. In order to keep to schedule as much as possible, a revised STS 51-D mission was quickly assembled, to be launched a mere six weeks later. The new payload would be the Telesat 1 (Anik C) satellite from STS 51-E, and the Hughes Communications Syncom IV-3 Navy communications satellite originally scheduled for STS 51-D.

All but one of the original crew members would fly: French payload specialist Patrick Baudry was replaced by Charles Walker from McDonnell Douglas as a result of payload requirements. This was the second flight for Walker, his first having been mission 41-D, about ten months earlier. The commander, Karol "Bo" Bobko, would be making his second flight as well. He first flew as the pilot on STS-6, the maiden voyage of *Challenger*. Navy Commander Donald E. Williams, the pilot for STS 51-D, was making his first trip into space. The mission specialists chosen for the flight were also space rookies: M. Rhea Seddon, M.D.; Navy Captain S. David Griggs; and Jeffrey A. Hoffman, Ph.D.

One of the other notable elements of this flight was that it would be the first mission to have a nonscientist, non-pilot crew member, Senator Edwin Jacob "Jake" Garn of Utah. As the chairman of the Housing and Urban Development and Independent Agencies Subcommittee, Senator Garn had control over funding for the space program. Much was made of the senator's participation, considered by some to be a public relations stunt, but he participated in valuable medical experiments, particularly in the area of motion sickness. Motion sickness commonly plagues most astronauts, causing many to have violent fits of nausea lasting up to two days. Because of the costs and short duration of the missions, the last thing needed is a sick crew, so extreme care was taken to avoid the malady. Since Garn had no active roll in the payload deployment or piloting of the spacecraft, he was the ideal candidate to get sick deliberately, offering the crew physicians important data without jeopardizing the mission.

Discovery got off the ground on April 12, 1985, at 8:59 A.M., four years to the day after STS-1. The crew successfully launched the Telesat communications satellite at about nine and a half hours into the mission.

Early on the second day the Syncom satellite was deployed successfully, using the "Frisbee" technique. Normally the satellites were seated vertically in the shuttle's payload bay, spun up like tops (the technique used for the Telesat), and shot straight out. The Syncom was much larger, however, being a squat cylinder 5 meters in diameter, 4 meters high, and weighing about 6,800 kilograms. It was therefore stowed on its side and practically "rolled" out of the spacecraft such that it resembled a giant Frisbee being thrown. While the physical deployment was on time and apparently correct, the "omni" antenna on top of the satellite did not erect itself, as it was supposed to do about a minute after release. Furthermore, the kick motor used to boost the spacecraft into its final orbit did not fire after deployment.

Immediately a special analysis team was formed; it decided that the on/off lever on the side of the satellite had not been switched to "on" at the time of deployment. This lever was supposed to activate an internal sequencing timer that in turn would deploy the antenna and fire the engine.

Several options were discussed, the main ones calling for an unplanned Extravehicular Activity (EVA) to throw the switch manually. Two EVA crewmen would attach snaring devices on the end of the Remote Manipulator System (RMS) arm. Afterward, the snares would be used in an attempt to trip the lever once the astronauts were safely inside.

Because of the change in plans, the mission was extended by two days, while ground crews put the finishing touches on the EVA schedule and designed the snaring devices. The Capsule Communicator (CapCom) radioed up descriptions of the snares, one termed a "flyswatter" and the other a "lacrosse stick," for the objects they resembled. The crew constructed them out of tape, plastic pages from their onboard documents, and "swizzle sticks" used by the crew to reach distant switches while strapped into their seats. Dr. Seddon used her onboard surgical equipment to construct the devices, assisted by Senator Garn.

The flyswatter consisted of a long plastic sheet with three large square holes in a vertical row, looking much like a small ladder with three rungs a few centimeters wide. The object would be to snag the lever with one of these rungs; if the flyswatter missed or broke a rung, the lever would likely hit a second. A single stick would attach it to the RMS. The lacrosse stick looked much like the swatter, except that it had

a loop of wire across the top and was supported by two sticks on either side so as to give it slightly more strength.

Although most flights do not have spacewalks scheduled, flight rules dictate that there always be two crew members trained for emergency operations. Such emergencies might include repair of the delicate tiles, releasing a satellite stuck in the payload bay, or manually cranking the payload-bay doors closed. On this mission, astronauts Hoffman and Griggs were the two trained in EVA operations. Rhea Seddon would assist them in manipulating the RMS as required. The EVA took place on the fifth day of the flight and lasted slightly more than three hours. Once the EVA was completed, preparations were begun for the following day's rendezvous with Syncom and snaring of the switch.

Visual sighting first took place while the Syncom was about 70 kilometers distant in the early morning hours of April 17. At a distance of 15 meters, the crew could clearly see the separation lever and that it was apparently thrown as it should have been originally. This meant either that there was an internal failure of the timer or that the switch needed only a tiny bit of motion to trip it, since the internal microswitches were activated in only the last few degrees of movement.

At 5 days, 13 minutes into the flight, Seddon was given the go-ahead to snare the lever. She slowly moved the arm toward the center of the satellite, exactly where the lever would be when it passed by the arm. At one time the lever could be seen slicing one of the rungs. Another time, the arm appeared to bump up against the Syncom. Six minutes later the attempts were halted and the crew reported that they got "a hard physical contact on at least two occasions." With that, *Discovery* separated from the satellite for the last time and maneuvered to an attitude from which the crew could observe the firing of the kick motor. No such firing was seen and the satellite was left in a 300-by-416- kilometer orbit for the possibility of a rescue mission on another flight.

Landing took place on Friday morning, April 19, 1985, at the Kennedy Space Center. As if enough had not happened during the flight, the landing was plagued by more troubles. First, the weather delayed reentry by one revolution. Next, during landing, crosswinds gusting to 15 knots blew the vehicle off the centerline of the runway. Bobko tried to correct this with "moderate" braking. As a result, both right brakes locked and the inboard tire blew out, rendering all tires unusable. Later investigation revealed that a tile had fallen off during ascent, exposing bare aluminum to the heat of reentry. More than 120 other tiles were damaged and required replacement.

The STS 51-B mission, carrying the Spacelab 3 scientific payload, was the seventeenth flight in the shuttle series. The primary areas of investigation for this mission were the life sciences and microgravity (the weightless environment and its effects on humans, animals, and properties of matter).

The STS 51-B commander was Robert F. Overmyer, who had previously flown on the fifth shuttle mission as pilot in

1982. His pilot on STS 51-B was to be Frederick D. Gregory, who was on his first shuttle mission. Three NASA mission specialists were named: Don L. Lind, a physicist, was making his first flight; Norman E. Thagard, M.D., who had flown previously on STS-7; and William E. Thornton, M.D., who was on his second flight, having flown on STS-8, and who, at age fifty-six, was the oldest person to date to have flown in space. The two payload specialists for the mission were selected from a group representing the scientific interests of the mission: Taylor G. Wang, a fluid mechanics expert and a principal investigator in one of the Spacelab 3 investigations, and Lodewijk van den Berg, a materials science expert. During the mission, these crew members were to work in twelve-hour shifts: The "gold" (day) shift would consist of Overmyer, Lind, Thornton, and Wang. The "silver" (night) shift would be Gregory, Thagard and van den Berg. Though no spacewalk (EVA) was planned, Gregory and Thagard were trained as contingency EVA crew members.

The payload and area of investigation for Spacelab 3 had been finalized several years before launch, and as a result of development difficulties with Spacelab 2, as well as several launch delays within the program, the manifest placed the launch of Spacelab 3 (which had been designated the first "operational" flight of the Spacelab series) before the Spacelab 2 flight. The countdown for 51-B proceeded without major incident, and its life science payload — two monkeys and twenty-four rats — was installed in the holding facility twenty-four hours before launch, allowing the animals to become accustomed to their new home. At 12:02 P.M. eastern standard time, *Challenger* lifted off and followed a nominal ascent profile to orbit.

The crew's first major activity was to deploy two small satellites from Get-Away Special (GAS) canisters in the payload bay. In order to ensure successful deployment and efficient battery power, it had been decided to deploy these satellites on flight day 1. The Northern Utah State Satellite (Nusat) was deployed by spring ejection 4 hours and 14 minutes into the mission; the Global Low-Orbiting Message Relay (GLOMR) satellite, which was to have been deployed some 14 minutes later, did not eject from its canister, though the GAS door did open. It was returned to Earth for reassignment to a later flight.

Challenger was then maneuvered into a gravity-gradient stabilization attitude with the tail toward Earth and *Challenger*'s port wing pointing in the direction of flight. The rest of the first day of the flight was spent in activating Spacelab and its experiments. For the next few days, the crew of STS 51-B worked hard to gather as much information as possible from their experiments.

Difficulties with the French Very Wide Field Camera (VWFC) were encountered some eight hours into the flight, when Lind had trouble using Spacelab's air lock in connection with the camera. A bent handle in the air lock forced ground controllers to decide against using the system for the rest of the flight, and the VWFC had to be abandoned with almost no data obtained.

On the whole, the crew members kept to their flight plan well, but only with much effort. Problems with the toilet system plagued them, and van den Berg experienced difficulties with his crystal-growing experiments during the second day of the flight. The animals, however, adapted well to their strange new environment. As time in orbit increased, thoughts of extending the mission were aired, a move that would enable the crew to complete research that had been slowed by the need to repair faulty equipment: Van den Berg had to shut down his experiment while he tried to repair its equipment, which he succeeded in doing during flight day 3, and Thagard had to repair the urine-monitoring system experiment and the Atmospheric Trace Molecules Spectroscopy (ATMOS) experiment. These efforts were worth the trouble. The ATMOS experiment yielded some spectacular results despite operating for only three days and obtaining only 19 of 60 planned data takes. Throughout his twelve-hour shifts in the Spacelab, Lind was able to obtain excellent photographs and record precise visual descriptions of auroral displays, which had been of scientific interest to him for years, especially after a period of intense solar activity on April 30.

A fine demonstration of the value of the human presence in space appeared in the untiring efforts of Wang, whose drop dynamics fluid mechanics experiment failed to work early in the flight because of an electrical malfunction. Determined not to return home unsuccessful, Wang almost completely rewired the machine himself, finally getting the experiment to function after several days' work. Another example of the importance of the "human factor" occurred when one of the two monkeys refused to eat for several days, apparently suffering from a bout of spacesickness; Thornton was able to encourage the animal to eat by hand-feeding him, probably saving his life.

The last full day in space was spent in stowing experiments and equipment in preparation for reentry into Earth's atmosphere. *Challenger* touched down on runway 17 at Edwards Air Force Base after an uneventful descent at 09:12:03, Pacific daylight time, May 6, 1985, and rolled to a wheel stop 47 seconds later. The shuttle had traveled more than 4 million kilometers in 7 days, 9 minutes, and 46 seconds, and had completed 110 orbits of Earth.

Despite a lightning strike on the 24-meter-tall lightning mast the night before, the fifth flight of *Discovery* on mission STS 51-G began with a perfect lift-off at 7:33 A.M., eastern daylight time, on June 17, 1985. Some 8 minutes and 46 seconds after leaving Kennedy Space Center's Launch Pad 39A, *Discovery* was in orbit.

Upon reaching the desired orbit, the crew began preparing for a seven-day mission that would see the release of a record four satellites, the retrieval of one of them, and a variety of experiments ranging from studies of human biology to measurements of the emissions of distant galaxies. The crew of NASA astronauts included the commander, Daniel C. Brandenstein; the pilot, John O. Creighton; and mission specialists John M. Fabian, Shannon W. Lucid, and Steven R. Nagel. While the crew was preparing the orbiter for its stay in space, Patrick Baudry began a series of measurements to study the flow of blood in his body. A payload specialist, Baudry was a French space traveler who had served as backup for his countryman Jean-Loup Chrétien on Soyuz T-6. The French Echocardiograph Experiment (FEE) that Baudry was using was very similar to the equipment Chrétien had used on Salyut 7.

Both Patrick Baudry and the other payload specialist, Sultan Salman Abdelazize Al-Saud, were on board to conduct specific experiments from scientists in their countries. Al-Saud's participation was sponsored by Saudi Arabia and was permitted because the twenty-two-member Arab Satellite Communications Organization was hiring NASA to deploy a satellite from the space shuttle. NASA policy allowed customers to include a passenger in such cases. Neither Baudry nor Al-Saud had any responsibilities directly related to the principal goals of the flight.

The first important goal in orbit was to release the Morelos -A communications satellite belonging to Mexico. Morelos was ejected into space only 8 hours and 5 minutes after *Discovery* itself was launched. *Discovery* fired its maneuvering rockets to move away from the free-floating satellite, and after 45 minutes a timer on Morelos activated its rocket engine to propel it to the desired orbit.

Although the Arabsat 1B release was scheduled for the next day, that satellite demanded attention when ground controllers received a signal indicating that one of its solar panels (used to convert sunlight into electricity for the satellite) had partially opened under the sunshield. The crew could see no evidence of this defect, so the next day the Remote Manipulator System (RMS) arm was used to get a closer view of the satellite. A camera at the "wrist" of the RMS allowed a detailed inspection of the solar array, and it was determined that the indicator was faulty. The array was correctly tucked against the side of the satellite. The release of the Arabsat took place on schedule and without difficulty on the second day of the mission, at 9:36 A.M. The successful deployment of the Telstar 3D occurred at 7:20 A.M. on June 19 and completed the delivery of communications satellites to orbit.

With the three satellites on their way, the crew had time to turn to other activities. One of these was the High-Precision Tracking Experiment (HPTE), an unclassified test in collaboration with the Strategic Defense Initiative Office (SDIO). The purpose was to evaluate various schemes for training a laser on an Earth target over a period of time. Earth's turbulent atmosphere makes this very difficult. The plan called for the astronauts to install a special 22-centimeter mirror in one of the windows. When a low-power laser at Mount Haleakala in Maui, Hawaii, was directed toward

Discovery, the mirror would send some of the light back to the ground station, which could then measure how accurately the laser had "hit" the orbiter.

The astronauts entered the location of the ground-based laser into the computers so that the orbiter could orient itself correctly to ensure that the mirror would reflect the laser back to the ground test facility. Shortly before the test, the astronauts reported that the window was not pointing toward Hawaii, but was pointing into space instead. The astronauts were able to see the blue-green light from the laser (it covered an area on the spacecraft about 9 meters in diameter), but the mirror was not in a position to return any of the light. It was soon determined that the location of the laser transmitter was entered into the computer incorrectly. Thus the computer miscalculated the laser's location. By the time this error was discovered, it was too late to correct it.

Before another attempt to perform the tracking experiment was made, the last satellite release had to be undertaken. The Shuttle-Pointed Autonomous Research Tool for Astronomy (SPARTAN) was designed to conduct observations independently of the space shuttle for as long as two days. This first test flight had instruments for measuring the X-rays emanating from the center of the Milky Way galaxy and from the galaxies clustered in a group known as Perseus. Shannon Lucid used the RMS to remove SPARTAN 1 from its cradle in the payload bay and point it so that its own systems would be directed toward the Sun and the star Vega. This orientation would allow the satellite to establish and maintain its own direction in space so that it could aim its instruments at the preprogrammed targets. The satellite, about the size of a telephone booth, was gently released at 12:03 P.M. on June 20. In order to minimize the cost and complexity of the SPARTAN, no communications equipment was built into it, so, to inform the crew that all of its systems had passed a self-test, the free-flying spacecraft performed a "pirouette." Satisfied that everything was working as planned, the astronauts directed *Discovery* away from the satellite to allow SPARTAN's observations to proceed without interference from the orbiter. The distance between the two craft would slowly grow to more than 190 kilometers.

On June 21, the laser tracking experiment was performed successfully. The next day, the crew members began a twenty-hour, thirteen-orbit chase to recover the SPARTAN. When they were fewer than 50 meters away from the satellite, they could see that it was not in the correct orientation. It was stable, but the fixture that the RMS needed to attach itself to was not on the side facing *Discovery*. The orbiter closed to within 10 meters of the craft, and John Fabian manipulated the RMS to reach behind the SPARTAN and pluck it from orbit. He then returned the satellite to its resting place in the payload bay so that scientists and engineers on Earth could evaluate both the data from its astronomical observations and the overall performance of the new, reusable satellite.

On June 24, Commander Brandenstein landed *Discovery*

on the dry lake bed at Edwards Air Force Base. The spacecraft rolled for 40 seconds before stopping, the final part of its 4.7-million-kilometer journey marred only by the left main landing gear digging 15 centimeters into the ground as *Discovery* came to a halt. The orbiter incurred minimal damage, however, and STS 51-G concluded with the same level of success it had demonstrated since it began.

Knowledge Gained

Although neither the U.S. Air Force nor NASA has given specific information about the military payloads flown aboard Department of Defense missions, it is generally believed that each satellite was placed into geosynchronous orbit and performed as designed. The IUS, which had proved troublesome on the STS-6 mission, worked as planned. There were other experiments flown aboard the flights, and some information about them has been released.

One of the experiments on mission STS 51-C, known as the aggregation of red blood cells experiment, was flown in two canisters in the middeck of the orbiter. A computer controlled the experiment, which involved the passing of eight donors' blood between two glass sheets so that cameras and a digital data system could assess how the blood aggregates in the near-weightless conditions of orbital spaceflight. Since blood aggregates differently in persons with disease from the way it aggregates in those who are free from disease, researchers can understand better how blood circulates by viewing its activities under controlled conditions.

Many of the shuttle missions are primarily targeted toward delivery of satellites; some have few if any major experiments. That was the case with STS 51-D. Vital experience was gained in on-the-spot planning for the unscheduled EVA, verifying that the current systems of training and mission planning did work.

On board the shuttle were two Get-Away Special canisters, (GAS cans, used for low-cost experimental units requiring little crew intervention). G035 was an experiment to test the surface tension, viscosity of fluids, and solids and alloy furnaces. A malfunction prevented G471, the capillary pump loop priming experiment, from being conducted.

Payload specialist Charles Walker operated the McDonnell Douglas Continuous Flow Electrophoresis System (CFES), which was used to develop techniques required for generating a pure pharmaceutical material. McDonnell Douglas reported that "all samples were processed and no contamination of the product was observed." What exactly the product was, however, was kept secret.

On the lighter side was the Toys in Space project. This served to demonstrate the behavior of more than thirty miniature mechanical systems in zero gravity. Films were taken of the crew that would later be used in classroom science discussions. Common toys, such as Slinkies, paper airplanes, jacks, yo-yos and paddleballs were used.

The scientific objectives of the STS 51-B Spacelab 3 mission could be divided into four areas: materials processing, environmental observations, fluid mechanics, and life sciences. Out of fifteen experiments, the crew had gained useful data from fourteen, resulting in 250 billion bits of scientific data. Preliminary data from the Spacelab 3 experiments were described as excellent by Mission Scientist George Fichti.

In the materials science experiments, a mercury iodide crystal about the size of a sugar cube was grown from a seed crystal in the vapor crystal growth system over a period of 104 hours, using a vapor transport technique. A fluid transport technique was used to grow two triglycerine sulfate crystals in the fluid experiment system. The vapor crystal growth and fluid experiment systems provided the first opportunity to observe in detail crystal growth in a microgravity environment and to determine the difference between growth on Earth and in orbit.

A total of 102 hours of geophysical fluid flow cell operations were completed on the flight to understand convection on the Sun, in planetary atmospheres, in Earth's oceans, and in basic fluid physics. The first experimental data on the behavior of a free-floating fluid in a microgravity environment were obtained in the drop dynamics module. For the first time, both solid and liquid samples were acoustically positioned and maneuvered in weightlessness; drops of varying sizes and viscosities were formed, rotated, and oscillated. In addition, studies were made on compound drops (drops formed within drops).

The ATMOS experiment obtained nineteen sequences of more than 150 independent atmospheric spectra, each containing more than 100,000 individual measurements used to analyze Earth's stratosphere and mesosphere at altitudes between 10 and 150 kilometers. In addition, high-resolution infrared spectra of the Sun were obtained during five calibrations and provided some surprising evidence about the molecular constituents of the Sun.

Mounted outside the pressurized module, the ionization of solar and galactic cosmic ray heavy nuclei experiment (IONS) recorded data on high-energy particles emitted from the Sun and other, more distant galactic sources. The life science experiments revealed that all the animals could successfully adjust to spaceflight conditions, although one of the monkeys did suffer from space adaptation syndrome, recovering in a manner similar to human recovery.

The flight of SPARTAN 1, during STS 51-G, yielded important scientific and technical results. The primary objective of mapping the X-ray emissions from the central region of the Milky Way and the complete Perseus cluster of galaxies was achieved.

The French experiments on the adaptation to spaceflight and the readaptation to normal gravity completed all the planned tests. The results of Baudry's measurements of the cardiovascular system and other systems in the body added to a small but growing data base on the effects of gravity (or its absence) on living organisms.

One of the other experiments conducted throughout the mission was designed to shed light on the process of convection in the melting of certain materials. Operated automatically while the astronauts slept (so that their movements about the orbiter would not cause disturbances in the sensitive experiment), it melted samples for later study to examine to what extent the absence of gravity reduced convection. The postflight analysis revealed the surprising fact that convection plays a very important role in solidification even when gravitational effects are greatly reduced.

Context

STS 51-C, which had been delayed in getting off the ground because of cold weather, landed a year and a day before the next flight to be seriously affected by the cold, STS 51-L. Ellison Onizuka would be on that tragic mission, too. After that cold morning in January, 1986, the space shuttle program would forever change. Never again would a flight be attempted in freezing weather.

The Department of Defense would fly only one more "secret" mission aboard the space shuttle — STS 51-J in October, 1985. After the *Challenger* accident in January, 1986, the military opted to puts its launch efforts back into expendable launch vehicles.

The flexibility of the space shuttle system was decisively demonstrated in the first-ever unplanned EVA on mission 51-D. The ability of ground personnel combined with the contingency training of the flight crew made for a successful operation even though the satellite was not activated. The space program was seen to have reached a new level of maturity. Senator Garn's presence served to emphasize that space travel was becoming increasingly routine.

STS 51-I was launched on August 27, 1985, to "hot-wire" the Syncom deployed by the *Discovery* crew on STS 51-D. (The physical design of the Syncom precluded any return to Earth.) Four days later, astronauts James "Ox" van Hoften and William F. Fisher left *Discovery* to snare the satellite manually and return it the payload bay. There Fisher installed the Hughes-designed Spun Bypass Unit to bypass the failed sequencer. Once the unit was installed, Fisher threw four switches, bringing the $90 million machine back to life. Using handles that he had installed in the side of Syncom,

van Hoften spun it up by hand to two revolutions per minute and "launched" it back into its own orbit.

The success of this flight, along with STS 51-D, deflated the arguments of those who claimed that manned spaceflight was a mere luxury and that robots could do all that was necessary.

The flight of STS 51-B provided scientists with their first real opportunity to investigate the science of fluid mechanics on an American manned spaceflight since the Skylab missions of 1973 and 1974. The growing of crystals in space has significant applications in the world of microelectronics, military systems, telescopes, cameras, and infrared monitors. The important link between the crew and ground teams was demonstrated in the work of Wang, who for the first time operated and repaired his own equipment in space, recovering the experimental facility and achieving almost all preflight goals, despite a late start in the mission.

Results from the round-the-clock operations on STS 51-B provided vast amounts of new data to complement the results obtained prior to the mission. For example, the ATMOS experiment gathered more data on one flight than previously obtained in decades of similar research with balloon-borne high-altitude sorties. The flying of principal investigators allowed them to refine their own hardware and to experience at first hand the operation of their experiments. What they learned was of great importance in the development of follow-up experiments. An understanding of the effects of spaceflight on hardware and the difficulties and advantages of the microgravity environment allowed Wang and van den Berg to make plans for more efficient and productive experimental hardware for later flights and use by other crews.

The three communications satellites sent into orbit by *Discovery*'s crew on STS 51-G reached their appointed slots in geosynchronous orbit. The Morelos satellite helps bring educational television, commercial programs, telephone and facsimile, and data and business transmissions to virtually every area in Mexico. SPARTAN's flight demonstrated the capabilities of a new family of free-flying spacecraft.

The high-precision tracking experiment provided experimenters with about 2.5 minutes (more than twice as long as needed) of data on the performance of different tracking techniques. It was the first SDIO experiment conducted with the space shuttle, and more were planned to follow.

Bibliography

Furniss, Tim. *Manned Spaceflight Log.* Rev. ed. London: Jane's Publishing Co., 1986. This reference provides a nontechnical overview of manned spaceflight since its beginning. Covers the flights of American astronauts, Soviet cosmonauts, and space travelers from other nations who have flown with them. Accounts of each flight are accompanied by black-and-white photographs.

———. *Space Shuttle Log.* London: Jane's Publishing Co., 1986. From STS-1 through STS 61-A, the space shuttles are covered in depth. Furniss discusses the design concepts for the Space Transportation System and provides a concise account of each shuttle flight. Includes black-and-white photographs from the missions.

Gurney, Gene, and Jeff Forte. *Space Shuttle Log: The First Twenty-five Flights.* Blue Ridge Summit, Pa.: Tab Books, 1988. A collection of chapters summarizing the first twenty-five missions of the Space Transportation System program, covering the operational flight time of April, 1981, to January, 1986, the first five years of the flight program. Each entry, from STS-1 through STS-25, lists information on crews, payloads, launch preparations, launches, orbital operations, and landing phases. A mission summary and text are accompanied by a selection of black-and-white photographs for each mission.

Kerrod, Robin. *Space Shuttle.* New York: Gallery Books, 1984. Most valuable for its beautiful color photographs of the space shuttle, this volume conveys the essence of the Space Transportation System vehicles and the people who fly them. Highlights of the first dozen missions are presented, as well as a fanciful look at the future of space exploration.

National Aeronautics and Space Administration. *Spacelab 3.* NASA EP-203. Washington, D.C.: Government Printing Office, 1984. This NASA publication provides a preflight overview of the 51-B mission and the scientific investigations planned for Spacelab 3. Background chapters on the flight crew, mission planning, and each of the major areas of scientific research are included, along with diagrams and color photographs, providing a handy summary of preflight objectives.

Otto, Dixon P. *On Orbit: Bringing on the Space Shuttle.* Athens, Ohio: Main Stage Publications, 1986. Early designs of the space shuttle are discussed. Also presents an account of each of the first twenty-five flights, including the crew, the payloads, and the flight objectives. Illustrated in black and white.

Smith, Melvyn. *An Illustrated History of Space Shuttle: X-15 to Orbiter.* Newbury Park, Calif.: Haynes Publications, 1986. A concise overview of the space shuttle and the experimental aircraft that led to its design. Spanning the period from 1959 through 1985, the book was written for the general reader. Illustrated with many photographs of the early lifting bodies, which help to show how the shuttle orbiter came to look as it does today. Arranged chronologically.

Wilson, Andrew. *Space Shuttle Story.* New York: Crescent Books, 1986. Traces the history of the space shuttle from the early days of rocketry to the *Challenger* accident. Furnished with more than one hundred color photographs, this volume provides little detail but emphasizes the men and women who fly the spaceplane to and from orbit.

Yenne, Bill. *Space Shuttle.* New York: Gallery Books, 1986. A large-format picture book which covers the shuttle in the most general fashion, from construction to operations. Sections are devoted to history, manufacture, and mission profiles. A brief flight log is included in the back.

Russell R. Tobias (STS 51-C)
Michael Smithwick (STS 51-D)
David J. Shayler (STS 51-B)
Marc D. Rayman (STS 51-G)

SPACE SHUTTLE FLIGHTS
JULY –DECEMBER, 1985

Date: January 24 to June 24, 1985
Type of mission: Manned Earth-orbiting spaceflight

During STS 51-F, astronauts conducted experiments in solar and space plasma physics and astrophysics. STS 51-I deployed three communications satellites, and crew members performed in-flight maintenance on a malfunctioning satellite. STS 51-J, the maiden flight of Atlantis, *featured the deployment of two military communications satellites during the second classified American manned spaceflight. STS 61-A was the first American spaceflight to have its control, once the shuttle was in orbit, centered outside the United States. Spacelab D-1 was an international mission dedicated to various scientific and technological investigations. During STS 61-B, astronauts launched three communications satellites and practiced assembling large structures in space, in preparation for the assembly of space stations.*

PRINCIPAL PERSONAGES

C. GORDON FULLERTON, STS 51-F Commander
ROY D. BRIDGES, STS 51-F Pilot
F. STORY MUSGRAVE, STS 51-F Mission Specialist-1
ANTHONY W. ENGLAND, STS 51-F
    Mission Specialist-2
KARL G. HENIZE, STS 51-F Mission Specialist-3
LOREN W. ACTON, STS 51-F Payload Specialist-1
JOHN-DAVID F. BARTOE, STS 51-F
    Payload Specialist-2
JOE H. ENGLE, STS 51-I Commander
RICHARD O. "DICK" COVEY, STS 51-I Pilot
JAMES D. VAN HOFTEN, STS 51-I Mission Specialist-1
JOHN M. LOUNGE, STS 51-I Mission Specialist-2
WILLIAM F. FISHER, STS 51-I Mission Specialist-3
KAROL J. BOBKO, STS 51-J Commander
RONALD J. GRABE, STS 51-J Pilot
ROBERT L. STEWART, STS 51-J Mission Specialist-1
DAVID C. HILMERS, STS 51-J Mission Specialist-2
WILLIAM A. PALES, STS 51-J Mission Specialist-3
HENRY W. HARTSFIELD, STS 61-A Commander
STEVEN R. NAGEL, USAF, STS 61-A Pilot
JAMES F. BUCHLI, USMC, STS 61-A Mission
    Specialist-1
GUION S. BLUFORD, STS 61-A Mission Specialist-2
BONNIE J. DUNBAR, STS 61-A Mission Specialist-3
REINHARD FURRER, STS 61-A Payload Specialist-1
ERNST MESSERSCHMID, STS 61-A Payload
    Specialist-2

WUBBO OCKELS, STS 61-A Payload Specialist-3
BREWSTER H. SHAW, STS 61-B Commander
BRYAN D. O'CONNOR, STS 61-B Pilot
MARY L. CLEAVE, STS 61-B Mission Specialist-1
SHERWOOD C. SPRING, STS 61-B Mission Specialist-2
JERRY L. ROSS, STS 61-B Mission Specialist-3
RUDOLFO NERI VELA, STS 61-B Payload Specialist-1
CHARLES D. WALKER, STS 61-B Payload Specialist-2

Summary of the Missions

The nineteenth space shuttle mission returned important data on the Sun, the stars, and the space environment with an advanced array of sophisticated instruments. The mission also demonstrated the Instrument Pointing System (IPS), designed as part of the Spacelab science system for the shuttle. Although the flight itself was designated STS 51-F, it is best known by its primary payload, Spacelab 2. Its scientific objectives were to scan the sky and to analyze the near-space environment around Earth.

The first attempt to launch *Challenger* on mission 51-F was aborted on the launch pad. Apparently, an engine had been slow in starting only three seconds before lift-off on July 12, 1985. The launch was reset for July 29.

After a ninety-minute delay caused by flight computer problems, *Challenger* lifted off at 5:00 P.M., eastern daylight time. Ascent was normal until 5 minutes and 45 seconds after lift-off, when the center engine was automatically shut down by its computer. That forced the crew to implement a procedure known as "abort to orbit." The remaining two engines would use the rest of the propellant to burn one minute longer in order to achieve orbit. Although called an "abort," this maneuver actually allows the mission to proceed by placing the spacecraft in an orbit lower than the one that had been planned. *Challenger's* engines were, in fact, performing as designed, but a pair of temperature sensors on a turbopump gave erroneous high readings and the computer deactivated the engine. A flight controller saw that all other engine readings were normal. When sensors on a second engine functioned the same way, the crew was instructed to override the computer's automatic command to deactivate this second engine.

Challenger was successfully inserted into an orbit 322 kilometers (instead of 385 kilometers) high. Because energy was conserved during the flight, Mission Control could extend the mission by one day to allow the solar science team to recover part of the experiment time lost when the instrument pointing system was malfunctioning. The mission ended at 12:45 P.M. Pacific daylight time, on August 6, when *Challenger* landed at Edwards Air Force Base in California.

Instruments in the Spacelab payload were designed to return data in the fields of astrophysics and solar and space plasma physics; experiments in the areas of atmospheric physics, technology development, and the life sciences were also on board. These were assembled on three U-shaped Spacelab pallets and on a special structure carried in *Challenger*'s payload bay.

The solar and atmospheric physics instruments were attached to an instrument pointing system designed to aim telescopes at the Sun, the stars, Earth, and other targets as the shuttle flew through space. Spacelab 2 had been delayed for several years because of problems in developing this highly sophisticated system.

Four solar instruments were mounted on the forward pallet of the instrument pointing system: the High-Resolution Telescope and Spectrograph (HRTS), the Solar Optical Universal Polarimeter (SOUP), the Coronal Helium Abundance Experiment (CHASE), and the Solar Ultraviolet Spectral Irradiance Monitor (SUSIM). Only the first three were considered to be true solar physics instruments, since the SUSIM was designed to support studies of Earth's atmosphere. The SUSIM had flown earlier on the STS-3 mission in 1982.

The HRTS and the SOUP were complementary instruments designed to return data on active regions of the visible surface of the Sun. It was hoped that the SOUP would help produce photographs of individual magnetic field activities within granules (convective eddies rising to the Sun's surface) during their 5- to 20-minute lives and within supergranules during their 20- to 40-hour lives. The HRTS would aid scientists in their study of the outer layers of the solar atmosphere, especially the transitional region between the chromosphere and the corona.

Astrophysics instruments were mounted on pallets and structures through the remainder of the payload bay. These were the X-Ray Telescope (XRT), the small, helium-cooled Infrared Telescope (IRT), and the elemental composition and energy spectra of Cosmic-Ray Nuclei (CRNE) instrument. The XRT actually consisted of two telescopes of similar design; each used a pinhole mask to project images of the sky in high-energy X-rays. The IRT used liquid helium to cool the detectors at the focal plane of a 15-centimeter telescope in order that cold objects in the 1- to 120-micron wavelength range, such as stellar nurseries and nebulae, could be observed. The telescope itself scanned at right angles to the shuttle's line of flight to construct images line by line. The CRNE, also called the "Chicago egg" because of its shape and its origin (The University of Chicago), was the largest cosmic-ray instrument placed in orbit. Its design allowed for

detection of extremely heavy cosmic rays at energies between 400 and 4,000 gigaelectronvolts. Cosmic rays are not actually rays but atomic and subatomic particles released after the explosion of stars and other such violent events.

Two plasma physics instruments shared pallet space with the IRT: the ejectable Plasma Diagnostics Package (PDP) and the Vehicle-Charging And Potential experiment (VCAP). Both the PDP and the VCAP had been flown on STS-3 in 1982. The PDP carried several instruments in an ejectable package that the shuttle was to circulate in order to help scientists determine the effects of large vehicles on the space environment. The VCAP comprised complementary instruments, including an electron gun to probe the plasma environment while the PDP measured responses. A third investigation, designed to return data on plasma depletion for ionospheric and radio astronomical studies, had no special equipment aboard *Challenger*. Instead, it used the shuttle's thrusters to burn "holes" in the ionosphere.

One technology experiment, whose subject was the properties of superfluid helium in zero-gravity, was carried on the pallet. Superfluid helium, in which electrical resistance disappears, behaves according to the laws of quantum mechanics. Two life sciences experiments were carried in the shuttle cabin to measure the effects of weightlessness on life; the first focused on the normal cycles in human bones, and the second focused on the production of lignin (a tough cellulose) in mung bean and pine seedlings.

The PDP was retrieved by the robot arm on the third mission day. It was then released so the shuttle could retreat to a point a few kilometers away and maneuver around it for six hours. Some of these activities required expert flying by the crew to place the shuttle and the PDP on the same magnetic field lines. Four of the eight in-orbit rocket firings planned for the plasma depletion experiments were canceled because of the fuel expenditure made during ascent.

STS 51-I was, perhaps, the most ambitious flight of the year. Plans for the mission included the deployment of three communications satellites. The first two were smaller HS-376 satellites, while the third was the rather large Syncom IV-4 (Synchronous Communications Satellite IV-4) set to roll out of *Discovery*'s payload bay on the third day of the mission. But the portion of the flight destined to get the most attention was the on-orbit repair of Leasat IV-3 (also known as Syncom IV-3). The satellite, a near duplicate of Syncom IV-4, had been deposited into orbit by the crew of STS 51-D four months earlier. The apparent failure of a triggering latch to activate resulted in the failure of the satellite's booster rocket to fire. Two astronauts on this flight would patch the electronics and send Syncom on its way.

On August 27, 1985, *Discovery* and its five-member crew were launched at 6:58:01 A.M. eastern daylight time from Pad 39A. The ascent was normal and *Discovery* was placed in a nearly circular orbit 306 kilometers above Earth.

The commander of the flight was Air Force Colonel Joe

Engle. He made his first trip into space as commander of STS-2 in 1981. Air Force Lieutenant Colonel Dick Covey, was making his first flight into space. The mission specialists on the flight were James D. van Hoften, Ph.D., John M. Lounge, and William F. Fisher, M.D. The only space-experienced mission specialist was van Hoften, who flew on STS 41-C and participated in the Solar Maximum Mission satellite repairs.

Six and a half hours after liftoff, the Aussat-1 (Australian Satellite-1) was deployed and later achieved its proper geosynchronous orbit with the aid of the Payload Assist Module, Delta class (PAM-D). A second satellite, American Satellite Company 1 (ASC-1), was deployed five hours later. The third satellite, Syncom IV-4, which is leased to the Department of Defense by its builder, the Hughes Co., was deployed as scheduled on flight day three. All three achieved geosynchronous orbit and became operational.

Discovery rendezvoused with the ailing Syncom IV-3 on day five of the mission, and it was grappled by the remote manipulator system arm. It was then lowered into position in the payload bay for the repairs. Astronauts van Hoften and Fisher performed two spacewalks (on days five and six) for a total of 11 hours and 27 minutes. During this time they replaced parts in the satellite needed to fire the satellite's booster rocket. After the activation lever was repaired, van Hoften grabbed the satellite and gave it a gentle spin while releasing it. Eventually, commands were sent to Leasat and it rode its booster rocket to the proper geosynchronous orbit.

The Physical Vapor Transport of Organic Solids (PVTOS) operated during the sleep periods on the mission. *Discovery* landed at Edwards Air Force Base in California, at 6:16 A.M., Pacific daylight time, September 3, 1985. The mission lasted 7 days, 2 hours, and 18 minutes.

Atlantis, the fourth of NASA's shuttle orbiters, was almost identical to its sister ship *Discovery*. It was named for the Woods Hole Oceanographic Institute research ship used from 1930 to 1966, which was the first American-operated vessel designed especially for oceanic research.

The commander of STS 51-J was Karol Bobko, a veteran of two previous shuttle flights. He was the pilot of *Challenger* for STS-6 and the commander of *Discovery* for STS 51-D. The remainder of the crew included Pilot Ronald Grabe, Mission Specialists David Hilmers and Robert Stewart, and Department of Defense (DOD) Payload Specialist William Pailes. Stewart was the only other veteran, having test-flown the manned maneuvering unit on STS 41- B.

The security surrounding STS 51-J was tighter than that surrounding STS 51-C; the DOD was not going to tolerate the leaking of any information regarding the payload. NASA imposed the same restrictions about flight details that it had on 51-C. The cargo for the flight was reported to be two military communications satellites known as Defense Satellite Communications System-3 (DSCS-3).

The twenty-first space shuttle mission was launched at 11:15:30 A.M., eastern daylight time, on October 3, 1985,

with very little notice. All systems worked as planned during the launch phase, and *Atlantis* was placed in a shuttle record-high orbit of 469 by 476 kilometers. Almost immediately, its secret payload was deployed. NASA remained silent about the progress of the flight until twenty-four hours before the planned landing. STS 51-J ended successfully on October 7, 1985, as *Atlantis* touched down on runway 23 at Edwards Air Force Base. The mission had lasted 4 days, 1 hour, 44 minutes, and 38 seconds.

The year 1985 was busy for shuttle operations, with nine manned missions flown. By the time STS 61-A took to the air, the media interest in reporting shuttle missions had dropped considerably; thus, this mission carrying the Spacelab D-1 payload was notable not only for the largest crew ever launched into space by one vehicle but also for the low-key coverage the general media devoted to the science mission.

The crew was commanded by shuttle veteran Henry W. Hartsfield, who had flown on the STS-4 in 1982 and STS 41-D in 1984. His pilot was Steven R. Nagel, who had flown on STS 51-G in 1985. Mission specialists for STS 61-A included Bonnie J. Dunbar, who had degrees in ceramic and biomedical engineering and who was on her first spaceflight, James F. Buchli, who had flown on STS 51-C earlier in 1985, and Guion S. Bluford, the first African American in space, who had flown on STS-8 in 1983. In addition, three rookie European payload specialists were members of the crew: Ernst W. Messerschmid and Reinhard Furrer, both from West Germany, and Wubbo J. Ockels, a Dutch national from the European Space Agency (ESA).

As with all Spacelab missions, the crew members were to alternate in twelve-hour shifts to operate the experiments in the Spacelab module. The blue shift was led by Nagel with Dunbar and Furrer; the red shift was led by Buchli with Bluford and Messerschmid. Hartsfield and Ockels were not assigned to a team and worked with either team as required. Buchli and Dunbar trained as contingency Extravehicular Activity (EVA) crew members.

Despite a few minor problems, the countdown for STS 61-A was one of the smoothest and most trouble-free of the program to date. With eight astronauts aboard, *Challenger* left the pad exactly on time, at noon, eastern standard time, on October 30, 1985. All stages of the ascent were nominal. Once orbit had been achieved, the payload bay doors were opened and a week of orbital science experiments began for the crew.

Three hours after launch, the crew floated into Spacelab to activate the experiments and equipment. Two hours later, this task was completed, and the control of payload operations was transferred from NASA Mission Control in Johnson Space Center, Houston, Texas, to the West German Space Operations Center in Oberpfaffenhofen, near Munich, a facility operated by the Deutsche Forschungs-und Versuchsanstalt für Luft-und Raumfahrt (German Federal Ministry of Research and Technology).

As soon as Spacelab was activated, the crew split into its two shifts, one team settling down for a sleep period, the other beginning the round-the-clock work in the science module. Once orbital operations had begun, the coverage of the mission began to decrease, an indication of how routine shuttle flights had become. Meanwhile, the crew successfully deployed the Global Low-Orbiting Message Relay (GLOMR) satellite, which had failed to deploy on STS 51-B five months before. During the week in space, seventy-three experiments out of a projected seventy-six were successfully activated, and their data were recorded.

Several of the experiments were conducted as forerunners to a planned Spacelab D-2 and the U.S. International Space Station, then planned for the early 1990's. One experiment by Ockels investigated a new sleep restraint designed to alleviate the sensation of floating in space and therefore disturbing sleep cycles. Tubes in the restraint were inflated to apply pressure to the body. More important was a demonstration of the capability of a crew of eight to work together in the confined environment of the Spacelab module and shuttle flight and middeck areas.

After a normal descent, *Challenger* landed on Runway 17 at Edwards Air Force Base just before 9:45 A.M., Pacific standard time, and began its rollout down the runway to a wheel-stop. During the 2,560-meter rollout, Hartsfield completed the last experiment of the flight with the nose wheel steering test. Crews of several previous missions had experienced difficulty in controlling the orbiter during the runway rollout and instrumentation had been fitted to *Challenger* for this flight to investigate the nose wheel steering inputs the pilot conducts during the rollout. Hartsfield successfully moved *Challenger* first to the left of the central line, then to the right, before moving back to the central line for a wheelstop.

Challenger had logged 7 days and 45 minutes in space and traveled more than 4 million kilometers in 111 orbits, landing during the 112th. Despite the signs of wear on its thermal protection system that were noticed after the landing of STS 61-A, mission planners were looking forward to seeing *Challenger* fly at least five more times in 1986.

On November 26, 1985, STS 61-B began with the lift-off, at 7:29 P.M., eastern standard time, of the orbiter *Atlantis*. During the week-long flight, the second one for *Atlantis*, the crew deployed three communications satellites and demonstrated the techniques needed for building large structures in space. The three satellites were the American Satcom 2, the Mexican Morelos 2, and the Australian Aussat-2.

Brewster H. Shaw commanded the mission. Bryan D. O'Connor was the pilot. Sherwood C. Spring, Jerry L. Ross, and Mary L. Cleave were mission specialists on the flight. They were accompanied by McDonnell Douglas engineer Charles D. Walker and Mexican engineer Rudolfo Neri Vela, payload specialists. Vela was present to observe the deployment of the Morelos communications satellite, to operate several Mexican-built medical experiments, and to photo-

graph areas of Mexico from space.

In addition to the trio of satellites, *Atlantis* carried several payloads in its crew compartment. Among these was the Continuous Flow Electrophoresis System (CFES), a commercial payload built by McDonnell Douglas. Electrophoresis is a process for separating cells using weak electrical charges. All living cells have a small negative charge on their surfaces; different types of cells have different charges, and in solution it is possible to separate them because of these differences. On Earth, such separation is difficult because the charges are extremely small and gravity causes sedimentation and convection. In the microgravity environment of orbital flight, however, such separation can be more easily accomplished. STS 61-B was the seventh flight to carry the CFES. Charles Walker, the payload specialist selected by McDonnell Douglas to operate the experiment, was on his third flight into space with the payload. Walker also operated a handheld protein growth experiment, a device to study the feasibility of crystallizing enzymes, hormones, and other proteins. Again, trying to crystallize such materials on Earth is extremely difficult. Nevertheless, successful crystallization permits the study of their three-dimensional atomic structure — important knowledge for enhancing or inhibiting certain functions of the proteins in the development of improved pharmaceuticals.

Other payloads in the crew compartment included the Diffusion Mixing of Organic Solution (DMOS), an experiment built by the 3M Corporation to try to grow large organic crystals for optical and electrical uses. Scientists at 3M wanted to see if larger, perfect crystals could be grown in space. Such crystals could be used for optical switches and computers that process information with light rather than electricity. This device also had cells to observe the mixing of fluids in weightlessness.

In the payload bay, *Atlantis* carried several attached payloads. One of these was an IMAX camera. IMAX, a Canadian large-format camera, had flown in the crew compartment on three previous space shuttle missions. The footage collected from these was used for a film first shown at the National Air and Space Museum in Washington, D.C. *Atlantis* also carried a small, self-contained payload, or Get-Away Special experiment, for Telesat of Canada. The result of a national competition among high school students in Canada, this experiment sought to fabricate better mirrors than those made on Earth by placing gold, silver, and aluminum coatings on quartz plates.

Atlantis flew a direct-ascent trajectory into its 352-kilometer-high initial orbit. The first satellite deployment of the mission, the Morelos 2, came only seven hours after launch, at 2:47 A.M., eastern standard time. About forty-five minutes later, the Payload Assist Module (PAM) attached to the satellite fired and moved it on a trajectory toward its eventual geosynchronous orbit. Later that same day, at 8:20 P.M., the crew released Aussat-2. The third and final satellite release of the mission, Satcom 2, occurred on Thursday, November 28.

After the satellite deployments, the crew prepared for the

next major mission activities, a pair of extravehicular activities (EVAs). During the spacewalks, Spring and Ross would build a 13.7-meter-tall tower and a 3.7-meter-wide tetrahedron in *Atlantis*'s payload bay. These two structures represented the culmination of nearly a decade of research on large space structures.

The tower, designed and fabricated by engineers at the Langley Research Center in Hampton, Virginia, was called Assembly Concept for Construction of Erectable Space Structures, or ACCESS. The other structure, developed jointly by the Marshall Space Flight Center and the Massachusetts Institute of Technology, was named Experimental Assembly of Structures in Extravehicular Activity, or EASE.

Each structure required a different assembly technique. EASE was a geometric structure resembling an inverted pyramid and comprised a few large beams and connecting nodes. ACCESS, by comparison, was a high-rise tower consisting of many small struts and nodes. ACCESS could be assembled from a fixed workstation in the payload bay, while EASE required the astronauts to move about the structure during assembly. Both were anchored on a special pallet which bridged the payload bay.

The first EVA began at 4:45 P.M. on November 29. For five and a half hours, Spring and Ross practiced assembling the structures. They first built the ACCESS tower. During preflight underwater simulations, they had taken an average of 58 minutes to build the tower. For the orbital EVA, mission planners had allotted two hours for the task. After only fifty-five minutes, however, they were finished. Spring and Ross disassembled the tower. They stowed the ACCESS components away and began working with EASE.

The pair assembled and disassembled the tetrahedron eight times during the first EVA. For the first four times, Ross acted as low man, handing beams to Spring and later putting them back in their storage rack. After the third assembly, Spring indicated that his hands were tired and his fingers were getting numb. By the fourth time, fatigue was beginning to set in, so Spring and Ross traded places for the remaining assemblies. Ross was supposed to perform only two assemblies from the upper position, but the process went fast enough that he had time for four. The first EVA provided fundamental information on space construction.

The second EVA, on December 1, explored specific space station assembly tasks and evaluated the use of the shuttle's manipulator arm in these operations. For the arm tests, the astronauts attached a portable foot restraint to the arm. Spring and Ross assembled the ACCESS tower. Then, while Ross stood on the foot-restraint platform, Cleave moved the arm to various work locations from inside *Atlantis*. While attached to the arm, Ross assembled one bay of the ACCESS tower's struts and nodes. Then, as Cleave maneuvered him along the tower, he attached a simulated electrical cable to its length, demonstrating a common space station assembly task. For the next test, Spring released the tower from its assembly jig in the payload bay, and Ross then maneuvered it by hand,

demonstrating manual movement and positioning of large space structures. After Ross put the ACCESS tower back in its jig, he and Spring exchanged positions on the end of the arm. Spring then practiced removing and replacing tower struts. He also moved the 13.7-meter tower by hand.

The astronauts then turned their attention to the EASE pyramid and assembled it while Spring was on the end of the arm, a different assembly technique from the one used during the first EVA. Ross then exchanged positions with Spring and tried moving the completed tetrahedron by hand. Following the conclusion of these tests, the astronauts disassembled the structure, stowed the components, and reentered *Atlantis*'s air lock. The EVA had lasted six and a half hours.

On December 3, the crew brought *Atlantis* back to Earth. Mission Commander Shaw landed *Atlantis* at Edwards Air Force Base, California, at 1:33 P.M., Pacific standard time. Following the landing, Shaw applied only light braking, allowing the orbiter to roll 3,279 meters.

Knowledge Gained

Although most of the mission objectives were achieved, the scientific results from Spacelab 2 were not all useful. Many of the HRTS images were fogged by overheating when the payload bay liner reflected more sunlight than expected. Nevertheless, scientists have called some of the images remarkable. They have discovered jets or explosive events in the solar corona, possibly where the solar magnetic field is perpendicular to the surface. Magnetic activity may also have been observed in superspicules, jets of hot gas rising to 15,000 kilometers above the solar surface and lasting three to five minutes. The HRTS also observed smaller spicules that match known spicules observed for some time in white light.

The CHASE instrument did not return useful data on helium ratios because of time constraints and because of internal instrument problems. Although the SOUP film was shot quickly, the experiment yielded more than six thousand striking, high-resolution images showing unusual evolution of granular structures. As the SOUP helped scientists discover, granules explode or break into bright rings that fade or they are destroyed by interaction with granules that have exploded. Also, the granules were found to be absent where solar magnetic activity is most intense.

The twin XRTs produced images of the center of the Galaxy and other stellar objects in a broader energy range than had previously been observed. The IRT provided useful data at short and long infrared wavelengths, including images of the center of the Galaxy. Data collected in the mid-range, though, were useless. The CRNE detected millions of low-energy events and some ten thousand high-energy events of interest. A few registered as high as 10 teraelectron volts, and a gamma-ray burst was detected by the secondary particles it created when it hit the detector shell.

Results from the plasma experiments were substantial. The most striking was the opening of a "hole" in the ionosphere through which the radio telescope at Hobart, Tasmania, could observe stars on wavelengths that normally are reflected by the electrified upper atmosphere. The "hole" was actually an area depleted of electrons and ions and thus transparent to those wavelengths for a few minutes, until the ionosphere regenerated itself. A similar "hole" was generated over New England and persisted for only fifteen minutes. The plasma wake left by the shuttle was found to be complex and turbulent. Thruster firings, water dumps, gas leaks, and other emissions from the shuttle generated a large cloud of neutral gas that expanded around the vehicle and altered the ionosphere. Water ions not normally present at the shuttle's orbiting altitude were detected in large quantities to a distance of several hundred meters from the shuttle, especially in the plasma wake. The PDP fly-around activity placed the shuttle and PDP on the same magnetic field line four times and showed that electrostatic noise (first detected on STS-3) extends far "downstream" from the shuttle but only a short distance "upstream."

STS 51-I successfully delivered its trio of satellites to their deployment spots and, once again, demonstrated the role that astronauts play in assuring mission success. A stuck sunshield covering the Aussat-1 satellite would have resulted in an aborted launch had a crew not been there to evaluate and correct the problem. More important, astronauts were required to accomplish the capture, repair, and redeployment of the Syncom IV-3.

The on-site service of Syncom IV-3 utilized extravehicular activity techniques developed on previous spacewalks. The astronauts were able to modify their procedures when required. The experience gained on the flight would be put to use on future repair missions and for the construction of the international space station.

A NASA experiment on STS 51-J called Bios, designed to study the damage to biological materials from high-energy cosmic rays, was positioned in the middeck area of *Atlantis*. A solid-state dosimeter was used to survey the interior of the orbiter to identify the areas most likely to be affected by such radiation. The experiments help to ensure that astronauts (and other biological passengers) are not exposed to potentially hazardous radiation from the Sun and other celestial bodies. In addition, the data are useful in the design of the space station and future space vehicles.

The Spacelab D-1/STS 61-A mission flew a scientific package consisting of seventy-six investigations involving fluid physics, solidification, biology, medicine, space-time interaction, GLOMR satellite deployment, flight test maneuvers during descent, and the steering wheel test during rollout. In the fields of biological and life sciences, the ground-based scientists were able to evaluate their data almost in real time, to confirm their findings and determine the success or failure of each experiment. The samples from the materials science experiments were evaluated over a longer period of several months after the conclusion of the mission.

From the materials science double rack with the isothermal heating facility, mirror heating facility, gradient heating facility with its quenching device, and fluid physics module, a total of 75 to 125 percent positive runs and stored data flows were obtained. From the process chamber with the holographic interferometric apparatus and the Marangini convection experiment in an open boat, a recorded level of 90 percent positive runs was achieved. Following early operational difficulties, from the Material science Experiment Double rack for Experiment modules and Apparatus (MEDEA) payload element carrying the monoellipsoid heating facility and the gradient furnace with a quenching device, a 110 percent success rate was achieved during the flight. From the life sciences experiments, success levels of 95 and 100 percent were recorded for the investigations. In addition, the vestibular sled, which was flown for the first time on this flight, achieved 120 percent test-run success. Among the navigation experiments, the clock synchronization and one-way distance measurement experiments were 100 percent successful.

In all, Spacelab D-1 provided a vast wealth of scientific data from experiments and investigations that had been years in the making; these data made planning for the D-2 mission much easier. The success of the data gathering of Spacelab D-1 was even more remarkable since only 40 percent transmission time could be achieved from the spacecraft to Earth because only one Tracking and Data-Relay Satellite was operational instead of the planned two.

The STS 61-B mission introduced a new satellite upper stage, the PAM-D2. All three satellites used the PAM, as had many of the satellites released during previous shuttle missions. The 61-B satellites, however, used an improved, more powerful version of the motor. PAM-D2 could propel satellites weighing 1,900 kilograms to geosynchronous orbit, while earlier PAM-D motors had only a 1,270-kilogram capacity. In fact, Satcom 2, with a weight of 1,860 kilograms, was the heaviest payload ever propelled by a PAM.

Nevertheless, the most significant results of the mission were from the two EVAs. During twelve hours outside *Atlantis*, Spring and Ross had demonstrated many of the techniques needed to build space stations and other large space structures. With ACCESS, they showed how to construct a long, thin tower structure from a fixed workstation with an assembly jig. EASE, on the other hand, required one of the crew to be free-floating as they assembled the large, pyramid-shaped structure.

During the second EVA, Spring, Ross, and Cleave showed that astronauts inside the spacecraft could work in concert with astronauts outside the vehicle. Cleave maneuvered the two EVA astronauts on the manipulator arm along the lengths of both EASE and ACCESS. She positioned her fellow crewmembers precisely at predetermined work locations. Working with Cleave, Ross attached a length of rope along the tower. This demonstration showed that it was possible to construct a structural framework in space, then route cables and electrical leads along it. Ross and Spring also showed that a spacesuit-clad astronaut could move large structures manually and could position them precisely by hand.

Context

Like Spacelab 1 and 3, Spacelab 2 advanced space science in several areas and demonstrated that the shuttle/Spacelab combination is an effective platform for conducting space science experiments. Observations made with the cluster of solar instruments marked the first time since Skylab's Apollo Telescope Mount in 1973-1974 that a manned solar observatory had been operated in space.

Despite the problems with the IPS, the data from the HRTS and SOUP telescopes were outstanding. Results from both experiments revealed details of solar activity that had been suspected but unobserved because Earth's atmosphere blocks the view. The images from the SOUP will provide a better understanding of the evolution and importance of granules (discovered only two centuries ago) in transporting energy to the solar surface. The HRTS images of super-spicules show that particular spicule phenomenon is larger than previously believed and plays a greater role in the transport of mass and energy from the solar surface into the transitional region where temperatures rise rapidly. A raster survey of 25 percent of the solar disk would help establish global properties of the fine structures of the solar surface. The SUSIM provided measurements of the solar ultraviolet output, with an accuracy of 6 to 10 percent. The SUSIM flown on Spacelab 2 was the beginning of a long-term program to collect data with respect to the influence of the Sun on the terrestrial environment, including the ozone hole. The value of these data will become known as SUSIM-type instruments are re-flown over the next few decades. X-ray and infrared images of the skies, and cosmic-ray data, are adding to scientists' understanding of celestial objects.

The shuttle's utility as a scientific platform was demonstrated with mixed results. The "abort to orbit" demonstrated a need for performance margins for experiments and for less intense mission planning. After Spacelab 2, infrared observations aboard the shuttle were perceived as risky at best, based on the IRT results. The plasma experiments, however, provided a wealth of data, as scientists were able to disturb the plasma environment in a controlled way and observe the effects with a nearby craft. Although the scientific results were generally good, the operation of the setup would need refining.

The capability to perform on-orbit repairs to the shuttle, its payloads and previously deployed payloads had been demonstrated on flights prior to 51-I. The Syncom IV-3 repair mission built on this experience and refined the techniques that would be used in later flights. It also gave NASA confidence in the ability of astronauts to work with large orbiting objects, carrying large amounts of explosive propellants. If a space station was in NASA's future, it would have to be constructed by astronauts working in close proximity to extremely large and potentially dangerous objects.

Atlantis came back from the STS 51-J mission virtually unscathed, a far cry from the various damaged parts and non-working systems the others brought back. More than anything, experience gained from the previous twenty missions contributed to the success of *Atlantis*. Major changes included improved construction materials and electronic hardware, lighter thermal protection systems, and provisions for carrying a Centaur high-energy booster. The Centaur, a liquid propellant vehicle, was the upper stage of the Atlas and Delta launch vehicles. It was to be used to boost several large deep space probes from the shuttle, including the Galileo probe to Jupiter. Like the orbiter's main engines, Centaur used liquid hydrogen and liquid oxygen for propellants. After the *Challenger* accident, the Centaur program was canceled, because the vehicle was considered too hazardous to be carried in the shuttle's payload bay.

The flight of Spacelab D-1 on STS 61-A provided a valuable link for the Americans between the early biomedical and materials experiments carried out on the earlier Apollo and Skylab missions in the 1970's and the planned flights on the space station in the 1990's. The Spacelab D-1/STS 61-A mission continued the scientific investigations carried out on the Spacelab 1 and 3 long module missions in 1983 and earlier in 1985.

From the human point of view, the flight of a crew of mixed sexes, races, and nationalities pointed the way to international cooperation on the space station and talks on the need of a united program of exploration of Mars in the next century. The effective use of the confined habitable quarters of the shuttle during D-1 allowed spacecraft designers to determine the most efficient, pleasing, and functional interior designs of the space station and future spacecraft. The Russians have gone a long way toward spacecraft habitability with their Salyut and Mir series. As crews increase in size and missions increase in duration, with the added complications of mixed sexes, religions, and races, suitable internal designs of spacecraft are an important element in planning. Spacelab D-1 represented a milestone in this ongoing aspect of the space program.

STS 61-B was the twenty-third flight of the space shuttle and the fifty-fourth American manned spaceflight. It came at a time when NASA managers were selecting the configuration for the International Space Station, which they planned for the mid-1990's. Validating the concept of EVA construction by in-flight experience was important as they made their decisions.

The EVAs were the culmination of nearly a decade's development and testing. During this period of hardware development, personnel aboard shuttle missions were demonstrating an amazing capability for EVA operations. The first space shuttle EVA was performed during STS-6 in April, 1983. One year later, astronauts repaired the ailing Solar Maximum Mission satellite during the STS 41-C mission. In November, 1984, two communications satellites that had been placed in incorrect orbits were retrieved and returned to Earth for refurbishment and relaunch. In August, 1985, the Leasat 3 satellite, which failed just after being released from the orbiter *Discovery* some four months earlier, was jumpstarted in space. Thus, by the time the EASE and ACCESS experiments flew on *Atlantis*, American astronauts had considerable experience in on-orbit satellite servicing and retrieval. The STS 61-B mission added experience with in-space construction.

Bibliography

Furniss, Tim. *Manned Spaceflight Log.* Rev. ed. London: Jane's Publishing Co., 1986. This updated version of the 1983 first edition covers in a minihistory the world's manned spaceflights in chronological order, from Yuri Gagarin's historic first flight in April, 1961, to the *Challenger* accident and the triumph of Mir in 1986, twenty-five years later. Presented in launch order, 115 manned spaceflights from the United States and the Soviet Union are described, along with thirteen X-15 Astro flights of the American research aircraft of the 1960's.

———, ed. *Space Shuttle Log.* London: Jane's Publishing Co., 1986. A collection of highly readable reports on the first twenty-two shuttle flights, from April, 1981, to the *Challenger* mission of October/November, 1985. The text provides a useful mission summary and data on each flight in sequence, as well as a collection of biographical sketches of shuttle astronauts up to 1985. Suitable for general readers.

Gurney, Gene, and Jeff Forte. *Space Shuttle Log: The First Twenty-five Flights.* Blue Ridge Summit, Pa.: Tab Books, 1988. A collection of chapters summarizing the first twenty-five missions of the space shuttle program, covering the operational flights from April, 1981, to the loss of *Challenger* in January, 1986. The entry on each mission covers data on crew and payload and flight records; the main text describes launch preparations and experiments and investigations on each flight. A collection of black-and-white photos from all the missions accompanies the text.

Kerrod, Robin. *Space Shuttle.* New York: Gallery Books, 1984. Most valuable for its beautiful color photographs of the space shuttle, this volume conveys the essence of the Space Transportation System vehicles and the people who fly them. Highlights of the first dozen missions are presented, as well as a fanciful look at the future of space exploration.

National Aeronautics and Space Administration. Marshall Space Flight Center. *Spacelab 2.* NASA EP-217. Washington, D.C.: Government Printing Office, 1985. This educational publication describes the instruments aboard Spacelab 2 and the planned scientific experiments. Written for reporters covering the mission.

Otto, Dixon P. *On Orbit: Bringing on the Space Shuttle.* Athens, Ohio: Main Stage Publications, 1986. Early designs of the space shuttle are discussed. Also presents an account of each of the first twenty-five flights, including the crew, the payloads, and the flight objectives. Illustrated in black and white.

Shayler, David J. *Shuttle* Challenger: *Aviation Fact File.* London: Salamander Books, 1987. A book devoted to the career and achievements of space shuttle orbiter OV-099, *Challenger*. This comprehensive text covers the role of the *Challenger* in the shuttle program, the construction and components of the vehicle, and its missions — including an account of the STS 51-L accident and summaries of all the astronauts and payloads *Challenger* carried into space on its ten missions. A selection of tables logging accumulated time and hardware data completes the work. Includes a selection of color photographs. A large-format book, this commemorative work on *Challenger* is aimed at a general readership.

Smith, David H., and Thornton L. Page. "Spacelab 2: Science in Orbit." *Sky and Telescope* 72 (November, 1986): 438-445. An extensive survey of scientific results from the Spacelab 2 mission. Well written and well illustrated with color photographs and charts. For the educated reader interested in astronomy.

Smith, Melvyn. *An Illustrated History of Space Shuttle: X-15 to Orbiter.* Newbury Park, Calif.: Haynes Publications, 1986. A concise overview of the space shuttle and the experimental aircraft that led to its design. Spanning the period from 1959 through 1985, the book was written for the general reader. Illustrated with many photographs of the early lifting bodies, which help to show how the shuttle orbiter came to look as it does today. Arranged chronologically.

Wilson, Andrew. *Space Shuttle Story.* New York: Crescent Books, 1986. Traces the history of the space shuttle from the early days of rocketry to the *Challenger* accident. Furnished with more than one hundred color photographs, this volume provides little detail but emphasizes the men and women who fly the spaceplane to and from orbit.

Yenne, Bill. *The Astronauts.* New York: Exeter Books, 1986. Presents an overview of the Soviet and American space programs and tells of the international passengers carried on various missions. Illustrated with several hundred photographs taken in both countries.

Dave Dooling (STS 51-F)
Dennis Chamberland (STS 51-I)
Russell R. Tobias (STS 51-J)
David J. Shayler (STS 61-A)

Space Shuttle Mission
STS 51-I

Date: August 27 to September 3, 1985
Type of mission: Manned Earth-orbiting spaceflight

STS 51-I was the twentieth flight of the United States' space shuttle program. In addition to deploying three communications satellites, crew members captured a malfunctioning satellite (which had been launched by a previous shuttle), repaired it successfully, and returned it to orbit. It was the second such in-orbit satellite repair in history.

Principal personages

A. D. Aldrich, Manager, Space Transportation System
Joe H. Engle, Mission Commander
Richard O. Covey, Pilot
James D. van Hoften,
John M. Lounge, and
William F. Fisher, Mission Specialists

Summary of the Mission

STS mission 51-I marked the sixth spaceflight of the orbiter *Discovery*. The seven-day mission proved the extraordinary versatility of the space shuttle as a manned delivery and repair platform.

The first launch date scheduled for mission 51-I was August 24, 1985, from Kennedy Space Center's Launch Complex 39A. The launch was rescheduled for the next day because of thunderstorms. During the second attempt, on August 25, 1985, an on-board computer malfunctioned during the countdown. The malfunction was described by National Aeronautics and Space Administration (NASA) engineers as a "GPC 5 byte fault." GPC 5 is general purpose computer number 5, located inside the space shuttle orbiter. This computer contained backup flight system software essential to the shuttle's launch. Engineers reinitialized the software after the error was found, but the error appeared again only 11 minutes later. The flight was postponed again, this time for two days, so GPC 5 could be removed and replaced.

The third countdown, on August 27, 1985, proceeded smoothly. *Discovery* and its five-member crew were launched without significant delays at 6:58:01 A.M., eastern daylight time, from Pad 39A. The launch proceeded normally in all respects. The solid-fueled rocket boosters separated without any problems 2 minutes and 1 second after lift-off. The three main engines were shut down 6 minutes and 27 seconds later, 18 seconds before the large external tank was jettisoned. (It later burned in the atmosphere over the Indian Ocean.)

For many missions it is necessary to fire the Orbital Maneuvering System (OMS) rocket engines twice to refine the orbital parameters. In this mission, however, the ship was flown in an ascent mode called a "direct insertion ascent trajectory," which precluded the necessity to initiate the first scheduled OMS firing. Exactly 40 minutes and 21 seconds after lift-off, the OMS engines were fired for 3 minutes and 3 seconds, placing the orbiter in a nearly circular orbit approximately 306 by 306 kilometers above Earth. At 10:41 A.M. on August 27, 1985, *Discovery* was safely in orbit.

Immediately following the OMS burn that placed *Discovery* in its orbit, the orbiter's payload bay doors were opened, exposing the satellites in the bay. At 2 hours and 2 minutes after lift-off, the sunshield covering the Australian satellite (Aussat) in the payload bay was commanded to open so that the satellite's systems could be checked prior to deployment from *Discovery*. The sunshield, however, did not fully open. It was determined that the shield was probably binding on an antenna bracket located on top of the satellite. Two hours after the problem was discovered, ground engineers authorized the crew to use the Remote Manipulator System (RMS), sometimes called the robot arm or the Canada arm, to help push the sunshield open and expose the satellite.

During these operations, the RMS "elbow joint" did not respond to computer commands. Fortunately, the arm had a backup system, and that backup system was used for the remainder of the mission. Yet the failure caused the cancellation of some operations involving use of the RMS-mounted video cameras, such as the monitoring of satellite engine burns and a wastewater dump from the orbiter.

Aussat was finally deployed at 6.5 hours into the mission. Three days later, the satellite reached its station in orbit 35,800 kilometers above Earth, propelled by an engine attached to it called the Payload Assist Module, Delta class (PAM-D). Aussat is used to provide communications relay services for Australia and its offshore islands. A second satellite, American Satellite Company 1 (ASC 1), was deployed at 11 hours and 9 minutes into the mission. Its deployment was successful, and ASC 1 reached its orbit on August 31. ASC 1 is a communications satellite for American business and government agencies.

The second day on board *Discovery* was much more relaxed than the first. It was spent performing experiments with an experimental package called PVTOS, for Physical Vapor Transport of Organic Solids, a package sponsored by the 3M Company and designed to collect data from chemistry experiments conducted in the weightless environment of space. The second day was also used to check out the third satellite still in the payload bay and prepare it for deployment on the third day.

The third satellite, synchronous communications satellite IV-4 (Syncom 4, also called Leasat 4), was deployed as scheduled on August 30. It reached geosynchronous orbit, at 35,800 kilometers in altitude, successfully. At such an altitude, satellites rotate at the same speed as Earth so that they appear to remain stationary in the sky. Unfortunately, for unknown reasons, all communications with the satellite were later lost.

Meanwhile, *Discovery* was effecting orbital corrections to rendezvous with Leasat 3, which had been launched by STS 51-D (also from *Discovery*) some four and one-half months earlier. Leasat 3's booster rocket, which would have placed it in a high, geosynchronous orbit, had failed. The *Discovery* crew planned to maneuver the malfunctioning satellite into the payload bay and repair it.

The fourth day in space was spent preparing for the encounter with Leasat 3. Two rendezvous maneuvers were performed while the crew members tested their Extravehicular Mobility Units (EMUs, or spacesuits). That included charging their batteries and checking out the Remote Power Unit (RPU), whose batteries had been charged on the second day. They would use the RPU to repair the satellite.

The following day, *Discovery* maneuvered to within a few meters of the ailing satellite. Crew members William Fisher, a physician, and James van Hoften, a researcher, exited the orbiter by way of the air lock into the payload bay. With the help of the RMS and some muscle, they captured the satellite and locked it into place in the payload bay to begin the long task of repairing it. Fisher and van Hoften's Extravehicular Activity (EVA) set a record for the longest spacewalk of the shuttle program: 7 hours and 10 minutes. During that time, Fisher and van Hoften worked to repair the satellite by replacing the parts needed to fire the satellite's booster rocket.

The sixth day was used to finish the repair work on the satellite in an EVA lasting 4 hours and 20 minutes. The crew members reentered the *Discovery*, and Leasat 3 was deployed. Days later, ground controllers successfully fired its troublesome booster rocket; the satellite attained the proper geosynchronous orbit and began normal service.

The seventh day of the flight was used to prepare for reentry. During this time, the crew members pressurized the cabin to sea level pressure. They tested the forward thrusters and the flight control systems using an Auxiliary Power Unit (APU), which provides the flight control systems with power during reentry. They also dumped wastewater into space and stowed all loose items in the cabin for reentry.

Early on the eighth day of flight, the payload bay doors were closed. The crew fired their OMS engines for 4 minutes and 9 seconds to reduce the orbiter's speed. Thirty minutes later, *Discovery* had descended from 305 kilometers to 126 kilometers, where it encountered the "atmospheric interface," or the upper, relatively dense portion of Earth's atmosphere. *Discovery* reentered Earth's atmosphere at twenty-five times the speed of sound (Mach 25). Thirty minutes after reaching the atmosphere, the spacecraft touched down at Edwards Air Force Base in California, at 9:16 A.M., September 3, 1985. The mission had lasted 7 days, 2 hours, and 18 minutes.

Knowledge Gained

The success of STS 51-I underscores many of the general aims of the United States' space shuttle program. The system delivered multiple large satellites to orbit. As a manned system, it was able to correct relatively simple payload problems in space (the failure of the Aussat sunshield to open, for example) that probably would have resulted in the loss of an unmanned payload. Its own systematic problem (the failure of an RMS mode) was overcome because crew members were able to evaluate the situation. Also, perhaps most important, *Discovery* was able to perform an in-orbit repair.

The RMS system was used in a unique way on this mission, to assist in the capture and stowing of an in-orbit satellite. Since the RMS required the help of a backup system, the crew expanded the knowledge of RMS capabilities and just how far the RMS could be pushed beyond its design.

The capture and repair of Leasat 3 incorporated a body of knowledge into the Space Transportation System that would be used in future repair missions and even eventual space construction. In the weightlessness of space, the crew members were able to maneuver the massive satellite and its attached PAM-D booster, weighing thousands of kilograms on Earth, into the spacecraft's restraints in the cargo bay. They were then able to anchor it into place for the repair work, releasing it later for boosting into its final orbit. Using the crew's experience, researchers would be better equipped to design tools for the most efficient methods of construction in weightlessness.

The repair provided valuable knowledge about bypassing complex electronic systems with alternate circuits and externally modified systems. It required assessing the problem from ground telemetry, working up a probable scenario of the circuitry involved, and designing a system to bypass the troubled circuits. Ground researchers had accomplished this over the span of a few months, and the astronauts were trained and sent into orbit to effect the repairs. All these activities provided a baseline of experience and knowledge that could be used repeatedly as a successful example of how such in-orbit repair missions could be effected in the future.

The seven-hour EVA set a very important precedent for work activities in space. It proved the ability of man to work in space for long periods and established that spacesuits and life-support systems are functional under extremely rigorous conditions.

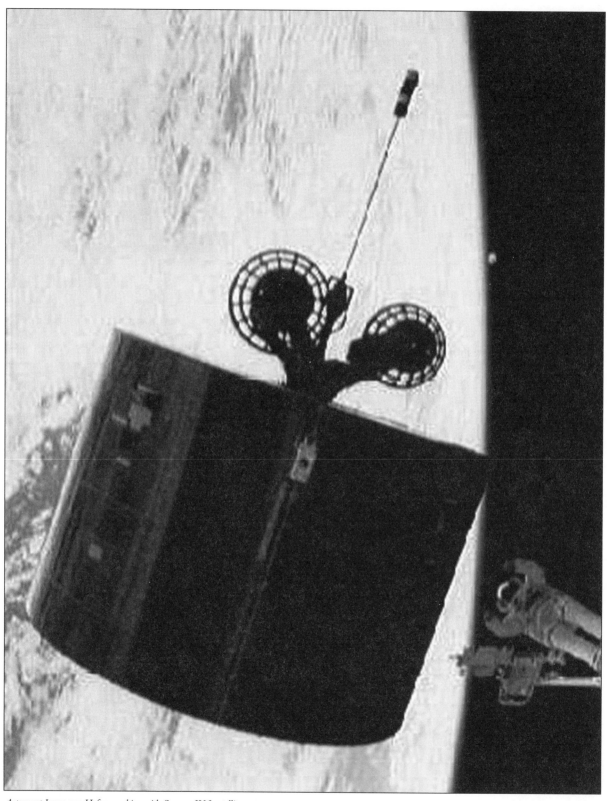

Astronaut James van Hoften working with Syncom IV-3 satellite. (NASA)

The PVTOS experiment provided knowledge of the transport of organic solids by vaporizing organic materials in what were called "reactor cells" within the experimental package. Data were obtained and stored in a special computer storage system that was a part of the PVTOS package itself. The exact data and parameters obtained were returned to the 3M Company as proprietary information; the data concerned chemical reactions that can only be performed in weightlessness.

Context

Prior to this *Discovery* mission, deployment of satellites from the shuttle system had become, for all practical purposes, commonplace. Twenty-four satellites had been deployed from the shuttle on previous missions. All deployments had been successful; yet, after leaving the shuttle payload bay, several had malfunctioned in orbit — which was not the fault of the shuttle delivery system.

The in-orbit repair of satellites, however, was not common at all. Although the single previous attempt, STS 41-C, had been successful, such repair missions incorporated many uncertainties. For example, the exact cause of the malfunctioning spacecraft could only be narrowed down to a list of possibilities, since the system could not be examined directly. Then the satellite engineers were required to manufacture a solution to cover the entire range of possibilities, plan how these could be installed in space by the crew, and assist the mission planners in training the astronauts to execute the repairs. The pilot and commander of the shuttle, meanwhile, were required to train in rendezvous maneuvers in simulators while other mission specialists trained in the use of the RMS, which would help maneuver the satellite into the payload bay. All of these were mere contingencies; the spectrum of the training program would also have to cover any unplanned events.

NASA engineers had been designing a system whereby an Orbital Maneuvering Vehicle (OMV) would ascend to higher orbits, retrieve malfunctioning spacecraft, and transport them down for repair in lower orbits. Yet that system was not available to mission 51-I.

As STS 51-I flew, the Soviet Union was at least two and one-half years away from the first launch of their shuttle system, and the Europeans had hardly released word of their planned shuttle-type system, Hermes. Hence, the United States was the only nation at the time to have such capability of launching and repairing spacecraft in orbit from a manned vehicle.

Unfortunately, the United States' shuttle system would fly only four more times before the tragic loss of the orbiter *Challenger*. Not only would the United States lose its lead in the operation of shuttle-type systems, but the program itself would emerge fundamentally changed after that paralyzing tragedy.

Bibliography

Allen, Joseph P., and Russell Martin. *Entering Space: An Astronaut's Odyssey*. New York: Stewart, Tabori and Chang, 1984. A copiously documented volume depicting the United States' space shuttle program. It describes the shuttle's movements from processing at the Kennedy Space Center to launch, orbital activities, and landing. The book is one of the most beautifully photographed and illustrated of all the books on the shuttle program.

Clarke, Arthur C. *Ascent to Orbit*. New York: John Wiley and Sons, 1984. This work is a compilation of many of Clarke's works from his early material (1930's) to his work of the mid-1980's. It is most effective in presenting the "history of conception" of space systems, from communications satellites to the distant future of space exploration. A mixture of technical and purely entertaining essays that can be appreciated by most readers with any interest in the space sciences. Illustrated.

Joels, Kerry M., Gregory P. Kennedy, and David Larkin. *The Space Shuttle Operator's Manual*. New York: Ballantine Books, 1982. This manual is a detailed space shuttle system reference work. It gives specifics on the space shuttle system, from weights and sizes to operational characteristics. It enumerates countdown procedures, emergency instructions, and standard operational modes. Written for a general audience, this book serves as an entertaining as well as an educational reference work. Illustrated.

Nova: Adventures in Science. Reading, Mass.: Addison-Wesley Publishing Co., 1982. A collection of essays and photographs from the public television series *Nova*. Includes several essays on space exploration, with details on the United States' space shuttle system. The book discusses the role of science in daily life, making it an especially valuable tool for referencing the space program and the shuttle's link with everyday existence. It is aimed toward the general reader.

O'Neill, Gerard K. *Two Thousand and Eighty-one: A Hopeful View of the Human Future*. New York: Simon & Schuster, 1982. Princeton physicist, founder of the Space Studies Institute, and "father" of the space colony, Gerard K. O'Neill has pieced together a thoughtful look at the year 2081. The book speculates on future shuttle systems in an insightful way that reveals the grand vision of today's missions. Illustrated and directed toward the general audience with an interest in the future and in space exploration.

Dennis Chamberland

SPACE SHUTTLE MISSION
STS 61-C

Date: January 11 to January 18, 1986
Type of mission: Manned Earth-orbiting spaceflight

During STS 61-C, the twenty-fourth flight of the space shuttle, astronauts aboard Columbia *launched a commercial communications satellite, tested a new payload carrier system, and photographed Halley's comet.*

PRINCIPAL PERSONAGES

ROBERT L. "HOOT" GIBSON, Mission Commander
CHARLES F. BOLDEN, JR., Pilot
GEORGE D. "PINKY" NELSON,
STEVEN A. HAWLEY, and
FRANKLIN R. CHANG-DIAZ, Mission Specialists
ROBERT J. CENKER and
C. WILLIAM "BILL" NELSON, Payload Specialists

Summary of the Mission

On January 11, 1986, Space Transportation System (STS) mission 61-C began with the predawn lift-off of the space shuttle *Columbia*. The 6:55 A.M., eastern standard time, launch was the twenty-fourth space shuttle mission and the seventeenth flight for *Columbia*. In its payload bay, *Columbia* carried the RCA Satcom K-1, thirteen Get-Away Special cannisters, the Hitchhiker payload carrier, a materials science laboratory, and an infrared imaging experiment. Inside the crew compartment, the astronauts operated the initial blood-storage experiment, the Comet Halley Active Monitoring Program, and three Shuttle Student Involvement Program experiments. *Columbia* also carried special flight instrumentation to determine more precisely orbiter aerodynamic and reentry heating characteristics.

Veteran astronaut Robert L. "Hoot" Gibson commanded *Columbia*. This was his second trip into space. The STS 61-C pilot was Charles F. Bolden, Jr., making his first spaceflight. The crew also included three mission specialists: Franklin R. Chang-Diaz, Steven A. Hawley, and George D. "Pinky" Nelson. RCA engineer Robert J. Cenker and Florida congressman C. William Nelson, served as payload specialists.

STS 61-C was the first mission for *Columbia* since the STS 9 flight in late 1983. Following that mission, National Aeronautics and Space Administration (NASA) managers returned *Columbia* to Rockwell International in Palmdale, California, for an eighteen-month overhaul. The hundreds of changes made to the first orbiter to fly in space included updating its navigation system, adding a cylindrical housing to its vertical stabilizer, and building a new nose cap to house the Shuttle Entry Air Data System (SEADS). For its first flight in two years, *Columbia* also carried instrumentation to sample air at its surface in the upper atmosphere and pressure transducers on the top and bottom sides of the wings to determine wing loading during ascent and reentry. *Columbia* originally had 90 wing load sensors. During its overhaul, engineers added 200 more.

The RCA Satcom was the only deployable payload aboard *Columbia* for the STS 61-C mission. The satellite cost $50 million, and RCA paid NASA $14.2 million to launch it from the shuttle. It was a Ku-band communications satellite to provide voice, television, facsimile, and data services to commercial customers throughout the forty-eight states. As with all earlier communications satellites carried aboard space shuttles, the RCA Satcom was attached to a booster motor that would propel it from low-Earth orbit to a higher, geosynchronous orbit. The motor attached to the Satcom was called the Payload Assist Module D2 (PAM-D2). Satellites that orbit at an altitude of 36,000 kilometers are in a geosynchronous orbit; that is, at that altitude, it takes twenty-four hours to complete one orbit. Thus, a satellite orbiting 36,000 kilometers above the equator will remain fixed in space with respect to an observer on the ground. Ground stations can receive broadcasts from geosynchronous satellites with fixed antennae.

This was the second of three planned vehicles for the RCA American Domestic Satellite System. The first RCA Satcom had been launched during the STS 61-B mission. RCA Satcom is a version of the RCA 4000 series of three-axis stabilized satellites. It carries sixteen operational transponders and six spares, each with an output of 45 watts. These are powerful enough to permit ground stations to receive their transmissions with antennae as small as one meter in diameter. The RCA Satcom system can provide direct-to-home television program distribution and television service to hotels, apartment houses, and other large institutions.

Early in the space shuttle program, NASA created the Get-Away Special (GAS) program. Get-Away Specials are small, self-contained payloads, carried in the orbiter's payload bay, which may be flown in space at a cost of as little as

$3,000. GAS experiments must be entirely self-contained. That is, they must have their own power and data-recording systems. All an astronaut will normally do with a GAS experiment is turn it on and off during the flight. They are flown on a space-available basis and are accessible to private individuals, foreign governments, and corporations. NASA provides standardized GAS containers for mounting in the payload bay. The containers are about 85 centimeters tall and 50 centimeters wide.

STS 61-C was the maiden flight of a new piece of GAS support equipment: the GAS bridge. The GAS bridge was an aluminum structure which spanned the width of the shuttle's cargo bay and could accommodate up to twelve GAS containers. A thirteenth GAS cannister was attached to the inside wall of the cargo bay near the GAS bridge. It contained instrumentation to measure the environment of the bridge during launch and landing. Prior to this flight, all GAS cannisters had been attached to the inside wall of the cargo bay. By the time Columbia flew the first space shuttle mission, NASA had sold more than two hundred GAS reservations. At the time of the STS 61-C flight, the backlog was even greater. Engineers at the Goddard Space Flight Center in Greenbelt, Maryland, devised the GAS bridge as a means of carrying large numbers of GAS payloads on individual shuttle missions to reduce the backlog.

Hitchhiker, a new payload carrier system, was also tested on STS 61-C. Like the GAS experiments, Hitchhiker was devised as a method of providing researchers with rapid and economical access to space. Hitchhiker can support scientific, technological, and commercial payloads. It has limited instrument pointing and data processing capabilities. Developed as a payload-of-opportunity carrier, Hitchhiker uses cargo space remaining after the space shuttle's primary payload has been accommodated. Unlike the autonomous GAS cannisters, Hitchhiker payloads are connected to the orbiter's communications and power systems. Communications with the payload are provided through a payload operations control center at Goddard, enabling real-time customer interaction and control.

NASA developed two separate Hitchhiker systems. Engineers at Goddard developed Hitchhiker-G, the type flown on STS 61-C. It was mounted on the front wall of the orbiter payload bay and could accommodate up to four experiments with a combined weight of up to 340 kilograms. The other system, Hitchhiker-M, was developed at the Marshall Space Flight Center in Huntsville, Alabama. It was a structure similar to the GAS bridge that could carry payloads heavier than those carried by Hitchhiker-G. Hitchhiker was created to support payloads too large for GAS and too small for the Spacelab carrier.

U.S. congressman Bill Nelson accompanied the crew as a payload specialist and congressional observer. He was the chairman of the House of Representatives' Space Science and Applications Subcommittee. Nelson represented the Eleventh Congressional District in Florida. During the mission, he operated the handheld protein crystal growth experiment. This experiment sought to use the weightless environment of space to produce protein crystals of sufficient size and quality to allow their nature and structure to be analyzed. Nelson also participated in detailed studies for NASA's Biomedical Research Institute. These studies provided additional data on the effects of spaceflight on the human body.

The flight of STS 61-C was canceled four times before the Columbia was launched on January 11, 1986. During the early part of the ascent, cockpit instruments indicated that one of Columbia's engines had a helium leak. The situation was serious enough to threaten a shutdown of one of the orbiter's three main engines. If such a malfunction had occurred at that point in the flight, the mission would have been aborted. Pilot Bolden took immediate action to correct the problem, and the mission proceeded according to schedule. Thirty seconds after lift-off, Columbia entered the area of maximum dynamic pressure, or "max Q." (This pressure is the product of air density times velocity squared.) After reaching a maximum during the first minute of flight, the aerodynamic forces on the vehicle decreased as the shuttle climbed higher.

The ascent profile flown by Columbia was deliberately selected to place greater stresses than ever before on the vehicle. Because actual stresses on the craft are greater than what is predicted based on wind-tunnel testing and other experiments, launch profiles that were less stressful (and therefore less capable of testing the shuttle's full payload capability) had been flown on past missions. STS 61-C was one of three flights planned to collect data that would explain the difference between actual flight results and wind-tunnel predictions. Once acquired, the new information could lead to a relaxation of ascent load constraints, allowing space shuttles to carry heavier payloads.

After the main engines finished their nine-minute burn, and after two firings by the Orbital Maneuvering System (OMS) engines, Columbia was in a 323-kilometer-high circular orbit. Nine hours after launch, during the seventh orbit, the crew opened the sunshield, which protected the Satcom in the cargo bay, and released the satellite from its launch cradle. Forty-five minutes later, the PAM-D2 motor fired and placed the satellite into a highly elliptical orbit which took it to 36,000 kilometers. On January 15, another rocket motor contained in Satcom fired and circularized the orbit at geosynchronous altitude.

During their first day in space, the crew also activated the Material Science Laboratory 2 (MSL 2). MSL 2 comprised three experiments in the cargo bay to study the behavior of materials in microgravity. Two of the experiments studied how melted materials solidify; the third observed liquid behavior in zero gravity. MSL 2 experiments continued throughout the mission.

The astronauts observed Halley's comet on the second day of the mission. The equipment used for this experiment, called the Comet Halley Active Monitoring Program, included a 35-millimeter handheld camera system provided by the University of Colorado. For this experiment, crew members photographed the comet using standard filters to obtain images and spectra. Unfortunately, an intensifier that boosted the light-gathering power of the camera malfunctioned, so the experiment returned only very limited results.

Another payload, an infrared imaging experiment, was operated by Cenker, the RCA payload specialist. Developed by RCA, this was one of two experiments aboard STS 61C that supported the Strategic Defense Initiative (SDI) and future space-borne surveillance systems. This experiment was mounted on the aft wall of the cargo bay. As he operated the setup, Cenker observed aircraft to measure their infrared signatures. The exact location and types of aircraft were classified. Cenker also used the system, which has possible applications for civilian remote-sensing systems, to observe such unclassified targets as cities and volcanoes.

The other SDI-related payload was carried by Hitchhiker. Developed by the United States Air Force, it was called the Particle-Analysis Camera System (PACS.) It comprised two 35-millimeter cameras and a strobe light to take photographs every 120 seconds, recording the amount and type of floating debris surrounding the space shuttle orbiter.

The STS 61-C mission was scheduled to land at the Kennedy Space Center (KSC) in Florida. This was the first landing scheduled for KSC since the STS 51-D mission in April, 1985. During that flight's landing, the orbiter *Discovery*'s right main landing tire experienced a blowout. After that landing, orbiters had landed on the dry lakebed of the Dryden Flight Research Center until mission 61-B. The STS 61-B flight concluded with a landing on the paved runway at Dryden. Following this successful landing, NASA managers opted for the shuttle to land at KSC. The weather at KSC, however, prevented a landing there on either January 16 or 17, so *Columbia* returned to Dryden on January 18 instead. The duration of the mission was 6 days, 2 hours, and 4 minutes. During reentry and atmospheric flight, infrared sensors in a housing on top of *Columbia*'s stabilizer measured the temperatures on the craft's upper surfaces. Other instruments in *Columbia* studied the composition of the upper atmosphere and provided precise measurements of the craft's flight attitude.

Knowledge Gained

The six-day STS 61-C mission demonstrated the utility and versatility of the space shuttle. On a single mission, NASA flew a diverse group of payloads. Major crew activities during the mission included deploying one commercial satellite, testing new payload support equipment, evaluating new space-based Earth imaging systems, and conducting materials processing experiments in space. *Columbia* also carried instrumentation that provided high-resolution infrared images of the top of the orbiter's left wing to create detailed maps of aerodynamic heating during reentry.

The one deployable payload was the RCA Satcom K-1. This was the second of three Ku-band communications satellites. Most previous communications had operated in the C-band frequency range, which can interfere with terrestrial microwave systems. Because the Ku-band frequencies are not shared with microwave traffic, antennae served by the RCA Satcoms can be located inside major metropolitan areas. Also, most C-band satellite transponders emit a signal strength of 12 to 30 watts. The RCA Satcom transponders transmit 45 watts of power. This makes direct reception from the satellites possible with antennae of less than 1 meter in diameter.

The Hitchhiker payload system, first demonstrated on this mission, promised to provide researchers with low-cost and rapid access to space. From its inception, the system was designed for simplicity and economy. It incorporated such features as standardized interfaces with orbiter systems and reusability to reduce hardware costs. In addition, with the introduction of the Hitchhiker, NASA reduced the level of paperwork and documentation normally required for shuttle payloads.

Another new piece of payload support hardware tested on this flight was the GAS bridge. Engineers at Goddard devised the GAS program as a means of providing researchers access to space at the lowest possible cost. Through this program, individuals and organizations, both public and private from all countries, have an opportunity to send experiments into space aboard the space shuttle.

This was also the second flight to have a congressional observer as a payload specialist. The first such flight of an elected official was the STS 51-D mission in April, 1985. On that flight, Senator Edwin Jacob "Jake" Garn flew aboard *Discovery*. Providing flight opportunities for appropriate congressional leaders gave them first-hand experience with spaceflight that they could use when evaluating proposed programs. In addition, they provided NASA physicians with an opportunity to evaluate the effects of spaceflight on individuals who were not career astronauts.

Context

The STS 61-C mission was the twenty-fourth flight of the Space Transportation System Program and the last space shuttle mission before the loss of *Challenger* in January, 1986. STS 61-C demonstrated the flexibility of shuttle payload scheduling by mixing deployable and attached payloads on *Columbia*.

Columbia was the first space shuttle orbiter to orbit Earth. On April 12, 1981, *Columbia* lifted off for the first time. Seven months later, it made its second voyage into space, becoming the first manned spacecraft to be reused. *Columbia* made four more spaceflights after that, then was temporarily removed from service for an overhaul. As the first operational orbiter, *Columbia* did not have many of the refinements that were built into subsequent vehicles. These included a "heads-up" display for the commander and pilot to use during landing, improvements in the thermal protection system, and structural changes. During its eighteen-month stay at the Rockwell International plant in Palmdale, California, engineers made these and hundreds of other modifications to *Columbia*.

While these modifications were under way, the other three orbiters made fourteen flights into space. Payloads included commercial satellites, classified Department of Defense experiments, scientific laboratories, and research satellites. The flight of STS 61-C combined many of these types of payloads into a single mission.

Bibliography

Couvalt, Craig. "Delays in Columbia Mission Complicate Shuttle Scheduling." *Aviation Week and Space Technology* 124 (January 20, 1986): 20-22. This article provides an overview of the launch and early flight results of the STS 61C mission. It also discusses the impact of this mission's launch delays on the flight schedule planned prior to the loss of *Challenger*.

Joels, Kerry M., and Gregory P. Kennedy. *The Space Shuttle Operator's Manual*. Rev. ed. New York: Ballantine Books, 1988. This book provides a description of how the space shuttle flies. It contains information on shuttle payload types, including Get-Away Specials and deployable satellites. It also contains data on the Payload Assist Module.

Microgravity Science and Applications Division, Office of Space Science and Applications. *Microgravity: A New Tool for Basic and Applied Research in Space*. NASA EP-212. Washington, D.C.: Government Printing Office, 1984. Written for a general audience, this document provides an overview of NASA's space materials processing programs.

Nelson, Bill. "Ascent." *Final Frontier* 1 (July/August, 1988): 18-21, 57. In this article, Congressman Nelson provides a firsthand account of the first 8.5 minutes of the STS 61-C mission.

Nelson, Bill, with Jamie Buckingham. *Mission*. New York: Harcourt Brace Jovanovich, 1988. This is a personal account by Congressman Bill Nelson of his flight aboard STS 61-C.

Gregory P. Kennedy

SPACE SHUTTLE MISSION
STS 51-L

Date: January 28, 1986
Type of mission: Manned Earth-orbiting spaceflight

STS 51-L, the twenty-fifth mission of the U.S. space shuttle, was to have launched the second Tracking and Data-Relay Satellite and a scientific mission called Spartan-Halley into Earth orbit; additionally, Teacher-in-Space Christa McAuliffe was to have broadcast a series of lessons to schoolchildren throughout America. STS 51-L exploded, however, only 73 seconds after launch, killing its crew and completely destroying the space shuttle Challenger *and its satellite cargo. In the wake of the STS 51-L disaster, the United States space program was severely set back, and the shuttle did not fly again for almost three years.*

PRINCIPAL PERSONAGES
> FRANCIS R. "DICK" SCOBEE, Mission Commander
> MICHAEL J. SMITH, Mission Pilot
> ELLISON S. ONIZUKA,
> RONALD E. MCNAIR, and
> JUDITH A. RESNIK, Mission Specialists
> GREGORY B. JARVIS, Payload Specialist
> S. CHRISTA MCAULIFFE, the first teacher in space
> WILLIAM P. ROGERS, the chairman of the Presidential
> Commission on the Space Shuttle *Challenger*
> Accident

Summary of the Mission

Preparations for the Space Transportation System's (STS) twenty-fifth mission, STS 51-L, began more than eighteen months before launch. When the flight crew was originally selected, on January 27, 1985, 51-L's launch was scheduled for the summer of 1985. Delays and a series of cargo changes, however, postponed the flight to mid-January, 1986. Because of these delays, both the detailed flight planning process and the crew's training were interrupted.

The major payloads carried on STS 51-L were the second National Aeronautics and Space Administration (NASA) Tracking and Data-Relay Satellite (TDRS) and the Spartan-Halley comet research observatory. In addition to these payloads, several small experiments were carried in the crew cabin, and the flight was to include the "teacher-in-space" activities of Christa McAuliffe.

According to preflight planning, mission 51-L was to last six days. During this time the crew would launch the TDRS satellite, activate and launch Spartan, conduct astronomical and medical experiments, recover Spartan from orbit, and broadcast lessons to students on the ground.

The planned 1986 shuttle launch schedule was very tight, and several very high-priority missions were to take place in the early part of the year. Within NASA, plans were discussed to skip 51-L if the launch date slipped beyond February 1 and proceed with the rest of the schedule. The purpose of this move would have been to clear the pad for the next launch (an important mission scheduled for March) and to begin readying *Challenger* for its planned launch of an international mission to explore Jupiter and the Sun.

An afternoon launch was originally planned for 51-L. Although scientists leading the Spartan project argued for retaining this time for scientific reasons, NASA mission planners insisted on changing the lift-off to mid-morning. NASA's reasoning for a morning launch was based on safety concerns. Were the vehicle to suffer an "engine-out" during its ascent from Cape Canaveral, it would have to glide to an emergency landing site at Casablanca on the west coast of Africa. Casablanca's runway was not equipped with lighting for night landings. It was decided therefore that the shuttle would be launched in the morning, eastern standard time, so that there would still be light in Casablanca, 6,400 kilometers to the east.

The countdown for STS 51-L began on January 24, but weather forecasts caused the launch to be postponed to January 27. The crew spent the extra time reviewing flight plans and watching the Super Bowl football game from their quarters. During this period, pressures within NASA to launch 51-L mounted. The Spartan satellite required a launch before January 31. With flights of even higher priority on NASA's schedule, the prospect of canceling 51-L became greater. Every effort was made to make sure the shuttle would be ready on the twenty-seventh, when the weather cleared.

On January 27, the vehicle was fueled and the crew had boarded when high wind conditions forced NASA to reschedule the launch once again, this time for January 28.

During the evening of January 27 and the early morning hours of the next day, a series of meetings were held among intermediate-level managers from NASA's Marshall Space Flight Center and the shuttle program's major industrial contractors, Rockwell, Morton Thiokol, and Martin-Marietta. Marshall had final responsibility for the Solid-fueled Rocket Boosters (SRBs). The purpose of these meetings was to assess the status of the launch. Such meetings take place during every countdown.

During these prelaunch meetings, concerns were expressed about the possible effects of a cold weather front approaching the Cape. Three concerns were expressed by some engineers from Thiokol, the manufacturer. In particular, Thiokol engineers Roger Boisjoly and Alan McDonald thought it possible that the booster's hot exhaust could "blow-by" (get past) its protective seals, called O-rings. The reason the O-rings might be bypassed was that they would become stiff in the cold. Once stiffened, the O-rings (which are supposed to be resilient) would not act to seal the SRBs. Without a good seal, exhaust would then leak from the booster's side, rather than from the nozzle, and the booster would be likely to explode or rupture — ending the mission in catastrophe. In one conversation, Alan McDonald went so far as to say that if anything happened, he would not want to have to explain it to a board of inquiry. McDonald noted that no shuttle had ever been launched at a temperature below 53 degrees Fahrenheit (12 degrees Celsius), and that even at that temperature the SRBs had experienced some exhaust blow-by.

The possibility of severe blow-by raised concerns and was the reason for many of the meetings that were held that night. Unfortunately, no actual tests of the boosters had ever been made at low temperatures; therefore, there was no clear case for what would happen. Thiokol's engineers recommended the launch be delayed until later in the day, or perhaps until January 29. NASA managers, however, feeling increasing pressure to launch, pressed for a firm decision from Thiokol. In testimony to the Rogers Commission, NASA managers later stated that the probability of an O-ring failure was believed to be low because each O-ring was backed up by another for increased protection, in case blow-by did occur.

Perhaps sensing the impatience of some NASA officials, Thiokol managers overruled their own engineers and signed a waiver form, stating that the SRBs were safe for launch. Without such a signature, the launch could not have occurred.

During the final hours before the launch of STS 51-L, all Challenger's mechanical and electrical systems were checked. The crew was awakened at 6:00 A.M. At 8:36 A.M. the astronaut crew arrived at the pad and boarded the shuttle. Because there was a buildup of ice on the launchpad and some delays in fueling the huge external tank, the launch was delayed first from 9:38 to 10:38, and then to 11:38. These intermittent delays were unusual in the STS program. Several of the onboard scientific experiments required launch times earlier than 10 A.M. on any given day in order to have the right lighting conditions in orbit. These experiments were sacrificed in order to get the flight launched that morning. The countdown proceeded.

As the final few minutes passed, Pilot Mike Smith powered up Challenger's turbines for flight. Only two minutes

before launch, Commander Scobee called to crew members McNair, Jarvis, and McAuliffe on the shuttle's lower passenger deck, "Two minutes, downstairs. Anybody keeping a watch running?"

At launch minus thirty seconds, Challenger's on-board computers took control. First, the orbiter's three powerful main engines were pressurized, then thousands of electronic checks were performed to verify the engines were ready to start. At minus 6.6 seconds, the main engines were ignited, one at a time, about a second apart. When 90 percent of flight-level thrust was reached on all three, the command was sent to ignite the SRBs. Both SRBs ignited simultaneously at exactly 11:38:01 A.M., and the vehicle rose from the pad. On board, astronaut Judy Resnik exclaimed, "Aaall riiight," as Challenger began its long-awaited push to orbit. In the launch control center three miles away, Thiokol engineers Boisjoly and McDonald relaxed a bit — apparently the O-rings had held. The ambient air temperature was 36 degrees Fahrenheit (about 2 degrees Celsius), 15 degrees colder than that of any previous shuttle launch.

Less than a half-second after the SRBs ignited, the first of eight small but ominous puffs of black smoke swirled from one of the lower joints in Challenger's right booster. These puffs were not obvious to onlookers (no one is allowed within five kilometers of the launchpad) but were recorded by cameras filming the launch. Later analysis by technical experts working with the Rogers Commission revealed that a primary O-ring had failed to seal in the right SRB and that its backup O-ring had failed as well. Blow-by had occurred. Engineering analyses have since indicated that either propellant residue or O-ring soot plugged this leak about two and one-half seconds into the flight.

For nearly a minute, the ascent went as planned. Challenger rolled to put itself on the proper flight path, thousands of electronic checks of on-board systems showed everything performing "nominally" (according to plan), and the vehicle properly throttled back its engines when aerodynamic forces increased.

About fifty-nine seconds into launch, however, trouble began. A review of film from ground cameras recording the launch detected flames coming from the right SRB. Challenger was being buffeted by a combination of high-altitude winds and the aerodynamic stresses of the launch. In response the vehicle flexed slightly and began to steer its engines to counteract the wind. In combination, these forces probably reopened the hole in the right SRB caused by the blow-by at ignition. Over the next five seconds the plume of flame grew and grew. By sixty-four seconds into the flight, a gaping hole was formed in the casing of the SRB. The thrust escaping through this hole exerted a force of 45,000 kilograms on the shuttle, greater than the thrust of many jetliners. Challenger's computers interpreted this force as unusually strong winds. To counteract the 45,000-kilogram side-force, the shuttle automatically swung its engines slightly to the left.

Crew members of the STS 51-L mission (top left); views of the liftoff of the Shuttle Challenger for STS 51-L mission. (NASA)

Inside the cockpit, the crew was jolted around by a combination of actual wind gusts, engine steering, and the thrust escaping from the breeched SRB. Pilot Mike Smith remarked, "Looks like we've got a lot of wind here today."

At 72 seconds into the flight, the searing exhaust from the right SRB either tore or burned loose the attachment strut between the SRB and the external tank. In the final second of flight, computers on board *Challenger* detected a fuel line break caused by the widening explosion and shut down each of the shuttle's three main engines. A moment later, the SRB slammed into and tore off *Challenger's* right wing, then careened into the tank, setting off a massive explosion that destroyed the orbiter. Simultaneously, the cockpit voice recorder taped the first and last indication that anyone on board knew of the serious trouble — Pilot Michael Smith, either seeing the SRB veering toward him through his window or noting the red main engine shutdown lights on his control panel said, "Uh oh." At an altitude of 48,000 feet (14,600 meters) and a speed of Mach 2 (twice the speed of sound), *Challenger* exploded.

On the ground, some spectators realized that the SRBs had separated too early. Others, unfamiliar with shuttle launches, thought this was the normal staging of SRBs. Soon,

however, it was clear to all that the shuttle was nowhere to be seen in a widening fireball, and that the SRBs were wildly spinning off on their own, still under thrust.

At Mission Control in Houston, telemetry signals suddenly stopped. At first, having seen no indications of trouble during the launch, flight controllers believed that either the tracking station or the shuttle's radios had failed. Within seconds, however, radar tracking detected hundreds of pieces of debris following *Challenger's* trajectory. Noting this, Flight Dynamics Officer Brian Perry, a veteran shuttle flight controller, confirmed the tracking data and reported the explosion to the flight director.

The explosion that destroyed *Challenger* (NASA's second and most experienced space shuttle) also destroyed the two satellites carried in its hull. Her crew, Commander Francis "Dick" Scobee, Pilot Michael Smith, Mission Specialists Ellison Onizuka, Judith Resnik, and Ronald McNair, Payload Specialist Gregory Jarvis, and spaceflight participant Christa McAuliffe, were all killed.

In the days that followed the *Challenger* accident the nation mourned. President Reagan eulogized the crew both in a nationally televised speech and at a memorial ceremony at NASA's Johnson Space Center in Houston, Texas.

Within hours of the explosion that destroyed the *Challenger*, calls were made for an official investigation. At NASA, Dr. Jesse Moore, the official in charge of the shuttle program, set up a task force to carry out a technical investigation of the cause, or causes, of the explosion. Moore's all-NASA team impounded all data relating to the flight and initiated a salvage effort to recover as much of the wreckage as possible from the ocean. The wreckage would provide physical evidence that would be available to help pinpoint the disaster's cause.

There were calls for a non-NASA investigation. Such an investigation, it was said, would more likely be freer of bias than any investigation carried out by NASA. Heeding these calls on February 3, 1986, President Reagan appointed a group of thirteen distinguished engineers, test pilots, and scientists to investigate the *Challenger* accident. This group was officially known as the Presidential Commission on the Space Shuttle *Challenger* Accident. The commission's chairman was William Rogers, a former secretary of state, former U.S. attorney general, and an accomplished lawyer. Other members of the commission included Neil Armstrong, the first man to walk on the moon; Sally Ride, an astrophysicist and the first American woman in space; and Richard Feynman, a physicist and Nobel Prize winner. Like the Apollo 1 investigating committee, the Rogers Commission was charged with carrying out a full assessment of all aspects of the accident and the shuttle program; unlike the Apollo investigation, the *Challenger* inquiry was performed publicly, by a basically non-NASA group.

The Rogers Commission took testimony from more than 160 individuals involved in the shuttle program and *Challenger's* last flight. More than twelve thousand pages of sworn testimony were taken, and more than sixty-three hundred documents relating to the accident were reviewed. More than six thousand engineers, scientists, technicians, and other individuals participated in the commission's work. On June 6, 1986, the Rogers Commission released its final report, fixing the immediate cause of the accident as well as discussing the contributing factors that had led to the decision to launch *Challenger* on January 28, 1986. The Commission's report also made recommendations to improve the design of the space shuttle and to prevent future accidents.

Using facts uncovered by the Rogers Commission, as well as supporting evidence and eyewitness accounts of the accident and the salvaged wreckage, it is possible to reconstruct the flight of STS 51-L and the crucial events that led to its ill-fated launch.

Context

The Rogers Commission made a methodical study of all the events leading up to the flight of STS 51-L. Also evaluated were flight records radioed to the ground, debris recovered from the ocean, and films of the flight taken by long-range cameras. Many individuals were interviewed, and a great number of technical studies were performed to test theories concerning the in-flight events of January 28.

The commission considered many things that could have caused *Challenger's* destruction. Possible causes that were investigated included a failure of the main engines, a rupture of the huge external fuel tank, a problem in one of the payload rockets (such as the ignition of the TDRS's upper stage), a failure in one of the SRBs, premature ignition of the shuttle's emergency destruct system, and sabotage. As the evidence mounted, many of the possible causes were eliminated from the list. By early February, only weeks after the launch, the investigators were already focusing their entire attention on the right SRB. Much of the reason for this early narrowing of the possibilities came about because films from automatic cameras developed after the flight clearly showed black smoke seeping from SRB joints at ignition. The films also depicted bright flames jetting from the rocket casings about fifty-eight seconds after launch. Engineers and technicians working for the commission considered propellant cracks, cracks in the rocket motor case, and O-ring seal problems as possible causes of the SRB failure.

In its final report, the commission pinpointed the cause of the accident and made several recommendations for improvements in the shuttle and its management. From a technical standpoint, the cause of the disaster was quite clear. The commission found that the cause of the accident was a failure of the O-ring pressure seal of the right solid rocket motor. In reaching this conclusion, many possible SRB failure modes had been evaluated. Once the O-ring was identified as the cause, the commission went on to determine what specifically caused the O-ring to fail.

Had the O-ring been improperly installed or tested? Had sand or water got into the O-ring joint to prevent it from sealing? Had the cold been to blame? Had the elastic putty used in the O-ring joint failed to seal? Again, more tests were performed, and the flight data and debris were reanalyzed. The commission did not, however, draw a definite conclusion about this cause of the accident. Too much of the evidence had been destroyed in the explosion. Although it was certain that the right SRB had experienced a failure in one of its joints, it was possible that one or more of the above causes were to blame. The commission did, however, conclude that the SRB design was prone to certain failures, including the one that destroyed *Challenger*.

The Rogers Commission's findings went far beyond a determination of the immediate cause of the accident. The commission also concluded that there had been "serious flaws in the decision-making process" leading to 51-L's launch. In particular, it concluded that exceptions to established rules had been granted "at the expense of flight safety" and that Morton Thiokol's management "reversed its position and recommended launch...contrary to the views of its engineers in order to accommodate a major customer."

It was found that previous ground tests and blow-by problems experienced on past flights should have alerted NASA and Thiokol to the serious deficiencies in the SRBs. The commission also found that the Marshall Space Flight Center had not properly passed evidence of SRB problems up the chain of command within the shuttle program but had instead "attempted to resolve them internally." The commission stated in its report to the president that this kind of management "is altogether at odds with...successful flight missions."

After analyzing the cause of the *Challenger* disaster, the Rogers Commission made a number of recommendations to NASA. These recommendations fell into several categories, including the design of the SRBs and shuttle management. The goal of the recommendations was to improve the reliability of the entire shuttle.

The commission recommended that the SRBs be redesigned and recertified to solve the numerous problems inherent in their O-ring joints. The redesign specifically called for an SRB that in future flights would "be insensitive to" environmental factors, including the cold and rain as well as "assembly procedures." Further, the commission called for a design that would have joints as strong as the rocket casings themselves. To verify the integrity of the new design, the commission recommended testing full-size boosters before the new SRBs were committed to actual flight. These tests began in the summer of 1987.

The commission also made specific recommendations relating to other potential problem areas in the shuttle. They insisted that the shuttle's brakes be improved (a long history of brake problems had occurred over many flights) and that a reevaluation of crew abort and escape mechanisms be undertaken to determine if launch and landing problems could be made "more survivable." Finally, the commission insisted that the rate of shuttle flights be controlled to maximize safety. Such a policy had not been implemented in the past, the commission said.

Beyond technical matters, the Rogers Commission also recommended a number of sweeping changes in the shuttle program's management structure. These were designed to prevent the problems that led to a "flawed decision-making process" regarding the launch of 51-L. These specific recommendations included the establishment of a safety panel, with broad powers, reporting directly to the manager of the shuttle program and the establishment of the Office of Safety, Reliability, and Quality Assurance within NASA reporting directly to the NASA Administrator, with broad powers to investigate and demand solutions to safety-related issues.

Additionally, it was recommended that a full review take place of all critical safety items in the space shuttle before the next flight. Finally, the commission recommended that astronauts be more fully involved in the shuttle program's management. This recommendation came in response to the anger expressed by some astronauts during the investigation that they, who were at greatest risk in each flight, had not been informed of the O-ring blow-by and erosion problems prior to the accident. In response to this call, NASA placed senior astronauts — Robert Crippen, Sally Ride, Paul Weitz, and others — in key advisory roles.

The *Challenger* disaster brought the United States space program to a halt. Within months of the accident, two unmanned launchers failed as well. With no way of launching satellites until either these rockets or the shuttle was recertified for flight, both NASA and the military were "pinned down." New research in space could not be conducted. Replacement military and weather satellites could not be launched. Planned space missions stagnated, awaiting the availability of a launcher. More than a dozen scientific payloads were canceled, and seventy more were delayed for years. Space planners were forced to buy dozens of expendable Titan and Delta rockets to supplement the grounded shuttle program.

Bibliography

Durant, Frederick C., III, ed. *Between Sputnik and the Shuttle: New Perspectives on American Astronautics*, 1957 – 1980. San Diego: Univelt, 1981. A comprehensive history of American manned spaceflight from 1957 to 1981. This book contains an excellent description of the origins of the Space Transportation System program.

Lewis, Richard S. *Challenger: The Final Voyage*. New York: Columbia University Press, 1988. A factual account of the *Challenger* disaster, this book relates the events of January 28, 1986. A popular version of the Rogers Commission Report.

Report of the Presidential Commission on the Space Shuttle Challenger Accident. Washington, D.C.: Government Printing Office, 1986. Contains the full text of the official report of the Rogers Commission. Technical in its content, this volume details both the immediate and root causes of the STS 51-L accident.

Stern, Alan. *The U.S. Space Program After Challenger.* New York: Franklin Watts, 1987. A detailed look at the *Challenger* disaster and its ramifications for the future of the United States space program.

Trento, Joseph J. *Prescription for Disaster: From the Glory of Apollo to the Betrayal of the Shuttle.* New York: Crown Publishers, 1987. Though somewhat biased, the author gives a detailed account of the *Challenger* accident and its root cause. Trento advances the theory that the events at Cape Canaveral on January 28, 1986, were the culmination of the decline of the U.S. space program since the early 1970's.

Washington Post Editorial Staff. *Challengers.* New York: Simon & Schuster, 1986. A touching and personal series of biographies of each of the seven crew members of the ill-fated STS 51-L mission.

Alan Stern

SPACE SHUTTLE MISSION
STS-26

Date: September 29 to October 3, 1988
Type of mission: Manned Earth-orbiting spaceflight

Space Transportation System (STS) 26 was the first shuttle flight following the catastrophic explosion of the space shuttle Challenger *thirty-two months before. STS-26 successfully returned the United States' manned space program to an active flight status. During the flight, a communications satellite vital to the U.S. space program was launched into geosynchronous orbit.*

PRINCIPAL PERSONAGES
  FREDERICK H. "RICK" HAUCK, Commander
  RICHARD O. "DICK" COVEY, Pilot
  GEORGE D. "PINKY" NELSON,
  DAVID C. HILMERS, and
  JOHN M. "MIKE" LOUNGE, Mission Specialists

Summary of the Mission

Space shuttle mission 26 was the first U.S. manned spaceflight following the loss of the space shuttle *Challenger* and its seven-member crew on January 28, 1986. Because of the protracted recovery time of thirty-two months, STS-26 became widely regarded as America's return to space. In fact, the National Aeronautics and Space Administration (NASA) officially designated the mission "Return to Flight."

Not since the fatal launchpad fire of Apollo 1 on January 27, 1967 (which killed astronauts Virgil Grissom, Edward White, and Roger Chaffee), had there been such an extensive reworking of an American spacecraft. Following the loss of *Challenger*, a thirteen-member investigative commission was appointed by President Ronald Reagan and headed by former Secretary of State William Rogers. Called the Rogers Commission, the panel issued its report to the president on June 9, 1986. More than four hundred changes in a $2.4-billion program were advocated to improve the shuttle and help ensure the safety of future flights. These changes included a complete redesign of the O-ring system connecting the solid-fueled rocket booster segments, the system that was blamed for the *Challenger* explosion.

Problems encountered after the initial launch date was set for February, 1988, caused a series of delays. Space planners wanted to be certain that the tests of the solid-fueled rocket boosters were completed successfully. Finally, on July 4, 1988, after the redesigned joints had been approved, the assembled vehicle was rolled from the vehicle assembly building at the Kennedy Space Center to Launchpad 39B. *Discovery* was poised for launch.

Even with the shuttle positioned on the pad, NASA officials waited before setting another launch date, ostensibly to evaluate the assembled system further and to receive the results from yet more booster tests. The media proved especially critical of the space agency in the wake of the *Challenger* disaster and these frequently shifting launch dates. After the final flight readiness review panel met, the launch date was set for Thursday, September 29, 1988.

The media gave as much attention to the upcoming launch as to any other space launch in American history, including the Apollo Moon flights. NASA had instituted a new launch management system to prevent a recurrence of what the Rogers Commission had called "a flawed decision-making process" for shuttle launches. Heading a launch management team was active astronaut Robert L. Crippen, pilot of the first shuttle, who would make the final launch decision. In the glare of world attention, NASA and the U.S. space program could ill afford a problematic launch or mission.

The countdown proceeded smoothly on September 29. Many predictions prior to the launch held that the new and untested safety and management systems were so bulky that there would likely be days of delays, holds, and reconsiderations before STS-26 could be launched. These predictions proved unfounded. After a delay of only one hour and thirty-eight minutes, caused by high-altitude winds that were lighter than predicted, *Discovery* was launched from Pad 39B at 11:37 A.M. eastern daylight time (EDT).

The launch was normal in every respect. The redesigned solid-fueled rocket boosters were jettisoned on schedule and dropped by parachute into the ocean off the coast of Florida. Subsequent inspection proved that they had come through the flight in pristine condition. The shuttle's three main engines performed well enough to preclude an initial burn of the Orbital Maneuvering System (OMS) engines to achieve stable orbit. Later, an OMS burn was performed in order to place *Discovery* in a 348-kilometer orbit above Earth.

Aside from the principal function of proving viability of the extensively reworked shuttle system, the four-day mission was to include eleven scientific experiments and the deployment of a vital NASA communications satellite. The satellite, a $100-million Tracking and Data-Relay Satellite (TDRS), was to replace the one lost on *Challenger*. With a mass of 2,268 kilograms, the TDRS was one of the largest communications satellites ever launched; it was so massive that it could fit only in the shuttle's payload bay. The TDRS was successfully deployed by a spring-loaded platform from *Discovery*'s payload bay six hours into the mission. After the astronauts maneuvered the orbiter to a safe 72 kilometers away, the TDRS automatically boosted itself 35,800 kilometers above Earth into geosynchronous orbit. (Satellites in geosynchronous orbit rotate at the same speed as Earth rotates and thus appear to remain at a fixed point in the sky.)

The purpose of the TDRS system is to provide a three-satellite communications constellation for NASA missions. The TDRS system would enable nearly constant communications between Earth and other orbiting spacecraft such as the orbiter fleet, the U.S. space station *Freedom*, and the Hubble Space Telescope.

One of the first problems the crew encountered on STS-26 was the partial failure of a cooling device called a flash evaporator, which is used to cool the crew cabin and the equipment. During lift-off, the evaporator became frozen with ice and operated at less than its optimal capacity. To assist in melting the ice, ground engineers allowed the temperature in *Discovery* to rise to 29 degrees Celsius for the first two days of the mission. The evaporator would malfunction again briefly during reentry, but it never endangered the crew or *Discovery*'s systems.

The Ku-band antenna, used to communicate with Earth-based stations, malfunctioned on the first day of the mission. The antenna failed to align itself properly for broadcast of signals to Earth, and it wobbled erratically on its mount. On the second day, the malfunctioning antenna was finally stowed back in the shuttle's payload bay; consequently, the flow of communications between *Discovery* and the ground for the remainder of the flight was reduced.

Eleven scientific experiments were carried on board the shuttle. Protein crystal growth experiments were conducted for use in the development of complex protein molecules that could be utilized in medicines and other chemical solutions. One of the crystals being examined was critical to the study of the Acquired Immunodeficiency Syndrome (AIDS) virus and how it replicates. Two experiments investigated medicinal properties unique to weightless space. Two other experiments were designed to investigate molten metal resolidification and crystallization for development of stronger metal alloys. Also on board was an experiment involving crystal growth on a semipermeable membrane (the results could ultimately help reduce medical X-radiation doses). Yet another was designed to produce thin films of organic material and study their properties.

Two meteorological investigations were conducted on STS-26. One photographed lightning and the other the glow of Earth's horizon near sunrise and sunset. This last experiment was remarkably similar to one of the first scientific investigations ever conducted by man in space.

At 11:34 A.M., EDT, on October 2, 1988, *Discovery*'s crew fired the OMS engines for two minutes and fifty seconds in the de-orbital maneuver. Forty minutes later, *Discovery* reentered Earth's atmosphere. Twenty-one minutes after reentry, at 12:37 A.M., EDT, the shuttle and its crew landed safely at Edwards Air Force Base in the high desert of Southern California. The duration of mission 26 had been four days and one hour.

Knowledge Gained

The primary goal of mission 26 was to prove the safety of the redesigned shuttle system and return American manned spaceflight to an active status. The successful return of the crew with all major mission objectives met fulfilled that goal. Precise details of the success of each redesigned element would take many weeks of detailed analysis, but all major redesigned components were proved sound.

The newly designed O-ring system appeared to perform flawlessly. According to engineers who examined the system after it was recovered from the ocean following the launch, the boosters appeared in better condition than any recovered from any other mission.

The three main engines had undergone forty significant design changes in a $100-million program that included strengthening of the engine's high-speed pump components. These components also appeared to have performed well, although one engine apparently experienced an oxidizer leak near the end of the main combustion chamber.

The orbiter's tires, brakes, and nosewheel steering system had all undergone extensive modifications. These systems absorb the monumental energies of landing when the shuttle orbiter touches down at speeds of more than 350 kilometers per hour. *Discovery*'s landing speed was said to have been 381 kilometers per hour. The vehicle required only 2,337 meters to stop, indicating a very good braking (energy absorption) efficiency when compared with that of past missions.

The performance of these critical systems indicates the viability of these redesigns. The ability of space planners to identify and produce the necessary redesigns gave NASA the confidence to incorporate future changes as they may be required in the shuttle program. One manifestation of this newfound capability was the redesign of a drogue chute braking system to be installed on all orbiters to absorb touchdown energies further and enhance the safety of landings.

Although overshadowed by the mission's main goal of returning the U.S. space program to flight status, the scientific experiments aboard STS-26 represented one of the primary reasons the space shuttle program was first conceived. The extensive study of protein crystal growth provided data for vital chemical studies that may have far-reaching implications. Crystal growth on membranes may one day reduce dosages of medical X-radiation, which account for the single largest source of ionizing radiation exposure for the American public. The experiment designed to study the AIDS virus also proved valuable. Information gathered from the experiment may lead to the development of medications.

Most important, STS-26 proved the ability of the space agency to recover from catastrophe and unprecedented tragedy in space to fly again. The system emerged from the flight of 26 fundamentally changed, with a better spacecraft and a baseline of knowledge that, unfortunately, might not have been obtained under other circumstances. From those hard-won lessons, the space program gained a clearer vision that would guide the United States into the next century.

Context

Mission STS-26 flew as the twenty-sixth space shuttle mission, the fifty-seventh United States manned mission, and the twenty-second manned mission of the operational shuttle system. During the thirty-two-month period of rebuilding, NASA had been heavily criticized by the Rogers Commission and the word media for a lack of proper management and a poor decision-making process. The space agency worked ceaselessly during this long hiatus to regain the confidence of the American public, the confidence they had enjoyed since their formation by Congress in 1958.

NASA took various measures to help ensure a successful 26R mission. The agency reviewed the cause of the *Challenger* accident and corrected the flawed O-ring system. It reviewed all shuttle systems, identifying and modifying any which could cause future problems.

Space planners at NASA also restructured the agency's management system, strengthening safety, quality, and reliability by making improvements in training and personnel. Finally, NASA reduced the planned flight rate and removed altogether the somewhat artificial dependence on civil payload manifests that had resulted in an overloaded system.

The crew selected to fly the 26 mission was an all-veteran crew; NASA wanted to maximize the experience available on this important return to space. Hauck and Nelson had both

flown twice before, and Covey, Hilmers, and Lounge had flown once before — making the STS-26 crew one of the most experienced ever to fly on a shuttle mission.

The space agency and the nation approached the launch with an apprehension that attracted much attention. The shuttle's systems were tested and retested as though *Discovery* had never flown before. The orbiter sat on the pad undergoing tests some thirteen weeks before launch, the longest launchpad preparation period since the earliest shuttle launches. Clearly, the launch was approached with maximum caution.

Just before the launch, crowds gathered at Kennedy Space Center. The atmosphere was reminiscent of that preceding the early Apollo Moon launches. Unlike the previous launch of mission STS 51-L, which was attended by relatively few and virtually dismissed by the press as "routine," the mission of 26 was anything but ordinary.

More than at any other time in history, the fate of the Western world's future in space was on the line. If STS-26 had failed, the manned space effort would have suffered another catastrophic setback, perhaps giving way to critics' calls for the abandonment of the manned program in favor of mechanical, robotic exploration. With the string of ongoing successes by the Soviets, another U.S. failure would have meant abdicating superiority in space exploration. The United States had only three shuttle orbiters left in its active fleet. The loss of another of these multibillion-dollar national resources would have been disastrous.

The success of STS-26 held an implicit significance for the continuance of manned spaceflight. Without question, it was the mission that had to succeed. Succeed it did, brilliantly restoring the United States to its competitive position in space. The tragedy of mission 51-L and all it entailed had been replaced by triumph. All space projects were suddenly possible again: the Hubble Space Telescope, Magellan to Venus, Galileo to Jupiter, and the space station *Freedom*. The United States had regained its foothold in space.

Bibliography

Allen, Joseph P., and Russell Martin. *Entering Space: An Astronaut's Odyssey*. New York: Stewart, Tabori and Chang, 1984. This heavily documented work describes the United States space shuttle program. It details the stages of a typical mission, from processing at the Kennedy Space Center to launch, orbital activities, and landing. The book is aimed toward all readers and is one of the most beautifully photographed and illustrated of all books on the shuttle.

Baker, David. *The History of Manned Space Flight*. New York: Crown Publishers, 1982. This book offers a precise chronology of the history of manned spaceflight, heavily oriented toward the United States' effort. It is a detailed work that chronicles the U.S. manned space effort from its beginning to the start of the space shuttle program. Includes analyses of U.S. spaceflights, beginning with the Mercury missions of the 1960's.

De Waard, E. John, and Nancy De Waard. *History of NASA: America's Voyage to the Stars*. New York: Exeter Books, 1984. A colorful pictorial essay on the history of NASA. It does not go into great detail on every flight but does include many color photographs of the various U.S. manned programs.

Joels, Kerry M., and Gregory P. Kennedy. *The Space Shuttle Operator's Manual*. Rev. ed. New York: Ballantine Books, 1987. This manual is a detailed space shuttle system reference work. It provides specific information about the space shuttle system, from weights and sizes to operational characteristics. It enumerates countdown procedures, emergency instructions, and standard operational modes. Thoroughly illustrated.

McConnell, Malcolm. *Challenger: A Major Malfunction*. Garden City, N.Y.: Doubleday and Co., 1987. This work offers the author's unflattering, unapologetic view of what happened to the U.S. space program from the closing days of the Apollo program to the Challenger disaster. It offers an often-bitter appraisal of the compromises that led to tragedy and the pressures that caused the NASA management system to fail. While some of the author's charges have not been entirely supported by formal investigations, the book does provide a baseline of information from which the reader may form his own opinions.

Dennis Chamberland

SPACE SHUTTLE FLIGHTS, 1989

Date: March 13 to November 27, 1989
Type of mission: Manned Earth-orbiting spaceflight

Space shuttle crews orbited the Earth for more than four weeks during 1989 as NASA successfully launched and landed five space shuttle missions. Highlights of the 1989 space shuttle flights included the boosting of the Magellan/Venus and the Galileo/Jupiter spacecrafts on their respective missions to explore the solar system.

PRINCIPAL PERSONAGES

> JAMES C. FLETCHER, Space Shuttle Program
>   Administrator
> DALE D. MYERS, Space Shuttle Program Deputy
>   Administrator
> MICHAEL L. COATS, STS-29 Commander
> JOHN E. BLAHA, STS-29 and STS-33 Pilot
> ROBERT C. SPRINGER,
> JAMES F. BUCHLI, and
> JAMES P. BAGIAN, STS-29 Mission Specialists
> DAVID M. WALKER, STS-30 Commander
> RONALD J. GRABE, STS-30 Pilot
> MARK C. LEE,
> NORMAN E. THAGARD, and
> MARY L. CLEAVE, STS-30 Mission Specialists
> BREWSTER H. SHAW, JR., STS-28 Commander
> RICHARD N. RICHARDS, STS-28 Pilot
> JAMES C. ADAMSON,
> DAVID C. LEETSMA, and
> MARK N. BROWN, STS-28 Mission Specialists
> DONALD E. WILLIAMS, STS-34 Commander
> MICHAEL J. McCULLEY, STS-34 Pilot
> SHANNON W. LUCID,
> FRANKLIN R. CHANG-DIAZ, and
> ELLEN S. BAKER, STS-34 Mission Specialists
> FREDERICK D. GREGORY, STS-33 Commander
> MANLEY L. CARTER,
> F. STORY MUSGRAVE, and
> KATHRYN C. THORNTON, STS-33 Mission Specialists

Summary of the Missions

After the successful conclusion of the STS-26 mission (October, 1989), the orbiter *Discovery* was returned from the Dryden Research Facility of the Kennedy Space Center for postflight inspection and reconstruction, a process which included the replacement of its three main engines. After the run-around activities were completed, *Discovery* was transferred from the Orbiter Processing Facility (OPF) to the Vehicle Assembly Building (VAB) where it was mated to an expendable External Tank (ET) and two Solid Rocket Boosters (SRB). On February 3, the assembled space shuttle orbiter mounted on the 47-meter-high ET and 45.46 meter SRBs was rolled aboard a mobile launcher platform 4.2 miles to Launch Pad 39-B. The payload for Mission STS-29 was installed, and the launch preparation countdown proceeded. On March 13, 1989, after a 1 hour, 50 minute delay owing to ground fog and upper winds, the 256,357 pound *Discovery* and cargo was launched at 9:57 A.M., eastern standard time (EST).

The primary objective of the STS-29 was to place a Tracking and Data Relay Satellite (TDRS-D; renamed TDRS-4 after deployment) into orbit. The TDRS-4, which was attached to a solid-propellant Boeing/U.S. Air Force Inertial Upper Stage (IUS), was deployed about six hours after liftoff. The IUS boosters were then fired, sending the TDRS communication satellite into its proper geosynchronous orbit. TDRS-4 was the third of its kind to be deployed from a shuttle. The successful deployment of TDRS-4 completed the constellation of on-orbit satellites for NASA's advanced space communication system.

During the remainder of this five-day mission, the crew — commanded by Brewster Shaw, Jr.— conducted chromosome and plant cell division experiments, studied the growth of protein crystals, performed two student involvement program experiments, and tested a space station heat pipe advanced radiator. After completing seventy-nine orbits and travelling approximately 2 million miles, the crew returned on March 18 with beautiful pictures of the Earth that were photographed with a hand-held IMAX camera.

The primary objective of the second mission of 1989, STS-30, was to deploy the Magellan Venus-exploration spacecraft. This mission received significant national media coverage, not only because the Magellan Project was the first planetary science mission attempted since 1978 but also because it was to be the first time that an interplanetary probe was to be deployed from a space shuttle. The proposed April 28 launch was scrubbed just 31 seconds before lift-off (T-31 seconds) owing to a problem with the liquid hydrogen recirculation pump. After repairs were made, the launch date was rescheduled for May 4. Cloud cover and high winds almost forced another postponement. Liftoff took place, however, at 2:47 P.M., eastern daylight time (EDT), about 5 minutes before the allowed 64-minute window opening expired. This was the fourth flight for *Atlantis*, an orbiter that flew its inaugural flight in 1985.

The five-member crew on this historic mission included David Walker, Commander; Ronald Grabe, Pilot; and Norman Thagard, Mary Cleave, and Mark Lee, Mission Specialists. Six hours and 14 minutes into flight, the crew deployed from the shuttle's cargo bay the Magellan/Venus radar mapper spacecraft along with its solid-fuel IUS booster. Shortly thereafter, the IUS boosters were fired, and the Magellan spacecraft began its fifteen-month journey around the Sun and to Venus. Before the completion of the Magellan mission in October, 1994, the Magellan space probe would radar map 98 percent of the planet's surface, and collect and transmit to NASA scientists high-resolution gravity data of Venus.

The four-day STS-30 mission produced another dramatic, yet unplanned moment. While in flight, one of the five General Purpose Computers (GPC) failed and had to be replaced with an on-board computer hardware spare. This was the first time a GPC was switched while a shuttle spacecraft was in orbit. Fortunately, the new computer was fully functional. After completing its sixty-four planned orbits, the *Atlantis* landed safely at Edwards Air Force Base in California on May 8.

On August 8, the first of two 1989 classified Department of Defense shuttle missions was launched on Pad 39-B at the Kennedy Space Center. This was the first mission in more than three and one-half years for the space shuttle *Columbia*, America's original orbiter, which flew its first flight in 1981. This mission, STS-28, lasted 5 days, 1 hour, and landed without incident on August 13. Three months later, the second classified mission of 1989 (and the fifth shuttle mission dedicated to the Department of Defense) was scheduled for the orbiter *Discovery*. The original November 20 launch date for this mission was delayed in order to allow time to replace suspect integrated electronics assemblies on the twin solid rocket boosters. On November 22 at 7:23 P.M., EST, the thirty-second U.S. space shuttle was launched for orbit. This was only the third shuttle mission launched at night, and the first night launching since the return of shuttle flights following the *Challenger* accident in 1986. The five-day mission, STS-33, completed seventy-nine orbits, traveled 2.1 million miles, and ended without complication on November 27.

Perhaps the most dramatic of the 1989 shuttle flights was the STS-34 mission. This mission was twice rescheduled, once owing to faulty main engine controller on the number two main engine, and the second time owing to weather conditions that scientists believed would prevent a return-to-launch-site landing at the Kennedy Space Center's Shuttle Landing Facility. After these delays, the five-member crew —

Donald Williams, Commander; Michael McCulley, Pilot; and Franklin Chang-Diaz, Shannon Lucid and Ellen Baker, Mission Specialists — was launched into orbit aboard the *Atlantis* at 12:54 P.M., EDT, on October 18. The primary task of the crew was to deploy the Galileo/Jupiter spacecraft and its attached IUS booster. This objective was accomplished six and one-half hours into flight. The IUS motors attached to the planetary orbiter and probe were fired, and Galileo was propelled on an interesting and historic trajectory to Jupiter.

Originally, the Galileo spacecraft had been designed for a direct flight to Jupiter, which would take about two and one-half years. Changes in the launch system after the *Challenger* accident, however, prevented this direct trajectory and forced engineers to design a new interplanetary flight path using several gravity-assisted swingbys, once past Venus and twice around the Earth. This Venus-Earth-Earth-Gravity Assist trajectory (known as VEEGA) would send the Galileo orbiter and probe on a six-year, 2.3 billion mile voyage to the mysterious Jupiter. The eight-fold objectives of the sophisticated and ambitious Galileo Project include: (1) to determine the temperature and pressure structure of Jupiter's atmosphere; (2) to determine the chemical composition of Jupiter; (3) to determine how many cloud layers exist, to find their location, and to characterize the cloud particles as to size and number density; (4) to measure the amount of helium relative to hydrogen on Jupiter to high accuracy; (5) to measure the winds in Jupiter's atmosphere and to determine how deep in the atmosphere the winds exist; (6) to measure how sunlight and energy coming from the deep interior are distributed in Jupiter's atmosphere; (7) to detect lightning if it occurs, to measure how energetic it is, and to observe the frequency of occurrence; and (8) to measure the characteristics of energetic protons and electrons trapped in Jupiter's magnetic field within a few Jovian radii from the planet.

After deploying the Galileo spacecraft, the STS-34 mission crew spent the remaining four days of their flight operating secondary payloads, conducting experiments, and taking IMAX photos of the Earth and space. This flight, which set in motion the historic Galileo mission, ended with the successful landing of the *Atlantis* on October 23 at Edwards Air Force Base.

A sixth mission, STS-32, was originally scheduled for 1989. The anticipated December 18, 1989 launch date for the mission, however, had to be postponed in order to allow time to complete and verify modifications to Pad A, a launch pad that had not been used since January, 1986. STS-32 ultimately was launched on January 9, 1990.

Knowledge Gained

In 1989, for the first time, NASA used shuttle orbiters to deploy interplanetary spacecraft designed to explore our solar system. Mission STS-30 launched the Magellan on a trajectory to Venus, the planet often called "the Earth's sister planet" because of its similar size and distance from the sun. Between

August, 1990, and October, 1994, the Magellan spacecraft orbited Venus. Because Venus is clouded by a dense, opaque atmosphere, conventional optical cameras could not be used to image its surface. The Magellan, however, using a sophisticated imaging radar, provided scientists with the most highly detailed maps of Venus ever captured. It also made global maps of Venus's gravity field. Craters depicted in the radar images tell scientists that Venus's surface is relatively young. Although Venus, like the Earth, was formed about 4.6 billion years ago, its surface was reformed by widespread volcanic eruptions only about 500 million years ago. Venus's present harsh environment has persisted since this resurfacing. Nothing was found to suggest the presence of oceans or lakes at any time in the planet's history. Nor does Venus appear to have plate tectonics — movements of crustal masses that result in earthquakes and continental drift. Unlike the Earth, Venus seems to lack an "asthenosphere," a buffer layer between the planet's mantle and crust. As a result, gravity fields on Venus are more affected by surface topography than they are on the Earth.

Mission STS-34, like STS-30, contributed to the success of a NASA unmanned planetary mission. The Galileo orbiter and probe, launched aboard space shuttle *Atlantis* on October 18, 1989, carried a total of sixteen scientific instruments. Knowledge gained and to be gained from this mission is voluminous. While en route, Galileo flew within 110,000 kilometers (68,000) of the moon, obtaining multispectral lunar images and data that will be useful in comparing our Moon with the Jovian satellites. Data from this lunar flyby suggests that the Moon has been more volcanically active and that the far side of the Moon has a thicker crust than researchers previously thought. Galileo also became the first spacecraft to fly closely by two asteroids, Gaspra and Ida. Data from these encounters confirmed and photographed a small moon, later named Dactyl, orbiting around Ida. In July, 1994, Galileo was the only observer in position to obtain images of the impact of fragments from Comet Shoemaker-Levy 9 on the far side of Jupiter. In December, 1995, Galileo became the first spacecraft to enter orbit around one of the outer planets of the solar system. On December 7, 1995, the Galileo probe

successfully entered the atmosphere of Jupiter. A radio link between the probe and the orbiter lasted for about fifty-seven minutes during the probe's descent. Data sent from the probe to the orbiter was stored into the orbiter's computer memory and its tape recorder. This data will be transmitted back to the Earth in the period from January, 1996, through May, 1996. The orbiter will then begin an intensive study from orbit of Jupiter's moons, magnetic field, radiation belts, and atmosphere. Scientists should continue to receive data from the Galileo orbiter at least through the year 1997.

Context

Two of the five completed missions of 1989 were classified Department of Defense flights, and little information is publicly available on these missions. Of the three remaining missions, however, two were highly publicized historic flights that initiated a new age of planetary space exploration. The Magellan and Galileo missions, both launched in 1989 from the shuttle orbiter *Atlantis*, have provided and will continue to provide scientists with a greater understanding of the composition and origins of the solar system. Galileo's probe of Jupiter's physical and orbital properties, in particular, offers spectacular opportunities to unlock the mysteries of the cosmos, since Jupiter and its four largest moons, which range in size from the diameter of the Earth's moon to the size of the planet Mercury, are in some ways analogous to a mini-solar system. By observing this miniature solar system, we will be better able to find new clues about how the Sun and the planets formed and about how they continue to interact and evolve.

Before 1989, the space shuttle — which takes off like a rocket, operates in orbit as a spacecraft, lands on Earth like an airplane, and then is refitted to fly again, perhaps for as many as eighty missions — had proven its effectiveness as a deployer and retriever of satellites and as a platform laboratory for conducting space research. In 1989, Missions STS-30 and STS-34 demonstrated the utility of the shuttle in another way. By serving as a base for the launching of the Magellan and Galileo space probes, the manned space shuttle *Atlantis* helped inaugurate a new era of unmanned interplanetary exploration.

Bibliography

Embury, Barbara, with Thomas D. Crouch. *The Dream is Alive*. New York: Harper & Row, 1990. An interesting account of the shuttle program intended for juvenile readers.

Lewis, Richard S. *The Voyages of Columbia: The First True Spaceship*. New York: Columbia University Press, 1984. An excellent description of space shuttle *Columbia* and its early missions. This early work, however, does not cover the *Columbia* missions of 1989.

NASA Space Shuttle Mission STS-29 Press Kit. March, 1989. Washington, D.C. A thirty-page NASA press release detailing among other things the objectives, countdown milestones, major activities, and landing and post-landing operations of the twenty-eighth space shuttle mission. This release, available in most research libraries on microfilm, is also available on the Internet at location: http://www.ksc.nasa.gov/shuttle missions/sts-29-press-kit.txt. For similar descriptions of the two other nonclassified missions of 1989, see *NASA Space Shuttle Mission STS-30 Press Kit* (April, 1989) and *NASA Space Shuttle Mission STS-34 Press Kit* (October, 1989). Other pertinent information concerning 1989 shuttle flights can be accessed on the Internet through the NASA Space Shuttle Launches Homepage located at http://www.ksc.nasa.gov/shuttle/missions/missions.html.

United States Congress. Senate Committee on Appropriations. *Space Shuttle and Galileo Mission: Hearing Before a Subcommittee on Appropriations.* Washington, D.C.: U.S. Government Printing Office, 1980. For a look into the politics of space in general, and into the process of funding for the Galileo Mission in particular, see this 132-page document published in 1980 by the United States Government Printing Office. Students of both the space program and the U.S. government will enjoy comparing this early document with the 1988 fact sheet prepared by the United States General Accounting Office for the Chair of the Subcommittee on Science, Technology, and Space. This thirty-two page document is entitled *Space Exploration: Cost, Schedule and Performance of NASA's Galileo Mission to Jupiter* (Washington, D.C., 1988).

"Variable Phenomena in Jovian Planetary Systems." *Journal of Geophysical Research* 98 (October 25, 1993): 18,727–18,876. A technical update that discusses some of the recent discoveries and theories concerning Jupiter and its moons. Intended for scientific audiences.

Young, Carolyn, ed. *The Magellan Venus Explorer's Guide.* Pasadena, California: NASA, Jet Propulsion Laboratory, 1990. This 197-page document produced by the California Institute of Technology provides a detailed description of the Magellan spacecraft, the Venus probes, and the space-based radar system used in this historic mission.

Terry D. Bilhartz

SPACE SHUTTLE FLIGHTS, 1990

Date: January 9 to December 11, 1990
Type of mission: Manned Earth-orbiting spaceflight

The National Aeronautics and Space Administration (NASA) flew six space shuttle flights in 1990, missions which totaled more than thirty-eight days in space. Among the highlights of 1990 was Mission STS-31, the flight that deployed the Hubble Space Telescope into an orbit from which it began collecting data on objects up to 14 billion light years away.

PRINCIPAL PERSONAGES

DANIEL C. BRANDENSTEIN, STS-32 Commander
JAMES D. WETHERBEE, STS-32 Pilot
BONNIE J. DUNBAR,
MARSHA S. IVINS, and
G. DAVID LOW, STS-32 Mission Specialists
JOHN O. CREIGHTON, STS-36 Commander
JOHN H. CASPER, STS-36 Pilot
DAVID C. HILMERS,
RICHARD M. MULLANE, and
PIERRE J. THUOT, STS-36 Mission Specialists
LOREN J. SHRIVER, STS-31 Commander
CHARLES F. BOLDEN, JR., STS-31 Pilot
BRUCE M. MCCANDLESS, II,
STEVEN A. HAWLEY, and
KATHRYN D. SULLIVAN, STS-31 Mission Specialists
RICHARD N. "DICK" RICHARDS, STS-41 Commander
ROBERT D. CABANA, STS-41 Pilot
BRUCE E. MELNICK,
WILLIAM M. SHEPHERD, and
THOMAS D. "TOM" AKERS, STS-41 Mission Specialists
RICHARD O. COVEY, STS-38 Commander
FRANK L. CULBERTSON, JR., STS-38 Pilot
CHARLES D. "SAM" GEMAR,
ROBERT C. SPRINGER, and
CARL J. MEADE, STS-38 Mission Specialists
VANCE D. BRAND, STS-35 Commander
GUY S. GARDNER, STS-35 Pilot
JEFFREY A. HOFFMAN,
JOHN M. "MIKE" LOUNGE, and
ROBERT ALLAN RIDLEY PARKER, STS-35 Mission
    Mpecialists
SAMUEL T. DURRANCE and
RONALD A. PARISE, STS-35 Payload Specialists

Summary of the Missions

The December 18, 1989, scheduled launch of Mission STS-32 was postponed to complete modifications to Pad 39-A, a launch site that had not been used since January, 1986. The rescheduled launch date of January 8, 1990 also was scrubbed due to poor weather conditions. Finally, on January 9, 7:35 A.M., eastern standard time (EST), the space shuttle *Columbia* was launched on what would be an 11-day, 172-orbit mission.

The crew on this thirty-third space shuttle flight performed a variety of deployment and retrieval tasks. Mission STS-32, for instance, deployed a communication satellite (named SYNCOM IV-F5 or LEASAT 5) that was designed to provide worldwide UHF communications between ships, planes, and fixed facilities on earth. This 20-foot long, 17,000, pound satellite with UHF and omnidirectional antennas was launched using the unique "frisbee" or roll-out method of deployment. On the fourth day of the flight, Mission STS-32 retrieved from orbit a Long Duration Exposure Facility (LDEF). The LDEF was a twelve-sided, open grid structure, thirty feet long and fourteen feet in diameter, which, since its deployment in April, 1984, had been exposing a variety of materials to the harsh space environment. Among the fifty-seven experiments conducted in the experiment trays of the LDEF between 1984 and 1990 were included an interstellar gas experiment (designed to provide insight into the formation of the Milky Way galaxy by capturing and analyzing its interstellar gas atoms), a cosmic radiation experiment (designed to investigate the evolution of the heavier elements in our galaxy), and a micrometeoroid experiment (designed to increase our understanding of the processes involved in the evolution of the solar system).

In addition to deploying and retrieving payloads, the crew of Mission STS-32 also conducted a number of experiments from the laboratories of the space shuttle. Mission Specialist Marsha Ivins, for example, operated an American Flight Echocardiograph (AFRE), a medical ultrasonic imaging system that, when attached to the skin of another crew member, provided in-flight measurements of the size and functioning of the astronaut's heart. Ivins, along with Mission Specialists Bonnie Dunbar and G. David Low, also conducted numerous Protein Crystal Growth experiments (PCG). Since protein crystals grown in space are purer and larger than crystals produced on Earth, these experiments performed in microgravity provide scientists with a better understanding of the three-dimensional structure of protein, information needed for the development of new drugs to combat cancer, AIDS, and many other diseases.

After nearly eleven days in orbit, the longest space shuttle-flight to date, STS-32 ended at 1:35 A.M., Pacific standard time (PST), on January 20. This was only the third night landing in shuttle flight history.

The second shuttle flight of 1990, STS-36, was the thirty-fourth U.S. shuttle mission and the sixth devoted to Department of Defense concerns. This flight, which originally was scheduled for launch on February 22, was postponed several times: first, because of Commander John Creighton's illness; second, because of a malfunction of the range safety computer; and third, because of unacceptable weather conditions. Mission STS-36 marked the first time since the Apollo 13 mission of 1970 that a manned space launch was affected by the illness of a crew member. After these delays, however, on February 28, at 2:50 A.M., EST, the five-member crew was launched aboard the space shuttle *Atlantis*. It was NASA's fourth nighttime launch. This classified mission ended after sixty-nine orbits with a successful landing on March 4.

The thirty-fifth space shuttle flight, STS-31, placed into orbit the Hubble Space Telescope (HST), a 2.4 meter reflecting telescope that was created in a cooperative effort by the European Space Agency (ESA) and NASA. The first attempted launch of this historic mission was scrubbed four minutes before the scheduled liftoff on April 10 owing to a faulty valve in the auxiliary power unit. Two weeks later, however, the *Discovery* and its crew were launched into orbit. Liftoff took place on April 24 at 8:34 A.M., eastern daylight time (EDT).

Filling most of the payload bay area of the *Discovery* was the HST, the largest space-based observatory ever built. On the second day of the five day mission, this twelve-ton, railroad-tank-car-sized observatory (43.5 feet long, 14 feet diameter) was deployed into a low-Earth (600 kilometer) orbit, a perch from which it would be able to image objects up to 14 billion light years away. The later discovery of a spherical aberration on the lens of the original Wide Field/Planetary Camera (WF/PC1) of the HST raised initial concerns about the functionality of the Hubble project. In December, 1993, however, the crew of another shuttle flight, STS-61, obviated the effects of the spherical aberration when it replaced the damaged WF/PC1 with an optically corrected spare WF/PC2 and installed corrective mirrors for the primary mirror.

In addition to delivering the HST, the STS-31 crew conducted a variety of mid-deck experiments, including protein crystal growth and polymer membrane processing studies. On April 29, after completing 76 orbits and traveling more than 2 million miles, the crew landed safely on Runway 22 of Edwards Air Force Base, California, at 6:50 A.M. Pacific daylight time (PDT).

After the completion of the successful STS-31 mission, NASA experienced six months of frustration before another shuttle was launched. The next two planned 1990 shuttle flights were the *Columbia*/STS-35 mission and the *Atlantis*/STS-38 mission. Launch dates for these flights originally were set, respectively, for May and July. Owing to a series of hardware problems, however, neither of these missions would be launched before the scheduled October 6 launching of the *Discovery*/STS-41 mission.

NASA's frustration began with the forced scrubbing of Mission STS-35 from its original May 16 launch in order to change out a faulty freon coolant loop proportional valve in *Columbia's* coolant system. Two weeks later a second launch was canceled during tanking owing to a hydrogen leak in the external tank/orbiter disconnect assembly. Since repairs could not be made at the pad, the *Columbia* was returned to the Vehicle Assembly Building (VAB), where it was demated from its external tanks. It was then transferred to the Orbiter Processing Facility (OPF) where it was fitted with new umbilical hardware.

With Mission STS-35 temporarily on hold, the *Atlantis* crew of Mission STS-38 prepared for its scheduled July launch. The hydrogen leak on *Columbia*, however, prompted NASA officials to conduct three precautionary tanking tests on the *Atlantis*. When these tests — performed June 29, July 13, and July 25 — confirmed a hydrogen fuel leak, the *Atlantis* also was returned to the VAB and to the OPF for demating and repair. Unfortunately, on the day that the *Columbia*/STS-35 was being transferred back to the pad for another launch preparation, the *Atlantis*, while parked outside the VAB, was caught in a hailstorm and suffered additional damage to its tiles.

Meanwhile, on August 9, the *Columbia*/STS-35 was rolled out to Pad A for a second time. Two days before the rescheduled September 1 launch, an avionics box on the payload equipment malfunctioned and had to be replaced and retested, forcing another postponement to September 6. During tanking for this launch, however, a high concentration of hydrogen was detected in the orbiter's aft compartment. NASA managers concluded that the *Columbia* had experienced separate hydrogen leaks from the beginning: one in the umbilical assembly, which subsequently had been replaced, and another in the aft compartment, which had resurfaced. Three hydrogen recirculation pumps in the aftcompartment were then replaced and retested, and the launch date was reset for September 18. When the leak in the aft compartment resurfaced again during tanking, NASA decided to put this mission on hold until the problem would be resolved by a special tiger team assigned by the Space Shuttle Director. *Columbia* again was transferred from Pad A to Pad B to make room for the *Atlantis*/Mission STS-36. On October 9, the Columbia orbiter was moved yet again, this time to the VAB to protect it from tropical storm Klaus.

Problems with the *Columbia* and *Atlantis* orbiters, howev-

LDEF grappled and positioned by RMS over OV-102's payload bay during STS-32. (NASA)

STS-31 Hubble Space Telescope is grappled by RMS. (NASA)

er, did not delay the scheduled October 6 launching of *Discovery*/STS-41 mission. The 7:47 A.M., EDT, liftoff occurred just twelve minutes after the two-and-one-half hour launch window opened at 7:35. This mission's launch weight of 259,593 pounds was the heaviest payload to date.

The primary objective of STS-41 was to send the Ulysses spacecraft on a five-year probe to explore the polar regions of the Sun. On the first day of the mission, the Ulysses was released from the cargo bay. Afterward, its two stage Inertial Upper Stage rockets were fired, thus sending the European Space Agency-built Ulysses spacecraft on a sixteen-month voyage to Jupiter. To reach the never-before-seen polar regions of the Sun, the Ulysses had to use the massive gravity of Jupiter to swing it up and perpendicular to the plane of the orbits of the Earth and all the other planets.

Secondary objectives of Mission STS-41 included studying the effects of atomic oxygen wear on solar panels; recording radiation levels in orbit; conducting a Chromosome and Plant Cell Division experiment designed to measure plant root growth patterns in microgravity; performing a physiological systems experiment designed to investigate how microgravity affects bone calcium, body mass, and immune cell functions; and completing a solid surface combustion experiment engineered to study flames in microgravity. Crew members on STS-41 also tested a new voice-actuated command system for future shuttle flights. After four days and sixty-five orbits, the *Discovery* landed on October 10, 6:57 A.M., PDT, at Edwards Air Force Base.

The *Atlantis*/STS-38 mission, which suffered yet another delay in October owing to damages caused by a falling platform beam during hoisting operations, finally received a favorable Flight Readiness Review (FRR) and was rescheduled with a launch date of November 9. Payload problems forced one more postponement, but at 6:48 P.M., EST, on November 15, the long-delayed mission was launched into orbit. This scheduled four-day flight, which was a classified Department of Defense mission, was extended for an additional day owing to unacceptable crosswinds at the original planned landing site at Edwards. Continued adverse conditions led to the decision to shift to the Kennedy Space Center landing site. The *Atlantis* landing, which took place at 4:43 P.M., EST, on November 20, was the first shuttle landing at Kennedy since April, 1985.

Mission STS-35, the other long-delayed 1990 flight, finally received the go-ahead for a launch on December 2. The 1:40 A.M., EST, launch of America's thirty-eighth shuttle mission was NASA's sixth night launch. The primary objective of this mission was to conduct round-the-clock observations of the celestial sphere. Working two twelve-hour shifts, the seven-member crew operated two sophisticated instruments: ASTRO, an ultraviolet astronomy observatory capable of making precise measurements of objects such as planets, stars, and galaxies in relatively small fields of view; and BBXRT, a Broad Band X-Ray Telescope designed to measure directly the amount of energy in electron volts of each X-ray detected from targets such as active galaxies, clusters of galaxies, supernova remnants, and stars. This mission, which had numerous problems before liftoff, also suffered in-flight troubles when the loss of the data display units used for pointing the telescopes impacted the crew-aiming procedures. The problem was partially solved by allowing ground teams at Marshall Space Flight Center to aim the ultraviolet telescopes. The crew on this troubled mission also experienced difficulties dumping waste water owing to a clogged drain, a problem managed by using spare containers for the waste. Fittingly, the mission was cut short by one day due to impending bad weather at the primary landing site. The *Columbia* and crew landed safely at Edwards on December 10 at 9:54 P.M., PST.

Knowledge Gained

Two shuttle missions in 1990, as in 1989, were classified Department of Defense flights, and for these missions, little information has been released. However, an impressive array of scientific intelligence gathered from the four nonclassified 1990 shuttle flights has been disseminated.

Mission STS-32 retrieved from space the LDEF, a facility that had been orbiting the earth since 1984 and at the time of retrieval was within one month of reentering the Earth's atmosphere. During the LDEF's 32,422 Earth orbits, it collected in its eighty-six trays a great variety of data about the space environment and its effects on various materials. Technical knowledge gained from this data led NASA management to conclude that: (1) space environments are hostile to spacecraft materials and coatings; (2) synergistic effects of all aspects of the Low Earth orbit environment must be considered in spacecraft design; (3) contamination should be a very significant consideration in design; (4) the pre-LDEF knowledge of space environmental effects on materials had major flaws; and (5) LDEF knowledge has forced the revision of environment-related test and qualification procedures. One of the many consequences of the LDEF findings was the decision to change the International Space Station radiator design from Teflon, a material that showed substantial deterioration due to atomic oxygen, to a Z-93 ceramic paint, a coating that proved to be very stable in the space environment.

In contrast to STS-32, a mission that retrieved scientific data that was many years in the collecting, Missions STS-31 and STS-41 launched into space vehicles that will be collecting and relaying information about the boundaries of the universe for years to come. For example, the Hubble Space Telescope (HST), which was deployed on STS-31, will enable us to measure stars 30 million light years away, and galaxies 100 million light years away, thus extending the volume of space we can survey to 100 times greater than that of the most advanced ground telescopes. By imaging the distant heavens, scientists hope to be able to determine more accurately the Hubble Constant (a calculation of the rate at which the universe is expanding) and the Deceleration Parameter (a measure of whether the distant galaxies are receding at a slower rate than nearby, newer galaxies). Moreover, by studying the motion of distant galaxies, the HST will collect data that will enable astronomers to infer the mass of galaxies, and from this data, to compute the mass of the universe as a whole. From this and other data, the HST will provide us with insights that will help scientists address some of the grandest questions the human mind has ever pondered, such as: How big is the universe? How old is it? Will it expand forever? How did struc-

ture (galaxies) arise from a fireball (big bang)?

The Ulysses spacecraft, launched from STS-41, also provides us with new measurements and insights about our own Sun. Some of the early findings taken from this probe suggest that in the Sun's polar region, solar wind (a very hot, ionized flow of gases and energetic particles emanating from the Sun) flows at a very high and steady velocity of about 750 kilometers per second (2 million miles per hour), twice the speed of solar winds at the Sun's lower latitudes. Prior to the Ulysses observations, scientists expected a continuous increase of velocity toward the poles. Moreover, many models of the solar magnetic field used prior to Ulysses assumed that the solar magnetic field was similar to that of a dipole. (For a dipole, the field strength over the poles is twice the strength over the equator.) Ulysses observations found, however, that the amount of outward magnetic flux in the solar wind did not vary greatly with latitude, thus indicating the importance of pressure forces near the Sun for evenly distributing the magnetic flux. Such technical knowledge of the Sun is of practical value for engineers since changes in the solar wind pressures and related magnetic disturbances impinge on the Earth's magnetosphere, often causing severe radio communication problems.

The final flight of 1990, STS-35, was perhaps the most disappointing mission of the year. The in-flight problems with ASTRO and BBXRT forced NASA management to abandon the entire preplanned, optimized mission timeline for a replanned, day-by-day schedule. The result was a lower-efficiency mission. In the end, BBXRT achieved a disappointing total of 185,000 seconds of observation time on cosmic X-ray sources. Even so, science teams at Marshall and Goddard Space Flight Centers estimated that nearly 70 percent of the planned science data was achieved on this mission.

Context

For many Americans, 1990 appeared to be a time of stupendous developments and rapid change. Americans greeted the year still celebrating the extraordinary opening of the Berlin Wall in late 1989. Later in the spring of 1990, U.S. President George Bush and Soviet Premier Mikhail Gorbachev signed a major trade agreement and accords on reducing strategic nuclear arsenals and ending the production of chemical weapons. By the year's end, the leaders of the two formerly hostile powers issued a joint statement condemning Iraq's invasion of Kuwait, agreed on the final settlement with respect to German reunification, and signed the Charter of Paris declaring an end to the military division of Europe. As Americans celebrated the ending of the Cold War, they also began to focus more of their attention on domestic concerns, in particular on the federal budget deficit. The 1990's would become a time when Americans would look for ways to reduce the federal government. In this new political environment, even the cherished U.S. space program would not be immune to increased budgetary scrutiny.

Unfortunately for space enthusiasts, during this time of

national reevaluation NASA did not enjoy a banner year. While there was no incident similar to the 1986 STS 51-L (*Challenger*) accident, the shuttle missions of 1990 experienced a number of flight postponements, in-flight troubles, and scientific disappointments. Even the excitement over the April deployment of the Hubble Space Telescope — a highly sophisticated piece of human engineering with the capacity to explore the very boundaries of the universe — dwindled in late June when it was learned that Hubble had a serious design flaw in one of its mirrors. This flaw, which incidentally would be rectified by a HST servicing crew on a 1993 shuttle flight, received more national media attention than the original HST deployment itself. The long-delayed STS-35 mission, and then the less than desirable scientific results obtained from it, bought this frustrating year to a close.

In hindsight, however, 1990 was a better year for NASA than it appeared to many Americans at the time. During 1990, the U.S. safely launched and landed six shuttle crews, conducted thousands of hours of microgravity observations and experiments, and launched the Hubble and the Ulysses projects, unmanned missions that ultimately would prove to be among the most successful space initiatives in U.S. aviation history.

Bibliography

Chaisson, Eric. *The Hubble Wars: Astrophysics Meets Astropolitics in the Two Billion Dollar Struggle Over the Hubble Space Telescope*. New York: HarperCollins Publishers, 1994. An entertaining 386-page treatise on the politics of space exploration.

Collins, Carolyn. *Hubble Vision: Astronomy with the Hubble Space Telescope*. Cambridge/New York: Cambridge University Press, 1995. A scholarly study on space astronomy and the Hubble program that will interest general readers as well as academics.

Embury, Barbara, with Thomas D. Crouch. *The Dream is Alive*. New York: Harper & Row, 1990. An interesting account of the shuttle program intended for juvenile readers.

NASA Space Shuttle Mission STS-32 Press Kit. December, 1989. Since few history monographs of shuttle missions since 1990 have been published, among the better sources of information on these missions are the official NASA press releases. These releases, generally prepared before scheduled liftoff, detail the objectives and planned timelines of the missions. The releases may be viewed on microfilm, or electronically on the Internet. For the other non-classified missions of 1990, see *NASA Space Shuttle Mission STS-31 Press Kit* (April, 1990), *NASA Space Shuttle Mission STS-41 Press Kit* (October, 1990), and *NASA Space Shuttle Mission STS-35 Press Kit* (December, 1990).

NASA. Jet Propulsion Laboratory. *Ulysses: A Voyage to the Sun*. Washington, D.C.: U.S. Government Printing Office, 1986. An early report prepared by the Jet Propulsion Lab at California Institute of Technology that outlines the objectives of the Ulysses project.

Page, D. Edgar and Edward J. Smith. "Reflecting on the Findings of the Ulysses Spacecraft." *Eros* 76 (July 25, 1995): 297-302. A technical update on the progress of the mission.

Terry D. Bilhartz

SPACE SHUTTLE FLIGHTS, 1991

Date: April 5 to December 1, 1991
Type of program: Manned Earth–orbiting spaceflight

Three space shuttle missions were flown in 1991 by the shuttle Atlantis, *two more by* Discovery, *and one by* Columbia. *Satellites to monitor the upper atmosphere, relay data from shuttles to the ground, and monitor missile launchings and nuclear explosions were deployed. Shuttle astronauts conducted experiments to determine the physiological effects of spaceflight and demonstrate techniques for construction of the space station.*

PRINCIPAL PERSONAGES

STEVEN R. NAGEL, STS-37 Commander
KENNETH D. CAMERON, STS-37 Pilot
LINDA M. GODWIN,
JERRY L. ROSS, and
JAY APT, STS-37 Mission Specialists
MICHAEL L. COATS, STS-39 Commander
L. BLAINE HAMMOND, STS-39 Pilot
GREGORY J. HARBAUGH,
DONALD R. "DON" McMONAGLE,
GUION S. BLUFORD, JR.,
C. LACY VEACH, and
RICHARD J. HIEB, STS-39 Mission Specialists
BRYAN D. O'CONNOR, STS-40 Commander
SIDNEY M. GUTIERREZ, STS-40 Pilot
JAMES P. BAGIAN,
TAMARA E. JERNIGAN, and
M. RHEA SEDDON, STS-40 Mission Specialists
FRANCIS A. "DREW" GAFFNEY and
MILLIE W. HUGHES-FULFORD, STS-40 Payload
    Specialists
JOHN E. BLAHA, STS-43 Commander
MICHAEL A. BAKER, STS-43 Pilot
SHANNON W. LUCID,
G. DAVID LOW, and
JAMES C. ADAMSON, STS-43 Mission sSpecialists
JOHN O. CREIGHTON, STS-48 Commander
KENNETH S. REIGHTLER, JR., STS-48 Pilot
CHARLES D. "SAM" GEMAR,
JAMES F. BUCHLI, and
MARK N. BROWN, STS-48 Mission Specialists
FREDERICK D. GREGORY, STS-44 Commander
TERENCE T. "TOM" HENRICKS, STS-44 Pilot
JAMES S. VOSS,

F. STORY MUSGRAVE, and
MARIO RUNCO, JR., STS-44 Mission Specialists
THOMAS P. HENNEN, STS-44 Payload Specialist

Summary of the Missions

The U.S. manned space program experienced a delay in early 1991, when cracked hinges were found on all three of the space shuttles: *Atlantis, Discovery,* and *Columbia.* These hinges allowed a hatch, which covers the entry port for hoses from the shuttle's external fuel tank, to close during reentry into the atmosphere. If this hatch did not close completely, the shuttle could burn up during reentry. The largest cracks were on the shuttle *Discovery,* and its launching, planned for February 28, was postponed for two months to permit repairs. Only one hairline crack was detected on *Atlantis,* so it made the first shuttle flight of 1991.

The space shuttle *Atlantis* lifted off from the NASA John F. Kennedy Space Center, Florida, at 9:23 A.M., eastern standard time, on April 5, 1991, under the command of United States Air Force colonel Steven R. Nagel, who was making his third trip into space. STS-37 mission was piloted by United State Marine Corp lieutenant colonel Kenneth D. Cameron, who was making his first spaceflight. *Atlantis* also carried three mission specialists, United States Air Force lieutenant colonel Jerry L. Ross, who was making his third spaceflight, and Jay Apt and Linda M. Godwin, both on their first shuttle flights.

On April 7, the crew deployed the seventeen-ton Compton Gamma Ray Observatory (GRO), the heaviest payload ever carried aboard the shuttle, into a nearly circular orbit 280 miles above the Earth's surface. The GRO was the second of four large astronomical observatories NASA planned to launch. The first was the Hubble Space Telescope launched in 1990. Unlike the Hubble, which observes astronomical objects in visible light, the GRO detects very high energy electro-magnetic radiation, called gamma rays. While the GRO was still attached to the robot arm used to remove it from the shuttle's cargo bay, the sixteen-foot antenna that it would use to communicate with Earth failed to unfold despite repeated commands transmitted from the NASA ground stations. Astronauts Ross and Apt, in a four-and-a-half-hour emergency spacewalk, were able to free the antenna.

Astronauts Ross and Apt conducted an even longer spacewalk, lasting six-and-a-half hours, on April 8. They assembled a forty-seven-foot long rail in the cargo bay, then tested prototypes of three carts, one manually propelled, one driven by an electric motor, and one using a mechanical pump, which might be used in the construction of NASA's space station.

After extending its flight for one day because of high winds in the landing area, *Atlantis* landed at Edwards Air Force Base in California at 6:55 A.M., Pacific daylight time, on April 11, 1991.

The space shuttle *Discovery* lifted off from the Kennedy Space Center in Florida at 7:33 A.M., eastern daylight time, on April 28, 1991, on a military mission to test systems developed for the Strategic Defense Initiative program. United States Navy captain Michael L. Coats, who was making his third shuttle flight, commanded the STS-39 mission, and United States Air Force lieutenant colonel L. Blaine Hammond, on his first flight on the shuttle, piloted *Discovery* into orbit. This was the first military mission of the space shuttle that was not cloaked in secrecy. On prior military flights the press was not even given advance notice of the launch or landing times. However, in a budget-cutting measure, the secret launch control room at the Kennedy Space Center and the secret flight control room at the Lyndon B. Johnson Space Center in Houston, Texas, had been closed in late 1990.

On April 29, the crew opened the cover on the Cryogenic Infrared Radiation Instrumentation for Shuttle (CIRRIS) instrument, designed to monitor the infrared radiation emitted by natural processes such as aurora. This experiment was designed to provide the information needed to construct instruments to distinguish between the heat emitted by enemy missiles and that emitted by naturally occurring processes.

On May 1, the crew released the $94 million shuttle pallet satellite into space. Sensors on this satellite monitored the emission from the rocket thrusters used to reposition the shuttle and compared this with emissions from vapor clouds sprayed from gas canisters aboard the shuttle. The objective was to determine how to distinguish actual rocket exhaust from vapor clouds released by enemy decoys.

On May 4, the crew overcame a problem with the data recorders on a third experiment, designed to distinguish between natural and human-induced x-ray sources. The objective was to monitor compliance with nuclear test ban treaties, since nuclear explosions release a burst of X rays.

Discovery had been scheduled to land at Edwards Air Force Base. However, high winds at the California site forced it to divert to the Kennedy Space Center in Florida, where the shuttle landed at 2:55 P.M., eastern daylight time, on May 6.

The space shuttle *Columbia* had been scheduled for a flight on May 21, 1991. However, this launching was postponed when *Columbia* developed three problems: electronic circuits linking the booster rockets to the orbiter failed, one of the five computers on the orbiter failed, and cracked welds

were found in sensors that monitored the temperature of the liquid hydrogen fuel lines. The weld problem, which had not been detected in earlier inspections, was the most serious. If a sensor broke free, it could be sucked into the engine's pumps causing an explosion. Further inspection found six cracks in welds on *Columbia*, as well as cracks in welds on *Atlantis* and *Discovery*.

Following the replacement of three temperature sensors, *Columbia* lifted off from the Kennedy Space Center at 9:25 A.M., eastern daylight time, on June 5, 1991, under the command of United States Marine Corps colonel Bryan D. O'Connor. The purpose of the nine-day STS-40 mission was to study the physiological effects of spaceflight on living organisms. The Spacelab module, a twenty-two-foot-long by sixteen-foot-diameter pressurized cylinder, was carried aloft in *Columbia's* cargo bay. It contained some of the world's most sophisticated medical instruments, as well as twenty-nine rats and 2,748 jellyfish. The jellyfish have gravity sensing mechanisms similar to those in the human ear, and the *Columbia* scientists studied how the jellyfish oriented themselves and adapted to the weightless environment. The rats were dissected after the flight to see how their bodies had changed in response to weightlessness. *Columbia's* flight ended on June 14, 1991, with a landing at Edwards Air Force Base at 8:39 A.M., Pacific daylight time.

The shuttle *Atlantis*, under the command of United States Air Force colonel John E. Blaha and piloted by United States Navy lieutenant commander Michael A. Baker, lifted off on mission STS-43 from the Kennedy Space Center at 11:02 A.M., eastern daylight time, on August 2, 1991. *Atlantis* carried a Tracking and Data Relay Satellite (TDRS) in its cargo bay. The 4,700 pound TDRS, built by TRW Inc., is a communications satellite designed to relay data from the space shuttle and other Earth orbiting satellites to ground stations. Since radio signals travel in straight lines, NASA had previously employed a worldwide network of ground stations to maintain communications with spacecraft in orbit. The use of TDRS relays allowed NASA to close many of these ground stations. This TDRS, deployed six hours after liftoff, joined three others, placed in orbit previously.

The second purpose of this mission was to evaluate equipment being considered for use on the planned space station. The crew tested a fiber-optic communications link, new cooling systems, and a method to minimize the effects of weightlessness by creating a low-pressure on the lower body. The crew, all but one of whom had flown previously, observed that the Earth was surrounded by an unusually thick haze that they speculated might have been caused by either the eruption of Mt. Pinatubo in the Philippines in June or by residue from the oil well fires set in Kuwait at the end of the Gulf War.

Atlantis landed at the Kennedy Space Center at 8:23 A.M., eastern daylight time, on August 11, 1991. This was the first scheduled landing of a shuttle in Florida since 1986.

The shuttle *Discovery*, commanded by United States Navy captain John O. Creighton and piloted by United States Navy commander Kenneth S. Reightler, Jr., lifted off from the Kennedy Space Center at 7:11 P.M., eastern daylight time, on September 12, 1991. STS-48 marked the beginning of NASA's Mission to Planet Earth, a decade-long effort to monitor the Earth's environment from space.

The Upper Atmosphere Research Satellite (UARS), weighing 14,500 pounds and costing $740 million, was released from the cargo bay on September 15. The UARS carried ten instruments to monitor the chemistry, winds, and heat distribution of the upper atmosphere of the Earth. One major objective was to study the ozone layer, which shields the Earth's surface from the ultraviolet rays that are harmful to biological organisms. Bad weather in Florida kept *Discovery* from making the first night landing on the Kennedy Space Center runway. The shuttle was diverted to Edwards Air Force Base, where it landed at 12:38 A.M., Pacific daylight time, on September 18.

The shuttle *Atlantis*, commanded by United States Air Force colonel Frederick D. Gregory and piloted by United States Air Force lieutenant colonel Terence T. Henricks, lifted off from the Kennedy Space Center, on the last shuttle mission of 1991, at 6:44 P.M., eastern standard time, on November 24, 1991. STS-44 was a military mission during which the crew observed ground sites to determine if human reconnaissance from space would be useful in times of world crises. They used binoculars, a small telescope, and a digital camera to observe twenty-five U.S. military sites around the globe.

The crew of *Atlantis* also deployed a 5,200 pound spy satellite, containing a large infrared telescope to detect missile launches and nuclear explosions. Earlier in 1991, similar satellites had helped track Scud missiles launched by Iraq during the Gulf War.

One of the three inertial measurement units, which monitor the velocity and orientation of the shuttle in space, failed during the flight, forcing *Atlantis* to return to Earth three days earlier than scheduled. *Atlantis* landed at Edwards Air Force Base at 2:35 P.M., Pacific standard time, on December 1.

Knowledge Gained

The Compton Gamma Ray Astronomy satellite, launched by the shuttle *Atlantis* on April 7, 1991, provided scientists with the opportunity to study some of the most violent astronomical events in the universe. Gamma rays, which are very high energy radiation, do not penetrate the Earth's atmosphere, and thus cannot be studied from the ground. Gamma ray detectors flown on balloons and smaller satellites had provided some information on these violent events, but the sensors on the Gamma Ray Observatory are ten to twenty times more sensitive than any previously flown, allowing detection of sources at greater distances from the Earth.

Vela satellites, whose primary mission is to detect nuclear-explosions on Earth, had detected bursts of gamma rays from space. One burst, detected on March 5, 1979, lasted only one-five-thousandth of a second, but gave off energy at a rate greater than the visible emission from the entire Milky Way galaxy. The Gamma Ray Observatory mapped the distribution of these gamma ray bursts across the sky and found they were apparently randomly distributed. This indicates to astronomers that these sources are not located within our galaxy, and that they must be even more energetic than previously assumed because they are located at large distances from the Earth.

Since the beginning of manned spaceflight, experts in space medicine had recognized that the human body reacted adversely to weightlessness. Studies of the physiological effects of weightlessness were a major objective of the long-duration missions on the U.S. Skylab space station, in the 1970's. The nine-day flight of *Columbia* carried three mission specialists, Dr. James P. Bagian, a surgeon, Dr. Francis A. Gaffney, a cardiologist, and Millie Hughes-Fulford, a cell biologist, to investigate these effects. The other *Columbia* astronauts were examined, and body fluid samples were collected throughout the flight to monitor the onset of the body's reaction to weightlessness.

On Earth, fluids accumulate in the legs due to gravity, but in weightlessness these fluids are redistributed more evenly throughout the body. The resulting increase in fluids in the torso causes the body to expel the perceived excess fluids. This fluid loss was observed to cause an increase in the heart rate. As the exposure to weightlessness continues, bones lose calcium and muscles atrophy. The physiological measurements on the *Columbia* crew provided data to support future long-duration space missions, including the space station and a human mission to Mars.

On its November flight, *Atlantis* crew members tested a rowing machine and a treadmill designed to determine if some of the physiological effects of weightlessness could be overcome by exercise.

Context

Mission to Planet Earth is NASA's contribution to the U.S. Global Change Research Program, an interagency effort to understand global change and the impact of human activity on the planet. This program is designed to distinguish changes caused by human activity from those resulting from natural processes. For example, both industrial processing and natural activity, such as volcanos, alter the chemical composition of the atmosphere. The Nimbus 7 weather satellite carried some instruments to monitor the upper atmosphere, but the Upper Atmosphere Research Satellite (UARS), launched by the shuttle *Discovery* in September, 1991, was the first satellite completely dedicated to this effort. The UARS monitored the day/night variation in the abundances of chemical species important in maintaining the ozone layer as well as measuring the abundances of nitrous oxides and halocarbons, important in maintaining the nitrogen and the chlorine cycles, in the upper atmosphere.

Each of the last two shuttle missions of 1991 had to take evasive action to avoid space debris. On September 15, *Discovery* maneuvered to avoid flying within 1.7 miles of the upper stage of the rocket that had placed Cosmos 955 into orbit in September, 1977. NASA officials indicated this was the first time one satellite had to maneuver to avoid a close approach with another. Because of the catastrophic consequences of high-speed collisions, NASA flight rules require that the shuttle avoid coming within five miles of orbiting space debris. On November 28, the Shuttle *Atlantis* maneuvered to avoid coming within 1.5 miles of part of the booster rocket of Cosmos 851, launched in 1976. These two incidents dramatically illustrated the increasing hazard to future spaceflight of man-made debris orbiting the Earth.

In April, 1991, the space shuttle *Endeavour*, the replacement for the shuttle *Challenger* lost in January, 1986, was unveiled at the Palmdale, California, facility of Rockwell International Corporation. *Endeavour* was built using many spare parts left over from earlier shuttles, but improved systems allow *Endeavour* to stay in orbit for up to twenty-eight days. Although NASA sought permission to build an additional shuttle, Vice President Dan Quayle, in his role as chair of the National Space Council, announced on July 24, 1991, that no additional space shuttles would be built. In the 1970's, the shuttle was envisioned as America's only access to space, but Quayle indicated the U.S. would shift to unmanned rockets to launch future commercial and military satellites. Quayle said: " The space shuttle, with its precious human lives, is just too valuable to use on missions that don't need its unique capabilities."

Bibliography

Covault, C. "USAF/SDI Space Shuttle Flight to Execute Complex Maneuvers." *Aviation Week and Space Technology* 134 (March 4, 1991): 46-69. A well-illustrated account of the April, 1991, flight of the shuttle *Discovery*, describing the instruments and experiments on this secret military mission.

Fox, Mary V. *Women Astronauts Aboard the Shuttle*. New York: Julian Messner. Provides biographical information and interviews with Dr. Margaret Seddon and Shannon Lucid, both of whom flew on shuttle missions in 1991.

Schuiling, R. L. and S. Young. "STS-37 Mission Report: Astronauts Give GRO a Helping Hand." *Spaceflight* 33 (June, 1991): 194-205. A detailed, well-illustrated account of the April, 1991, flight of *Discovery*, including a description of the Compton Gamma Ray Observatory and its scientific objectives.

Schuiling, R. L. and S. Young. "Life Sciences Get Important New Data from Spacelab Mission." *Spaceflight* 33 (December, 1991): 336-341. Description of the June, 1991, flight of *Columbia*, including a thorough account of the biomedical results obtained on this mission.

George J. Flynn

SPACE SHUTTLE FLIGHTS, 1992

Date: January 22 to December 9, 1992
Type of mission: Manned Earth-orbiting spaceflight

Eight space shuttle missions were flown in 1992. Several new manned spaceflight records were set during these missions, including the longest shuttle mission to that date and the most spacewalks in one mission until 1993. The fourth space shuttle orbiter, Endeavour, *was put into service, replacing the* Challenger, *which was destroyed during a launch accident in 1986.*

PRINCIPAL PERSONAGES:

EUGENE F. KRANZ, Director, Mission Operations,
    Johnson Space Center
ROBERT B. SIECK, Launch Director, Kennedy Space
    Center
RONALD J. GRABE, STS-42 Commander
STEPHEN S. OSWALD, STS-42 Pilot
NORMAN E. THAGARD,
WILLIAM F. READDY, and
DAVID C. HILMERS, STS-42 Mission Specialists
ROBERTA L. BONDAR and
ULF MERBOLD, STS-42 Payload Specialists
CHARLES F. BOLDEN, JR., STS-45 Commander
BRIAN DUFFY, STS-45 Pilot
KATHRYN D. SULLIVAN,
DAVID C. LEETSMA, and
C. MICHAEL FOALE, STS-45 Mission Specialists
DIRK D. FRIMOUT, and
BYRON K. LICHTENBERG, STS-45 Payload Specialists
DANIEL C. BRANDENSTEIN, STS-49 Commander
KEVIN P. CHILTON, STS-49 Pilot
RICHARD J. HIEB,
BRUCE E. MELNICK,
PIERRE J. THUOT,
KATHRYN C. THORNTON, and
THOMAS D. "TOM" AKERS, STS-49 Mission Specialists
RICHARD N. "DICK" RICHARDS, STS-50 Commander
KENNETH D. BOWERSOX, STS-50 Pilot
BONNIE J. DUNBAR,
ELLEN S. BAKER, and
CARL J. MEADE, STS-50 Mission Specialists
LAWRENCE J. DELUCAS and
EUGENE H. TRINH, STS-50 Payload Specialists
LOREN J. SHRIVER, STS-46 Commander
ANDREW M. ALLEN, STS-46 Pilot
CLAUDE NICOLLIER,

MARSHA S. IVINS,
JEFFERY A. HOFFMAN, and
FRANKLIN R. CHANG-DIAZ, STS-46 Mission
    Specialists
FRANCO MALERBA, STS-46 Payload Specialist
ROBERT L. "HOOT" GIBSON, STS-47 Commander
CURTIS L. "CURT" BROWN, JR., STS-47 Pilot
MARK C. LEE,
JAY ABT,
N. JAN DAVIS, and
MAE C. JEMISON, STS-47 Mission Specialists
MAMORU MOHRI, STS-47 Payload Specialist
JAMES D. WETHERBEE, STS-52 Commander
MICHAEL A. BAKER, STS-52 Pilot
C. LACY VEACH,
WILLIAM M. SHEPARD, and
TAMARA E. JERNIGAN, STS-52 Mission Specialists
STEVEN G. "STEVE" MACLEAN, STS-52 Payload
    Specialist
DAVID M. WALKER, STS-53 Commander
ROBERT D. CABANA, STS-53 Pilot
GUION S. BLUFORD,
JAMES S. VOSS, and
MICHAEL R. "RICH" CLIFFORD,
    STS-53 Mission Specialists

Summary of the Missions

The year 1992 was dubbed International Space Year. Eight space shuttle missions were flown during the year. The orbiter *Endeavour* (OV-105) was added to the fleet of active orbiter vehicles, joining *Columbia* (OV-102), *Discovery* (OV-103), and *Atlantis* (OV-104). Each of the four shuttles flew two missions during the year. Most of the missions during the year were science missions. In addition to the science missions, satellite deployment and retrieval missions were also flown, and the final flight of the year was a classified military mission.

The first space shuttle mission of the year was STS-42. The orbiter *Discovery* was launched January 22, 1992, at 9:53 A.M., eastern standard time, from Pad A, of Launch Complex 39, Kennedy Space Center. The mission was primarily a scientific mission with the International Microgravity Laboratory-1 (IML-1) flying in the shuttle's cargo bay.

STS-50 Columbia, *OV-102, landing with drag chute deploy at the Kennedy Space Center.* (NASA)

Landing was postponed for one day to allow for additional scientific investigations. The landing finally occurred at 8:07 A.M., Pacific standard time, on Runway 22, Edwards Air Force Base, in California.

The International Microgravity Laboratory-1 was designed to investigate the effects of weightlessness on living organisms and materials processing techniques. The term "microgravity" is used rather than the older terms "zero gravity" or "weightlessness" since all objects have some gravity, and thus the experiments experienced almost no gravity but were not really done in zero gravity conditions.

In addition to scientific experiments, astronauts also took film footage of the STS-42 mission with IMAX cameras for a film to be made in commemoration of the International Space Year. This film was later shown in Omniplex theaters around the country.

The second space shuttle mission of the year, STS-45, began with the launch of the Space Shuttle *Atlantis* from Pad A of Launch Complex 39 at the Kennedy Space Center at 8:14 A.M., eastern standard time, on March 24, 1992. The

launch was one day late due to a possible fuel leak, which had been detected during the filling of the shuttle's external fuel tank. This mission, like the first space shuttle mission of the year, was primarily a scientific mission. The primary payload of the *Atlantis* was the Atmospheric Laboratory for Applications and Science (ATLAS-1), which allowed astronauts to study the Earth's atmosphere from above. STS-45 ended with *Atlantis* landing on April 2, 1992, at 6:23 A.M. on Runway 33 at the Kennedy Space Center.

STS-45 was an international mission, with instruments in the ATLAS-1 investigations being provided by institutions and scientific agencies in seven different nations. ATLAS-1 allowed space scientists to study the composition and dynamics of the Earth's upper atmosphere from a unique perspective, from above. In addition to providing data for other ongoing satellite studies of the atmosphere, ATLAS-1 was designed to provide a set of comparison data for the later ATLAS missions, scheduled for space shuttle flights over the next few years.

The third space shuttle mission of 1992, STS-49, was the

maiden voyage of the Space Shuttle *Endeavour*. Launch was from Pad 39-B on May 7, 1992, at 7:40 P.M., eastern daylight time. The primary goal of this mission was the rescue of the INTELSAT VI (F-3) satellite, which had been stranded in a useless orbit by the malfunction of its Titan Rocket launch vehicle. Difficulties with the satellite rescue portions of the mission caused mission controllers to extend the mission two extra days to complete all of the flight's objectives. Landing occurred at 1:58 P.M., Pacific daylight time, on Runway 22 at Edwards Air Force Base, in California.

The primary objective of STS-49 was the rescue of the INTELSAT VI (F-3) satellite. This proved to be much more difficult than had been anticipated. The satellite had not been designed to be retrieved in space and thus did not have any good places for the astronauts to grab it with the shuttle's robot arm, the Remote Manipulator System (RMS). Difficulties in attaching a capture bar to the satellite for the RMS to grab forced a new, more daring plan to capture the satellite. Astronauts built a bridge structure across the shuttle's cargo bay and three astronauts, Pierre Thuot, Richard Hieb, and Thomas Akers, stood on the bridge and manually grasped the satellite to stabilize it so that the capture bar could be attached. A new kick motor was attached to the satellite to push it into a useful orbit, and after several unsuccessful attempts, the satellite was finally redeployed.

The fourth mission of the year, STS-50, began at 12:12 P.M. on June 25, 1992, with the launch of the Space Shuttle *Columbia* from Pad 39-A. STS-50 was another science mission, studying the effects of microgravity on materials processing and on living organisms. The primary payload was the U.S. Microgravity Laboratory-1 (USML-1). The USML-1 experiments were similar to those performed in IML-1 in the STS-42 mission earlier in the year. The orbiter *Columbia* had been extensively refurbished during the year since its last flight, enabling STS-50 to be an extended duration mission of more than thirteen days, marking the longest space shuttle mission to that time. Landing was at the Kennedy Space Center on Runway 33 at 7:42 A.M. on July 9, 1992.

Several problems developed with the experimental equipment; however, almost all of the mission objectives were met. A more serious problem developed on the second day of the flight when the new system installed on board the *Columbia* to remove carbon dioxide gas from the air on board the orbiter failed. A backup system consisting of lithium hydroxide canisters was used to absorb the carbon dioxide while the astronauts worked to repair the primary system. The regenerative carbon dioxide removal system was repaired by the sixth day of the mission, and the mission was able to continue as planned.

The fifth mission of the year was STS-46. The Space Shuttle *Atlantis* lifted off from Pad 39-B at 9:57 A.M., eastern daylight time, on July 31, 1992. This mission was both a science and a satellite deployment mission, requiring the deployment of two satellites. The European Retrieval Carrier

(EURECA) was deployed by STS-46 to be retrieved later in another shuttle mission. Another satellite scheduled for deployment was the Tethered Satellite System (TSS-1). Difficulties with deployment of both primary payloads forced an extension of the mission of one day. Landing was August 8, 1992, at 9:12 A.M., eastern daylight time, on Runway 33 at the Kennedy Space Center.

The first part of the mission involved deployment of the EURECA satellite. This satellite was designed to investigate the effects of microgravity and space radiation on materials and living organisms. This is a similar mission to the IML-1 and USML-1 missions flown aboard the space shuttle, but the investigations last for longer durations than a single space shuttle flight. EURECA was retrieved by the STS-57 mission nearly one year later.

The TSS-1 mission objectives were not met. TSS-1 was to have been deployed on a twelve-mile-long tether. As the satellite and its tether swept through space, scientists would have been able to study the Earth's magnetic field. The tether was to have been used as a test of generating electric power from the Earth's magnetic field, perhaps for a future orbiting space station.

On September 12, 1992, the Space Shuttle *Endeavour* lifted off from Pad 39-B at 10:23 A.M., eastern daylight time, to begin the sixth space shuttle mission of the year, STS-47. This mission was another science mission, studying the effects of microgravity. The primary payload was Spacelab-J, the first United States and Japan joint space shuttle mission. The mission was extended one day to allow for additional data to be gathered. Landing was on Runway 33 at the Kennedy Space Center at 8:53 A.M., eastern daylight time, on September 20, 1992.

Astronauts conducted forty-three materials science and life science experiments during the Spacelab-J mission. This mission had an unusually large number of life science experiments. Among the firsts of this mission was the Frog Embryology Experiment, which marked the first study of an animal's ovulation, fertilization, and development under microgravity conditions. Astronauts also studied the effectiveness of using biofeedback techniques to control motion sickness, which commonly affects space travelers.

The seventh space shuttle mission of 1992 was STS-52. The Space Shuttle *Columbia* began the mission October 22, 1992, at 1:10 P.M., eastern daylight time, from Pad 39-B at the Kennedy Space Center. The mission was a joint science mission and satellite deployment mission. Many of the science experiments were performed as part of the U.S. Microgravity Payload-1 (USMP-1) in the cargo bay. STS-52 also deployed the Laser Geodynamic Satellite II (LAGEOS II). Landing was at Kennedy Space Center on Runway 33 at 9:06 A.M., eastern standard time, on November 1, 1992.

The USMP-1 payload performed quite well. USMP-1 allowed scientists on the ground an opportunity to interact with their experiments in orbit using a new system for relaying telemetry from the experiments to the ground for imme-diate analysis. This allowed scientists to issue commands to the USMP-1 investigations while the experiments were under way and after initial data had already been examined.

LAGEOS II was deployed on the second day of the mission. A rocket motor on the satellite lifted it from the space shuttle's low Earth orbit to a higher circular orbit. The LAGEOS II satellite replaced the LAGEOS I satellite launched in 1976 on a Delta rocket. These satellites were designed to use laser-ranging systems to study motions of the Earth's crust, tides, and the Earth's rotation.

The eighth, and final, space shuttle mission of the year was a Department of Defense mission, STS-53. The mission began with the launch of *Discovery* from Pad 39-A at the Kennedy Space Center on December 2, 1992, at 8:24 A.M., eastern standard time. The primary mission was the deployment of a classified military payload. Landing was at 12:44 P.M., Pacific standard time, on Runway 22 at Edwards Air Force Base, in California, on December 9, 1992.

In addition to the classified primary payload, STS-53 astronauts experimented with a new system developed to determine the positions of targets on the Earth from orbit. The system is called the Hand-Held Earth-oriented, Real-time, Cooperative, User-friendly, Location-targeting and Environmental System (HERCULES). HERCULES allows astronauts to point a camera at a feature on Earth, take its picture, and determine its latitude and longitude to an accuracy of less than four kilometers.

Knowledge Gained

The materials science microgravity experiments performed in IML-1 (STS-42), USML-1 (STS-50), Spacelab-J (STS-47), and USMP-1 (STS-52) allowed scientists and engineers to learn how crystals grow in space. All crystals grown on Earth experience gravitational forces, and these forces are all in one direction during the formation of the crystal. Under microgravity conditions, such as aboard the shuttle, crystals are free to grow with minimal external influences. These crystals tend to be more pure and contain fewer defects than those grown on Earth. Crystals also tend to grow larger in orbit than they do on the Earth. The EURECA satellite also had materials science experiments aboard, but the results of these experiments would become known when the satellite was retrieved during the STS-57 mission in 1993.

In addition to crystal growth studies, one of the major investigations of the USMP-1 payload was the study of the lambda point transition of helium. This transition is the change of state that occurs in helium at 2.17 Kelvin (degrees above absolute zero). At this temperature, liquid helium changes from an ordinary liquid to a superfluid. Superfluids have properties different from ordinary liquids, including the ability to flow without viscosity. While this transition of helium from normal liquid to superfluid had been studied, it had never been studied in microgravity.

Numerous life science experiments were performed during almost every space shuttle mission. Every mission involved study of the astronauts themselves as they adapted to microgravity conditions. Some surprises indicated that seeds tended to sprout sooner and plants grow more rapidly in space than on Earth. This finding was particularly apparent during the IML-1 mission. The Spacelab-J mission studied the growth of insects in orbit. Experiments with frogs provided the first studies of the conception and early developmental stages of animals under microgravity conditions. These studies allow scientists to learn in part what development is due to genetics and what development in environmental.

The ATLAS-1 mission (STS-45) provided valuable data on the Earth's atmosphere. Many of these investigations were designed to establish a baseline from which to make comparisons for other atmospheric studies. Ozone levels in the upper atmosphere were studied along with other measurements of atmospheric composition. ATLAS-1 discovered residual aerosols from the huge volcanic eruption in 1991 of Mount Pinatubo in the Philippines.

The STS-49 mission to rescue a stranded satellite provided valuable experience in securing and servicing a satellite from the space shuttle. This experience was put to use over a year later during the STS-61 mission to service and repair the Hubble Space Telescope.

Not all the knowledge gained by the space shuttle flights of 1992 was gained by experiments in orbit. Some useful engineering data was gathered in the landing of the orbiter. *Endeavour* and the newly refurbished *Columbia* employed drag chutes upon landing. These chutes allowed for shorter stopping distances on landing, thus adding an extra margin of safety. Unfortunately, these chutes tended to sometimes cause the orbiter to veer slightly to the side of the runway centerline during landing. This may have been due to crosswinds at landing.

Context

The space shuttle flights of 1992 continued the saga of manned spaceflight with the setting of several new records. Among those records were a record number of four spacewalks, or Extravehicular Activities (EVAs). Two of the EVAs of the STS-49 mission became the longest EVAs conducted by astronauts, with one lasting 8 hours 29 minutes and the other lasting 7 hours 45 minutes. Both of these activities surpassed the previous record of 7 hours 37 minutes held by Apollo 17 astronauts on the surface of the Moon.

In addition to long spacewalks, the *Columbia* spent more than thirteen days in space. This marked a new endurance record for manned spacecraft. Only space station missions, such as Skylab or the Russian space station, Mir, have lasted longer than the STS-50 mission. The importance of long duration missions is that they allow time for additional experiments or for space construction activities, such as would be needed to build a space station in orbit.

One of the chief rationales for the space shuttle program given in the 1970's was that such a system would allow astronauts to deliver satellites to orbit or to retrieve satellites already in orbit. The space shuttle deployed EURECA and a defense payload. Attempts to deploy TSS-1 failed, yet valuable information was gained from the attempt.

Much more important, the STS-49 mission proved the importance of the space shuttle system. By retrieving and replacing the motor of the INTELSAT VI satellite, STS-49 saved millions of dollars that would have been lost replacing the satellite with another using yet another expendable launch vehicle. This rescue mission, in fact, required human ingenuity and could not have been done with an unmanned spacecraft.

The microgravity experiments were vital to our understanding of how materials and living organisms adapt to gravity or the lack of gravity. Future humans will be living in space, and it is important to know how long term exposure to the lack of gravity or to the radiation in space affects living organisms. The plant studies are vital to any attempt that may be made to try to grow food in space. The extended missions allowed further studies to be made than would have been possible with shorter duration flights, yet they also pointed to the need for a permanent manned space station to conduct even longer duration scientific investigations. The materials science investigations are essential to our understanding of crystal structure. Most of modern electronics depends upon solid state technology, and this technology utilizes the crystal structure of semiconductors. Better understanding of crystal and crystal growth helps in the development of better computer chips. The higher purity and greater perfection of crystals grown in space would tend to indicate that space-based manufacturing of these crystals may be in our future.

The ATLAS-1 mission was the first of ten scheduled ATLAS missions. These missions provide data on the Earth's atmosphere from above, a perspective impossible to ground-based scientists.

Bibliography

Branley, Franklyn M. *From Sputnik to Space Shuttles.* New York: Thomas Y. Crowell, 1986. This is a very good juvenile book on the development of space exploration, from unmanned satellites to the modern space shuttle. The book emphasizes the uses of the space shuttle more than the shuttle itself.

DeWaard, E. John, and Nancy DeWaard. *History of NASA: America's Voyage to the Stars.* New York: Exeter Books, 1988. This book is intended for the general audience. The last chapter of the book covers the beginnings of the space shuttle program. The need for a system such as the space shuttle is discussed. The book ends with the explosion that destroyed the *Challenger* in 1986.

Hawkes, Nigel. *Space Shuttle.* New York: Gloucester Press, 1983. This juvenile book includes a very nice description of the design and layout of the space shuttle. It also includes a very brief discussion of living in space and the sorts of payloads and missions available to the space shuttle.

Joels, Kerry M., Gregory P. Kennedy, and David Larkin. *The Space Shuttle Operator's Manual.* New York: Ballantine Books, 1982. This is the definitive description of the space shuttle itself available for general audiences. It includes diagrams and charts illustrating key parts of the space shuttle, and even includes fold-out sections showing instrument panels. It includes a very good description of living facilities on board the shuttle. There are even detailed timelines for both the launch and landing phases of the missions.

Rycroft, Michael, Ed. *The Cambridge Encyclopedia of Space.* New York: Cambridge, 1990. This general audience level work contains a section on space launch facilities, including the Kennedy Space Center, where space shuttle missions are launched. Further sections discuss living in space, with a very thorough description of the various types of microgravity experiments that can be conducted in space. The space shuttle system itself is given considerable coverage.

Shapland, David, and Michael Rycroft. *Spacelab: Research in Earth Orbit.* New York: Cambridge, 1984. Written in the early days of the space shuttle program, this general audience level book describes the types of scientific missions anticipated for the space shuttle. The book describes a spacelab, which is a living/working module that is flown in the space shuttle's cargo bay during science missions.

Torres, George J. *Space Shuttle: A Quantum Leap.* Novato, California: Presidio Press, 1986. This general audience level book traces the development of manned spaceflight, culminating with the space shuttle. Included are brief descriptions on many different types of shuttle missions. The book was written at the time of the *Challenger* explosion, and does not include any of the missions that occurred after the resumption of the shuttle program.

Raymond Benge

SPACE SHUTTLE MISSION
STS-49

Date: May 7, 1992, to May 16, 1992
Type of mission: Manned Earth-orbiting spaceflight

The newest space shuttle orbiter, Endeavour, *was built as replacement for the* Challenger, *lost during a launch accident in 1986. STS-49 was the maiden voyage of* Endeavour.

PRINCIPAL PERSONAGES

DANIEL C. BRANDENSTEIN, Mission Commander
KEVIN P. CHILTON, Pilot
RICHARD J. HIEB, Mission Specialist 1
BRUCE E. MELNICK, Mission Specialist 2
PIERRE J. THUOT, Mission Specialist 3
KATHRYN C. THORNTON, Mission Specialist 4
THOMAS D. "TOM" AKERS, Mission Specialist 5

Summary of the Mission

Every orbiter underwent an on-the-pad test firing of its main engines prior to first launch. *Endeavour's* Flight Readiness Firing (FRF), a twenty-two-second engine burn, was performed at 11:13 A.M. on April 6. At the post-firing press conference the launch team announced it was quite pleased with the test firing, and it reported no significant problems were encountered.

Endeavour's first flight was entrusted to Commander Daniel C. Brandenstein and Pilot Kevin P. Chilton. The rest of the STS-49 crew complement included five Mission Specialists: Richard J. Hieb, Bruce E. Melnick, Pierre J. Thuot, Kathryn C. Thornton, and Thomas D. Akers. With the exception of Chilton, each crew member already had shuttle flight experience. Brandenstein flew as Pilot on STS-8, and Commander of both STS 51-G and STS-32. Melnick flew on STS-41; Hieb on STS-39; Thornton on STS-33; and Thuot on STS-36.

Although not a STS-49 payload, the primary focus of STS-49 centered on the Intelsat 6 satellite. (Intelsat is an acronym for International Telecommunications Satellite Consortium, a group of 124 member nations formed in the second half of the 1960's.) Intelsat 6 was launched on March 14, 1990, atop a commercial Titan launch vehicle. A malfunction left Intelsat IV attached to a spent Titan second stage. As a result, the satellite's Perigee Kick Motor (PKM) was unable to fire to push the satellite toward geosynchronous orbit. In order to save Intelsat, controllers jettisoned its PKM and, using only small thrusters, managed to place the satellite into a stable orbit from which it might later be rescued.

STS-49 would include three back-to-back Extravehicular Activities (EVAs), a first for shuttle operations. During the first EVA, astronauts would attempt to attach a new PKM to Intelsat IV. This new PKM, developed by United Technologies, weighed 10,454 kilograms and was 235 centimeters in maximum diameter and 323 centimeters long. During the second and third EVAs, two teams of astronauts would demonstrate and assess several methods of space station assembly and maintenance. During these Assembly of Space Station by EVA Methods (ASEM) exercises, five different crew rescue device prototypes (a bi-stem pole, inflatable pole, crew propulsive device, telescopic pole, and astrorope) would be evaluated.

The National Aeronautics and Space Administration (NASA) had sponsored a contest among school children throughout the country to name the new orbiter. Representatives from the school submitting the winning entry — *Endeavour* — were present at the Kennedy Space Center (KSC) to witness launch on the evening of May 7, 1992. Problems with the master events controller and weather conditions at a trans-Atlantic abort landing strip delayed the countdown. *Endeavour* lifted off at 7:40 P.M., quickly rolling to the proper heading for a 28.35 degree-inclination orbit. *Endeavour* passed through cloud patches before, during, and after the point of maximum aerodynamic pressure on the vehicle. After the crew received the go-at-throttle-up call from Mission Control, skies over KSC were clear enough to provide an unusually splendid view of solid rocket booster separation. *Endeavour* entered orbit 9 minutes and 27 seconds after liftoff. The new orbiter was in space on the anniversary of its delivery to KSC.

Intelsat IV was 560 kilometers above Earth when *Endeavour* lifted off. *Endeavour* entered a 328-kilometer-high orbit, 13,500 kilometers away from the satellite. Controllers at Intelsat's Washington, D. C., control center began a series of maneuvers to actively engage their satellite in the rendezvous phase. Intelsat IV would perform four thruster firings over the next two days to lower its orbit closer to that of *Endeavour.*

The astronauts prepared for the Intelsat capture EVA. First they lowered the cabin pressure to 10.2 pounds per square inch (psi) to minimize the amount of time EVA astronauts would have to pre-breathe pure oxygen before leaving the safety of *Endeavour's* airlock. Second, the EVA teams checked out all their spacesuits, making sure they were ready to support the planned EVA. Third, the Remote Manipulator System (RMS) was unberthed and commanded through a set of preprogrammed maneuvers to verify it could execute required motions during the EVA.

Late in the evening of May 9, the astronauts were remote participants in activities being held at the Peabody Hotel in Orlando, Florida. Five of the original Mercury astronauts were being honored at the Give Kids the World annual banquet. The crew spoke to John Glenn, Donald Slayton, Gordon Cooper, Scott Carpenter, and Alan Shepard, and also with congressman Jim Bacchus.

Thornton received special Mother's Day (May 10) greetings from her three daughters. The wake-up music they selected for their mother was from the film version of Winnie The Pooh. *Endeavour* was in a 305-by-300 kilometer orbit when the crew awoke.

The astronauts got their first glimpse of Intelsat IV around 12:30 P.M. when the satellite was still a little more than 75 kilometers away. Thuot and Hieb exited the airlock at 4:42 P.M. when Intelsat was still 270 meters from *Endeavour.* Brandenstein moved cautiously closer to Intelsat at a 0.3 meters-per-second closing rate. Thuot positioned himself on the RMS arm's end effector, and he was moved up close to the satellite by Melnick, who operated the RMS from inside *Endeavour* at its aft flight station. Thuot reached out with a specially-designed capture bar to grasp the satellite and stop its rotation rate. Both the astronauts and Mission Control were surprised when Thuot failed on his first attempt. Interaction dynamics between the capture bar and Intelsat proved to be different from what Thuot had expected based upon his training experiences. He was amazed at how reactive the massive satellite was to his application of relatively small forces. The satellite began to wobble. Thuot's second capture attempt also failed, and resulted in an even greater coning motion, the cone angle estimated to be as much as 45°.

EVA1 was terminated, and Hieb and Thuot were recalled into *Endeavour's* payload bay. Mission Control ordered the flight crew to execute a separation maneuver that would put 37 kilometers between *Endeavour* and Intelsat. During the night, Intelsat controllers stabilized the satellite for a second EVA attempt to capture the errant satellite.

The astronauts were awakened around 9:30 A.M. on May 11. The EVA team exited the airlock around 5:10 P.M. to begin preparations for another try at Intelsat IV capture. Prior to going after the satellite itself, Thuot attached himself to the RMS end effector and allowed Melnick to maneuver him to the side of the orbiter. There, Thuot practiced bumping the capture bar against *Endeavour* in a test of the dynamics of the bar and capture procedure.

Thuot was nearly perfectly positioned for his second capture attempt. At 7:18 P.M. he inserted the part of the capture bar into the center of the satellite's underside, and began to fire the latches that should have rigidified the capture bar to the satellite. Instead of firing properly, Thuot's efforts gave the satellite a small rate away from him and induced further wobbling.

The astronauts spent the next several minutes discussing what had gone wrong during this latest capture attempt, and watched the satellite's motion to ascertain how best to next approach Intelsat. It was decided to try again using the original method of capture. The crew pulled Thuot back after this third failed capture attempt and then maneuvered him in even closer on the next try. Mission Control even asked the astronauts about the possibility of what might have seemed a rather desperate plan — Thuot grabbing the satellite by hand.

Try as he might, Thuot continued to fail. EVA2 was ordered terminated, having lasted 5 hours 30 minutes. Thuot attempted a total of five captures before he induced unacceptable wobbles into Intelsat IV's motion. NASA management took a hard look at the wisdom of a three-man EVA consisting of hand-grasping so large a satellite as Intelsat. During a brief televised conference, Brandenstein pushed gently for approval of such a move. Just before 8:00 P.M., Mission Control approved an adventuresome three-person EVA plan.

EVA preparations on May 13 were uneventful. After two previous EVAs the astronauts had become quite proficient in preparing their spacesuits and other EVA equipment, and running through the checklist for depressurization and payload bay entry.

Once outside again, Hieb put himself into his assigned foot restraint. He held the capture bar nearby. Thuot assisted Akers into his foot restraint and then Melnick moved Thuot on the RMS arm to a position 120 degrees away from each of his fellow spacewalkers. Hieb was stationed on the starboard longeron and Akers was in the middle of the payload bay on the ASEM strut.

Brandenstein had to level *Endeavour* and the three spacewalkers relative to the Intelsat's base. The astronauts grabbed Intelsat at its solar drum positioners, solid devices which couldn't be damaged by human contact, and slowly nulled the satellite's rotation. The astronauts spent nearly half an hour getting comfortable holding the huge satellite before going ahead to capture bar installation. Akers and Thuot secured the capture bar's left end to Intelsat's aft ring. With the satellite attached to the capture bar, Melnick was able to grapple the bar's fixture and maneuver Intelsat. He moved Intelsat over to the PKM's location for attachment. Latches were thrown, and all indications suggested a hard mating between Intelsat and its new PKM.

EVA3 ran into the early morning hours of May 14. Thornton and Melnick began preparations for Intelsat deployment. These two astronauts would command deployment from the aft flight deck. Astronauts in the airlock would stand ready to return to the payload bay if an anomaly pre-

Three Mission Specialist EVA manual INTELSAT capture. (NASA)

vented deployment of Intelsat. After commands failed twice, the third time was the charm. Intelsat slowly departed *Endeavour*'s bay with a slow rotation about its symmetry axis provided by a spring-loaded deployment mechanism referred to as a super-zip. Down in Mission Control the flight team applauded. Intelsat was deployed near orbital apogee. *Endeavour* then executed a separation maneuver that would place the orbiter a safe distance away from the satellite by the time of its PKM burn.

When it was all over, EVA3 set a number of impressive records. It was the longest spacewalk ever — 8 hours and 29 minutes — the first to include three spacesuited astronauts, and the one hundredth spacewalk in the history of manned spaceflight.

Mission Control informed the astronauts that STS-49 would be extended a full day in order to provide them an ample rest period following upcoming EVA4 activities and to give them plenty of time to prepare for reentry and landing operations.

Intelsat controllers in Washington, D.C., commanded a sequencer to ignite the satellite's new PKM at 1:25 P.M. The burn was nominal, and the astronauts had the added bonus of viewing it.

A problem threatened the viability of EVA4 — failure of Thornton's spacesuit display electronics. However, telemetry was being sent down to Mission Control, so the EVA was given a go to proceed under the provision that Mission Control would advise Thornton on her suit performance. Akers and Thornton left the airlock at 5:30 P.M. to perform abbreviated ASEM work, combining the most important aspects of the original plans for EVA2 and EVA3 into a single spacewalk. EVA4 ended after Thornton and Akers stowed the ASEM hardware. Its duration was 7 hours and 45 minutes.

May 15 provided a leisurely day to prepare to end the mission. One thruster developed a minor leak following a hot-fire test, but that ceased after thermal restabilization. A news conference was held from orbit. The astronauts were told that President Bush had called NASA Administrator Daniel Goldin to congratulate the agency on its successful capture/redeploy of Intelsat IV.

The crew's final sleep period began just prior to 1:00 A.M. on May 16. They were reawakened at 8:40 A.M. to prepare *Endeavour* for deorbit and landing. Runway 22 at Edwards Air Force Base was the selected landing site.

Deorbit was performed an hour before scheduled landing. Entry was nominal, and *Endeavour* touched down on the concrete runway at 4:57 P.M. ending the 8-day, 21-hour, 17-minute, and 38-second mission. *Endeavour*'s rollout took only 58 seconds under the braking effect of the red, white, and blue drag chute deployed from the base of *Endeavour*'s vertical stabilizer immediately following nose wheel touchdown.

Intelsat controllers executed an interim orbital maneuver on May 17 sending the satellite out of its PKM-provided 36,800 by 312 kilometer transfer orbit, leaving it in a circular geosynchronous one. After a week, the satellite was on station ready for initial pre-operational testing.

Knowledge Gained

STS-49 carried the latest version of the Protein Crystal Growth (PCG) facility, an experiment that had flown on twelve previous shuttle missions. On STS-49, the PCG was used to grow protein crystals of bovine insulin from solution. The Air Force Maui Optical System (AMOS) was tested again on STS-49. An electro-optical facility on Maui recorded orbiter signatures during Orbital Maneuvering System (OMS) and RCS thruster firings. Another part of the ambitious STS-49 flight plan was a space shuttle demonstration of space station Freedom Assembly of Station by EVA Methods (ASEM).

The ASEM truss had been partially constructed to support the EVA3 Intelsat grab. Thornton and Akers completed that 4.5-by-4.5-by-4.5-meter pyramid-shaped truss assembly. ASEM work involving evaluation of astronaut large mass handling capability was eliminated from EVA4. Thornton and Akers also skipped the part of the ASEM work devoted to evaluating the orbiter's nose as an assembly station for Freedom components. Some evaluations of Crew Self-Rescue aids were performed.

Context

Endeavour, orbiter OV-105, was authorized in late 1986 as the replacement for *Challenger* (OV-099), lost on mission STS 51-L during a launch accident. Rockwell International had to reassemble its space shuttle orbiter production line at the Palmdale plant. *Endeavour* was completed ahead of schedule and slightly under budget, a first for a shuttle orbiter. *Endeavour*'s design included technology updates and the ability to remain in orbit for up to 28 days. *Endeavour* was rolled out of Palmdale plant 42 on April 25, 1991 to the theme from the motion picture *2001: A Space Odyssey* as thousands of spectators cheered. *Endeavour* had cost $1.9 billion and took 2.5 million manufacturing hours to complete after assembly began in August, 1987. *Endeavour* assembly flow benefitted heavily from lessons learned during construction of previous shuttle orbiters. For example, *Columbia* required 55 hours per tile to complete installation of the 32,000 Thermal Protection System (TPS) tiles. *Endeavour*'s 24,000 TPS tiles were installed with only 11 hours per tile required. *Endeavour* was also 9,000 pounds lighter than *Columbia* (the first production orbiter).

Endeavour began its maiden voyage on the specified launch date, May 7, 1992, performing with fewer anomalies than any other shuttle orbiter experienced during its first mission. The flight crew, through extraordinary efforts, was able to fulfill all STS-49 mission objectives and set a few space flight records in the process: first three-man EVA and the first shuttle mission to support four EVAs.

STS-49 was Goldin's first mission as the new NASA

Administrator. Following STS-49 he formed a group of experts to investigate NASA pricing policies. Intelsat would easily generate over one billion dollars in revenue as a result of its salvage on STS-49, yet the Intelsat consortium paid NASA only $93 million in compensation for services rendered. It had also paid $46 million for the new PKM, but the total cost to Intelsat hardly covered the total expense of NASA training and flight operations for the STS-49 Intelsat rescue.

Bibliography

"*Endeavour's* Intelsat Rescue Sets EVA, Rendezvous Records." *Aviation Week and Space Technology* (May 18, 1992): 22-24. Describes the rendezvous sequence that led up to capture of Intelsat IV, and the EVA techniques used to attach a new Perigee Kick Motor to the satellite.

"EVAs to Influence Development of Space Station Hardware." *Aviation Week and Space Technology* (May 18, 1992): 25-26. Describes lessons learned during the STS-49 EVAs that are applicable to construction of a space station in low Earth orbit.

"Mission Control Saved Intelsat Rescue From Software, Checklist Problems." *Aviation Week and Space Technology* (May 25, 1992): 78-79. Describes problems encountered during the Intelsat rescue, and also the first entry and landing of *Endeavour*.

National Aeronautics and Space Administration. *STS-49 Press Kit*. May, 1992. Fully documents facts and figures concerning the flight activities, *Endeavour* space shuttle, and Intelsat satellite.

David G. Fisher

SPACE SHUTTLE FLIGHTS, 1993

Date: January 13 to December 13, 1993
Type of mission: Manned Earth-orbiting spaceflight

Seven space shuttle missions were flown in 1993. Several of the missions were science missions. There was also a mission to retrieve a satellite left in space a year earlier by another space shuttle as well as missions to deploy new satellites. The crown jewel of the 1993 space shuttle missions, though, was the service and repair mission of the Hubble Space Telescope in December by the STS-61 mission.

PRINCIPAL PERSONAGES

    EUGENE F. KRANZ, Director, Mission Operations,
        Johnson Space Center
    ROBERT B. SIECK, Launch Director,
        Kennedy Space Center
    JOHN H. CASPER, STS-54 Commander
    DONALD R. "DON" MCMONAGLE, STS-54 Pilot
    MARIO RUNCO, JR.,
    GREGORY J. HARBAUGH, and
    SUSAN J. HELMS, STS-54 Mission Specialists
    KENNETH D. CAMERON, STS-56 Commander
    STEPHEN S. OSWALD, STS-56 Pilot
    C. MICHAEL FOALE,
    KENNETH D. COCKRELL, and
    ELLEN OCHOA, STS-56 Mission Specialists
    STEVEN R. NAGEL, STS-55 Commander
    TERENCE T. "TOM" HENRICKS, STS-55 Pilot
    JERRY L. ROSS,
    CHARLES J. PRECOURT, and
    BERNARD A. HARRIS, JR., STS-55 Missions Specialists
    ULRICH WALTER and
    HANS WILHELM SCHLEGEL, STS-55 Payload Specialists
    RONALD J. GRABE, STS-57 Commander
    BRIAN DUFFY, STS-57 Pilot
    C. DAVID LOW,
    NANCY J. SHERLOCK,
    PETER J.K. "JEFF" WISOFF, and
    JANICE E. VOSS, STS-57 Mission Specialists
    FRANK L. CULBERTSON, JR., STS-51 Commander
    WILLIAM F. READDY, STS-51 Pilot
    JAMES H. NEWMAN,
    DANIEL W. "DAN" BURSCH, and
    CARL E. WALZ, STS-51 Mission Specialists
    JOHN E. BLAHA, STS-58 Commander
    RICHARD A. SEARFOSS, STS-58 Pilot
    M. RHEA SEDDON,

WILLIAM S. "BILL" MCARTHUR, JR.,
DAVID A. WOLF, and
SHANNON W. LUCID, STS-58 Mission Specialists
MARTIN J. FETTMAN, STS-58 Payload Specialist
RICHARD O. COVEY, STS-61 Commander
KENNETH D. BOWERSOX, STS-61 Pilot
KATHRYN C. THORNTON,
CLAUDE NICOLLIER,
JEFFREY A. HOFFMAN,
F. STORY MUSGRAVE, and
THOMAS D. "TOM" AKERS, STS-61 Mission Specialists

Summary of the Missions

Seven space shuttle missions were flown during 1993. Three of the four active space shuttle orbiters were involved in these missions. *Endeavour* (OV-105) flew three missions, while *Columbia* (OV-102) and *Discovery* (OV-103) each flew two missions. *Atlantis* (OV-104) spent most of the year being refurbished and did not fly again until 1994. All of the missions involved scientific investigations; however, only three of the missions were dedicated to science only. Three missions had as a major objective satellite deployment or retrieval. The final mission of the year was the long awaited Hubble Space Telescope servicing mission.

The first space shuttle mission of 1993 was designated STS-54. This mission began with the launch of the orbiter *Endeavour* from launch pad 39-B at the Kennedy Space Center on January 13, 1993, at 9:00 A.M., eastern standard time. The primary objective of this mission was the deployment of the fifth Tracking and Data Relay Satellite (TDRS-F). The *Endeavour* landed on January 19, 1993, on Runway 33 at the Kennedy Space Center at 8:38 A.M., eastern standard time.

The Tracking and Data Relay Satellites relay telemetry from space shuttles, the Hubble Space Telescope, and other satellites to controllers on the ground. These satellites are located in geosynchronous orbits. Geosynchronous orbits are orbits that are so high above the surface of the Earth that it takes exactly one day to orbit the Earth. This means that satellites in geosynchronous orbits seem to always stay above one part of the Earth, since they go around the Earth in the same length of time that it takes the Earth to rotate.

TDRS-F was deployed on the first day of the STS-54 mission. Since the space shuttle cannot fly high enough to put the satellite into a geosynchronous orbit, the satellite must be boosted into the proper orbit by a small rocket engine called the Inertial Upper Stage booster. The Inertial Upper Stage booster successfully boosted TDRS-F into its proper orbit.

Crew members aboard *Endeavour* spent the remainder of the mission performing various microgravity experiments on board the orbiter, and on the final full day of the mission they spent several hours practicing extravehicular activities in the space shuttle's cargo bay. These activities were designed to give astronauts practice at maneuvering in weightlessness and moving heavy objects, tasks that might be necessary in repairing satellites in orbit or assembling components of a space station.

The second mission of 1993, STS-56, began at 1:29 A.M., eastern daylight time, on April 8, 1993, with the launch of the space shuttle *Discovery* from Pad 39-B at the Kennedy Space Center. Launch had initially been scheduled for April 6, 1993; however, the launch had been canceled only eleven seconds from liftoff due to an indication that there may have been a problem with a fuel valve. This was only the sixth night launch of a space shuttle. This mission was primarily a science mission studying the Earth's atmosphere from above. Landing was on Runway 33 at the Kennedy Space Center at 7:37 A.M., eastern daylight time, on April 17, 1993. The landing had been delayed one day due to bad weather in Florida on April 16, 1993.

The primary payload of this mission was instrumentation for the Atmospheric Laboratory for Applications and Science-2 (ATLAS-2). The equipment was located in a Spacelab pallet in the cargo bay of the *Discovery*. A Spacelab pallet is a self-contained laboratory that can be loaded into a space shuttle. Astronauts climb into the Spacelab from the main cabin of the space shuttle to perform the experiments or make the measurements necessary for the mission.

ATLAS-2 was designed to study the Earth's atmosphere from above and to investigate changes in the Earth's atmosphere resulting from human activity and from interactions between the sun and the atmosphere. Among the studies undertaken were studies of ozone depletion in the atmosphere, the level of pollutants in the atmosphere, and the composition of the atmosphere. The data from ATLAS-2 were compared to data collected from the ATLAS-1 mission flown one year earlier aboard the STS-45 mission.

Astronauts aboard the *Discovery* also performed several astronomy, materials science, and life science experiments during the mission. In addition to the science aspects of the mission, astronauts communicated with school children around the world by short wave radio as they passed overhead. They also made radio contact with the Russian space station Mir using the same amateur radio equipment.

The third space shuttle mission of the year was also a science mission. Designated STS-55, this mission began with the launch of the *Columbia* at 10:50 A.M., eastern daylight time, on April 26, 1993, from Pad 39-A at the Kennedy Space Center. *Columbia* carried into orbit a German Spacelab module called Spacelab D-2. Landing was on May 6, 1993, at 7:30 A.M., Pacific daylight time on, Runway 22 of Edwards Air Force Base in California.

Many of the experiments conducted in the Spacelab D-2 were extensions on the earlier German Spacelab D-1, which flew aboard the space shuttle mission 61-A in 1985. These experiments investigated the effects of microgravity on fluids, materials processing, and living organisms. While in orbit, the space shuttle experiences almost gravitational effects, hence the term microgravity. Most scientists prefer the term "microgravity" to "zero gravity" in recognition of the fact that all mass has gravity, and thus the experiments in orbit, while away from the strong gravitational forces experienced on the Earth's surface, still experience tiny gravitational forces.

Among the materials science experiments, astronauts studied the effects of microgravity on crystal growth. Some of the life science experiments involved the astronauts themselves. The human body loses certain fluids and chemicals during spaceflight, and part of the experiments involved injections of saline solutions as an experiment investigating the body's response to direct replacement of fluid loss.

STS-57, the fourth space shuttle mission of the year, began June 21, 1993, with the launch of *Endeavour* from Pad 39-A at the Kennedy Space Center at 9:07 A.M., eastern daylight time. In addition to biomedical and materials science experiments, a primary goal of the STS-57 was the retrieval of the European Retrievable Carrier (EURECA), which had been left in orbit during the STS-46 mission one year earlier. Landing was delayed two days due to bad weather in Florida. Landing finally occurred at 8:52 A.M., eastern daylight time, on Runway 33 at the Kennedy Space Center. The EURECA satellite had been left in orbit a year earlier in order to study long term effects of microgravity and the radiation in space on materials and living organisms. The original plan for retrieval of the EURECA was for the satellite to be captured by the *Endeavour*'s robot arm, called the Remote Manipulator System (RMS), and for EURECA's antennas to be retracted using power from the space shuttle. A misaligned electrical connector on the RMS prevented EURECA from getting power from *Endeavour* to retract its antenna. Astronauts were forced to don spacesuits and manually close the antennas after the satellite had been pulled into the shuttle's cargo bay.

During the rest of the mission, astronauts studied crystal growth and performed other materials science experiments. Astronauts also practiced manipulating equipment that might one day be needed in the construction of an orbiting space station. Astronauts also studied the effects of extended weightlessness on human posture.

The fifth space shuttle mission of the year was STS-51. This mission had numerous delays launching. The mission was nearly two months late in launching, after twice having been stopped in the countdown mere seconds from launch. A further delay was caused by fears that the August 11, 1993, Perseid meteor shower might prove hazardous to spaceflight. Launch finally occurred at 7:45 A.M., eastern daylight time, September 12, 1993, when the Space Shuttle *Discovery* lifted off from Pad 39-B at the Kennedy Space Center. The primary objective of the STS-51 mission was the deployment of the Advanced Communications Technology Satellite (ACTS). Landing was at the Kennedy Space Center, on Runway 15, at 3:56 A.M., eastern daylight time, on September 22, 1993. This marked the first nighttime landing of a space shuttle at the Kennedy Space Center.

The deployment of ACTS was delayed through communication problems between the *Discovery* and mission controllers on the ground. An accident that occurred during deployment caused minor damage to the interior of the *Discovery*'s cargo bay. After deploying ACTS, astronauts performed various biomedical experiments and practiced extravehicular activities that would be needed later in the year to repair the Hubble Space Telescope.

The sixth space shuttle mission of the year, STS-58, was dedicated to life sciences research. The space shuttle *Columbia* lifted off at 10:53 A.M., eastern daylight time, on October 18, 1993, from Pad 39-B at the Kennedy Space Center. Landing was at Edwards Air Force Base in California on Runway 22 at 7:06 A.M., Pacific standard time, on November 1, 1993.

Most of the scientific work done on the STS-58 mission was biomedical research. The long duration of this mission permitted scientists to study the physiological adjustments that occur during weightlessness. Several experiments were performed to attempt to counter the negative effects that occur to astronauts' bodies in microgravity. In addition to studies involving the astronauts, forty-eight rats were also taken into space on board the *Columbia*.

The seventh and final space shuttle mission of 1993 was the long awaited repair and servicing mission for the Hubble Space Telescope. This mission, designated STS-61, involved a night launch and a night landing for the space shuttle *Endeavour*, both rare events for space shuttle missions. Launch was from Pad 39-B at the Kennedy Space Center on December 2, 1993, at 4:27 A.M., eastern standard time. Landing was on December 13, 1993, at 12:26 A.M., eastern standard time on Runway 33 at Kennedy Space Center.

The Hubble Space Telescope (HST) had been placed into orbit three years earlier with the anticipation that it would eventually need a servicing mission to update its instruments and to replace any equipment on board that had failed during the intervening time. With this in mind, most of the equipment on board HST had been designed to be removed and replaced while the satellite was in orbit. Perhaps the most publicized repair of the mission was the Corrective Optics Space Telescope Axial Replacement (COSTAR) system. COSTAR was designed as an optical correction for a flaw

that had been discovered in the telescope's primary mirror. This flaw prevented the telescope from properly focusing light, thus reducing its effectiveness for some of its scientific objectives. While correcting the telescope's optics got most of the press attention, this was only one of many tasks performed by the astronauts of the STS-61 mission.

The *Endeavour* spent two days carefully maneuvering so it could safely grab hold of HST with its RMS arm. The astronauts had to be very careful not to damage the telescope. During the first Extravehicular Activity (EVA), astronauts changed two of HST's six gyroscopes, which keep the telescope pointed in the desired direction. The second EVA was to replace HST's solar panels. The original solar panels had flexed whenever the space telescope passed in or out of the Earth's shadow. This flexing caused the telescope to lose a stable fix on any object that it tried to view at that time. The new solar panels were designed not to flex when undergoing large temperature changes. The original plan had been to safely stow the old solar panels in the *Endeavour*'s cargo bay and return them to Earth; however, one of the solar arrays had become twisted and could not be stowed aboard the shuttle. This array was thrown overboard to eventually burn up in the Earth's atmosphere as its orbit spirals down towards the Earth.

The third EVA had two objectives. Astronauts changed out HST's original Wide Field/Planetary Camera (WFPC) with a new and more sophisticated Wide Field/Planetary Camera-2 (WFPC-2). After installation of WFPC-2, astronauts changed two magnetometers on the HST. The magnetometers sense the Earth's magnetic field, allowing the space telescope to orient itself relative to the Earth's magnetic field.

The fourth EVA also had two separate objectives. First, astronauts replaced HST's high speed photometer with COSTAR. The high speed photometer had been one of the least-used instruments on the HST, so it was sacrificed to allow installation of COSTAR to correct the flaw in the telescope's primary mirror. A second objective of this EVA was the upgrade of HST's on-board computers. A fifth EVA replaced various other electronic components aboard HST. Then HST was released back into its orbit, thus successfully ending one of the most sophisticated space shuttle missions of all time.

Knowledge Gained

The ATLAS-2 investigations of STS-56 studied the composition of the Earth's atmosphere. Among the factors studied were the impact of both human-induced pollution and variations in the sun's energy output on the composition of the upper atmosphere. ATLAS-2 measured ozone concentrations in the upper atmosphere and concentrations of greenhouse gases in the lower atmosphere. While the information gathered by ATLAS-2 is important in its own right, it is even more important when compared with ATLAS-1, which flew aboard STS-45 one year earlier, and ATLAS missions to be flown later. Comparisons between the measurements made by all of the different ATLAS missions allow scientists to study changes in the atmosphere.

The crew of STS-58 took numerous infrared photographs of wildfires that were burning out of control in southern California during the mission. These photographs were made available to scientists after the mission was over in the hopes that some of the information in the photographs would help scientists understand the phenomena that led to those fires.

The EURECA satellite recovered by STS-57 permitted scientists to study the long term effects to materials and organisms of exposure to microgravity and radiation in space. Spacelab D-2 of the STS-55 mission studied similar effects but on a much shorter time scale. Much of this data is still being evaluated, but some initial findings are available. Crystals grown in orbit tend to be larger and more perfect than those grown on Earth.

Astronauts aboard STS-59 studied changes in human physiology that occur in microgravity. This information is being used to design better living and working areas on board future shuttle missions. The STS-58 mission had numerous life science findings. Astronauts studied bone tissue loss that occurs in astronauts who spend extended periods of time in microgravity. Calcium appears not to be maintained in bones as efficiently in microgravity as in the comparatively heavy gravity of Earth. Several different experiments were performed to try to minimize calcium loss in astronauts. Astronauts also kept logs of fluid and food intake as well as physical activity in an attempt to determine the nutritional and energy needs of astronauts in microgravity conditions. The nutrition needs of humans in microgravity differ from those of humans under normal gravity. For the first few days of the mission, astronauts also performed experiments aimed at trying to determine the mechanism for space motion sickness, a form of motion sickness that has plagued astronauts since the early days of manned spaceflight.

Throughout several missions during the year, astronauts conducted extravehicular activities that were designed to provide familiarity and practice with moving around and manipulating tools and materials under microgravity conditions from within the confines of a spacesuit. The experience gained during these EVA's was put into use during the HST servicing mission of STS-61 at the end of the year. This experience may also be put to use in future satellite retrieval and repair missions or in the construction of a possible space station in the future.

Context

The scientific investigations conducted during the year indicated the importance of space-based research. The atmospheric studies of ATLAS-2 could only be done from above the atmosphere. The studies of the atmosphere from above complement the studies done on Earth. Photographs and measurements of Earth from above give a much different perspective than those done on the Earth's surface. ATLAS-2 is the second of a series of ten ATLAS missions scheduled to study changes in the Earth's atmosphere over a decade.

The nearly weightless conditions aboard a spacecraft in orbit permit studies of materials and organism under microgravity conditions. Material science studies often center on the study of crystal growth in orbit. These sorts of studies are very important for future technology. Much of modern technology is reliant upon semiconductors. Semiconductor devices may be manufactured with greater precision and accuracy under microgravity conditions.

Long duration exposure to the microgravity conditions of space have demonstrated detrimental effects on the human body. Since astronauts will be exposed to long durations of microgravity in proposed missions to Mars or proposed space stations, it is important for scientists to understand the effects of microgravity on astronauts. To accomplish these studies, the space shuttle *Columbia's* STS-58 mission was the fourth-longest manned spaceflight in NASA's history.

The space shuttle demonstrated its versatility during the satellite deployment and retrieval missions of 1993. Only the space shuttle could have retrieved EURECA from orbit. Furthermore, the HST servicing mission also demonstrated the value of a manned space vehicle. The space telescope could not have been repaired or upgraded without the space shuttle's unique capabilities.

The HST servicing and repair mission was perhaps the most complicated space shuttle mission ever flown. To service the space telescope, astronauts were required to perform five EVA's during the mission. This was a new record for most extravehicular activities in a manned spaceflight.

Bibliography

Branley, Franklyn M. *From Sputnik to Space Shuttles.* New York: Thomas Y. Crowell, 1986. This is a very good juvenile book on the development of space exploration, from unmanned satellites to the modern space shuttle. The book emphasizes the uses of the space shuttle more than the shuttle itself.

Bruning, David. "Hubble: Better Than New." *Astronomy* 22 (April, 1994): 44-49. This general audience level article has little information but lots of pictures taken with the newly repaired Hubble Space Telescope. It gives an idea of how much improved images from the space telescope will be with the new corrective optics installed.

DeWaard, E. John, and Nancy DeWaard. *History of NASA: America's Voyage to the Stars.* New York: Exeter Books, 1988. This book is intended for the general audience. The last chapter of the book covers the beginnings of the space shuttle program. The need for a system such as the space shuttle is discussed. The book ends with the explosion that destroyed the *Challenger* in 1986.

Hawkes, Nigel. *Space Shuttle*. New York: Gloucester Press,1983. This juvenile book includes a very nice description of the design and layout of the space shuttle. It also includes a very brief discussion of living in space and the sorts of payloads and missions available to the space shuttle.

Hoffman, Jeffrey A. "How We'll Fix the Hubble Space Telescope: An Astronaut's Anticipations." *Sky and Telescope* 21 (December, 1993): 23-29. Although this article was written prior to the space telescope servicing mission, the mission went so smoothly that there were virtually no deviations from the plans outlined in the article. The article, written by one of the astronauts to fly on the mission, gives a detailed account of the various activities involved in fixing the Hubble Space Telescope.

Joels, Kerry M., Gregory P. Kennedy, and David Larkin. *The Space Shuttle Operator's Manual*. New York: Ballantine Books,1982. This is the definitive description of the space shuttle itself, available for general audiences. It includes diagrams and charts illustrating key parts of the space shuttle, and even includes fold-out sections showing instrument panels. It includes a very good description of living facilities on board the shuttle. There are even detailed timelines for both the launch and landing phases of the missions.

Rycroft, Michael, Ed. *The Cambridge Encyclopedia of Space*. New York: Cambridge, 1990. This general audience level work contains a section on space launch facilities, including the Kennedy Space Center, where space shuttle missions are launched. Further sections discuss living in space, with a very thorough description of the various types of microgravity experiments that can be conducted in space. The space shuttle system itself is given considerable coverage.

Shapland, David, and Michael Rycroft. *Spacelab: Research in Earth Orbit*. New York: Cambridge, 1984. Written in the early days of the space shuttle program, this general audience level book describes the types of scientific missions anticipated for the space shuttle. The book describes a Spacelab, which is a living/working module that is flown in the space shuttle's cargo bay during science missions.

Torres, George J. *Space Shuttle: A Quantum Leap*. Novato,Calif.: Presidio Press, 1986. This general audience level book traces the development of manned spaceflight, culminating with the space shuttle. Included are brief descriptions on many different types of shuttle missions. The book was written at the time of the *Challenger* explosion and thus does not include any of the missions that occurred after the resumption of the shuttle program.

Raymond Benge

SPACE SHUTTLE MISSION
STS-61

Date: December 2, 1993 to December 13, 1993
Type of mission: Manned Earth-orbiting spaceflight

The Hubble Space Telescope (HST), first in a series of the National Aeronautics and Space Administration's (NASA) Great Observatories, was launched in 1990 with a serious deficiency in its primary mirror. STS-61 was responsible for an impressive demonstration of the value of manned space flight — the orbital repair of HST.

PRINCIPAL PERSONAGES

RICHARD O. COVEY, STS-61 Commander
KENNETH D. BOWERSOX, STS-61 Pilot
KATHRYN C. THORNTON,
CLAUDE NICOLLIER,
JEFFREY A. HOFFMAN,
F. STORY MUSGRAVE, and
THOMAS D. "TOM" AKERS, STS-61 Mission Specialists
EDWARD J. WEILER, HST Project Scientist
DANIEL GOLDIN, NASA Administrator
LYMAN SPITZER, scientist who first proposed orbiting
     a Large Space Telescope
EDWIN HUBBLE, astronomer for whom the HST
     is named

Summary of the Mission

On April 23, 1993, NASA issued a press release concerning STS-61/HST mission operations. NASA was concerned, as the list of required repairs continued growing at an alarming rate during the second quarter of 1993. Thought of lowering the scope of this first HST repair mission was not well received. Considering the bad press surrounding HST's problems, shuttle managers decided it would be preferable to attempt a more ambitious schedule and possibly not complete everything, than to set their sights lower and find the repair-work easier than imagined.

Extravehicular Activity 1 (EVA1) would consist principally of getting set up for repairs to follow, establishing work stations and unpacking tools, and replacement of two Rate-Sensing Units (RSU). EVA2's primary task would be to remove and stow the old solar arrays and put new ones in their place. EVA3 would be dedicated to the installation of the Wide/Field Planetary Camera-II (WFPC-2). EVA4 was dedicated to the Corrective Optics Space Telescope Axial Replacement (COSTAR) installation. EVA5 would include

work on areas of HST that were not originally designed to be EVA-friendly. A second magnetometer would be installed, and a co-processor module added to the telescope's main computer.

NASA's cost estimates for the HST repair effort, a $692 million expenditure, included $429 million for STS-61, $12 million for new solar arrays from ESA, $86.3 million to characterize the HST primary mirror's spherical aberration, $49.9 million for COSTAR, and $23.8 million for WFPC-2.

Within press circles, STS-61 was touted as a make-or-break mission for NASA. Public support ran somewhat on the cynical side prior to launch but quickly turned to a sense of excitement and sincere hope that the shuttle *Endeavour* astronauts would succeed in their repair efforts. Dr. Edward J. Weiler cautioned the press not to expect a guaranteed success, emphasizing this was a risky and complex mission.

Weather was excellent on December 2, and *Endeavour* lifted off in total darkness at 4:27 A.M. Solid Rocket Booster (SRB) exhaust quickly turned darkness into brilliant light as the vehicle punched off the pad, and climbed along the proper flight azimuth. SRB separation was clean, and first stage performance nominal. Portions of ascent were visible for several hundred miles along the eastern sea coast.

Endeavour flew a direct ascent trajectory to the targeted 493-by-344 kilometer orbit. Later, Orbital Maneuvering System (OMS) engines were fired to alter the orbit and initiate rendezvous. That burn, called OMS-2, was commanded when *Endeavour* was 9,440 kilometers behind HST. Over the next two days a sequence of small thruster firings was performed to close in slowly on HST.

The astronauts were awakened at 7:00 P.M.. on their third day in space to prepare for the final phase of rendezvous and subsequent grapple and capture operations. Maneuvers over the course of the next eight hours, each designed to slow approach speed, ultimately brought *Endeavour* and HST to within grappling range for the Remote Manipulator System (RMS) end effector. By December 4, *Endeavour* had narrowed the gap between HST and itself from 9,600 kilometers at liftoff to only 12.8 kilometers by terminal phase initiation.

The final portion of the rendezvous was flown very slowly. Nicollier raised the RMS arm high above the payload bay. Four midcourse corrections were performed before HST sat 11 meters from the Orbiter's payload bay. The RMS was moved so its end effector camera's field of view was centered on the HST star tracker. *Endeavour's* Reaction Control System (RCS) thrusters were turned off so as not to interfere with grapple operations. Nicollier grabbed HST at 3:47 A.M. without inducing the slightest shaking motion in the telescope. Nicollier very slowly inched HST down for berthing on the Flight Support System (FSS) in *Endeavour's* payload bay.

The crew reported seeing more than just a bow in the starboard solar array. It had suffered a kink attributed to thermally induced flexing the array experienced on each orbit and to an anomalous event that occurred about eighteen months before the repair flight, the exact nature of which was unknown.

While Hoffman and Musgrave worked in the payload bay, Akers acted as choreographer, watching from the aft station inside *Endeavour* during EVA1. Covey and Bowersox were responsible for maintaining or changing *Endeavour's* attitude. Thornton documented the EVA using various cameras.

About one hour into EVA1, Hoffman was ready to open up HST's aft shroud area to work inside at the RSU replacement. RSU replacement provided no surprises, as EVA1 continued into December 5. Hoffman undid the latches, pulled out the units, handed them to Musgrave, and put in new ones.

The first difficulty of the mission surfaced as Hoffman could not simultaneously close both aft shroud doors at the top, bottom, and middle bolts. So as not to lose the advantage that had been gained from the ease of RSU replacement work, Hoffman was told to move on to changeout of the Electronic Assemblies (ECUs) for the RSUs. The astronauts then performed some work at the Solar Array Carrier (SAC), tasks that would help EVA2 work begin quickly. Hoffman closed out both ECUs and shut the Bay 10 door on HST at 3:10 A.M., moving on to the replacement of eight fuses about 4.5 hours into the EVA.

The astronauts installed a portable foot restraint on the right side of HST so that Musgrave could assist Hoffman in closing the aft shroud doors. It was decided to use a bungee cord between latch handles to maintain even tension. At 5:30 A.M. Hoffman and Musgrave finally got the door bolts engaged.

To retract the solar arrays, HST was rotated and pivoted on the FSS to the proper position as EVA1 concluded after 7 hours 54 minutes; it had been scheduled for only six hours.

For EVA2, Thornton was fixed to the RMS foot restraint while Akers floated freely. The airlock hatch was opened at 10:30 P.M. Before the EVA team exited the airlock, Mission Control rolled up the solar array that was slated for return to Earth. The damaged one would be discarded in orbit since it could not be retracted.

Thornton placed a handhold on the damaged solar array so that she could handle the array once it was disconnected from HST. Nicollier moved Thornton to the optimum position for holding the solar array once Akers completed the disconnection. Nicollier moved Thornton and the solar array

high above the payload bay at 11:40 P.M. to await orbital sunrise. Meanwhile, Akers started working on extracting the new solar arrays from their protective covers. Thornton simply let go of the array at 11:48 P.M. Nicollier pulled her quickly down into the bay away from the discarded array. The crew at the aft flight deck commanded the orbiter to move away from the solar array. When *Endeavour's* exhaust hit the array, it picked up twisting and flapping motions. One large RCS firing really clobbered the array with its shock wave. The array picked up both a spin, tumble, and twist.

Thornton and Akers now turned their attention to the new solar arrays, removing them from protective carriers near the front of the payload bay and installing one in place of the discarded array before tackling the other array's replacement. Thornton and Akers completed all work associated with the first array by 1:00 A.M., December 6.

By 1:30 A.M., the discarded solar array was 18 kilometers away from *Endeavour*. The first new solar array was in place and properly mated to HST. Inside the shuttle, the astronauts rotated HST 180 degrees to provide Thornton and Akers access to the second solar array. EVA2 concluded after Thornton and Akers did some EVA3 prep work, setting up for WFPC removal and replacement.

EVA3 began at 10:45 P.M. By the time Musgrave looked deep inside the WFPC cavity, it was past midnight on December 7. Hoffman put WFPC in its temporary stow position on *Endeavour's* left side, extending over the payload bay longeron. WFPC-2 installation was done even more slowly than WFPC removal. By 1:30 A.M. WFPC-2 was inserted, Hoffman tightened bolts and latches, and attached a grounding strap as Musgrave looked in to observe indicator lights from the fine guidance sensor bay; each light signaled the desired configuration.

For magnetometer attachment, both spacewalkers loaded an RMS caddy with tools by 3:00 A.M. and were lifted up together to the work site. A small piece of multilayer thermal insulation got away from the astronauts. It moved away at a few feet per second off to the left side of Endeavour. This was no significant concern, as *Endeavour* would reboost HST away from the current shuttle's orbit after HST repair operations were concluded.

EVA4 began at 10:15 P.M. The first EVA4 task to perform was removing the telephone booth-sized High Speed Photometer (HSP) from its aft shroud bay. HSP had the same mass-handling characteristics as COSTAR and was precisely the same size and shape. Thus it was a good surrogate on which to practice before COSTAR insertion.

Shortly after midnight December 8, Akers returned to the open cavity from which HSP had been removed. At 12:15 A.M. he was in position, ready to assist Thornton in moving COSTAR into proper position. COSTAR installation was completed by 1:00 A.M. Thornton and Akers then released HSP from its temporary parking position. Nicollier moved Thornton back over to the forward part of the payload bay, putting HSP in COSTAR's protective enclosure.

Next came attachment of a new computer co-processor. This was mounted right on top of HST's existing IBM-compatible 286 computer, using the threaded holes of previously used handles. Akers chuckled at how easy it was to make bolt contacts and electrical connector attachments. Co-processor work was completed by 3:40 A.M., approximately five hours and twenty minutes into EVA4.

EVA5 began at 10:30 A.M. EVA5's first task — changeout of the solar array drive electronics — was closed out halfway through revolution 104. The spacewalkers turned their attention next to getting the solar arrays down along the telescope's side. After that, the astronauts worked on the Goddard High Resolution Spectrometer (GHRS). A cabling kit bypassed a short circuit that had disabled one of GHRS's power supplies and essentially cut its capability in half. Installation of the redundancy kit was completed at 3:15 A.M.

The last action the astronauts took was to remove a small protective cover that they had placed over the telescope's low-gain antenna. HST was ready for redeployment except for unfurling the solar arrays and deployment of the high-gain antenna.

Hoffman entered the airlock, and Musgrave remained on the RMS while the first solar array deployment was commanded. Both arrays blossomed to full extension without kinks in the bi-stem. Once the arrays were deployed, HST had to be rotated to the position required for proper thermal control and even solar illumination until release. EVA5 ended at 5:05 A.M.

Difficulty encountered in initial attempts to communicate with the telescope's computer system delayed HST release. Nicollier grappled HST and then raised it off the berthing equipment. He held HST high above the bay before releasing the telescope.

Nicollier commanded the RMS end effector grapple wires to retract at 5:27 A.M. HST's attitude was stable as it began to fly freely again after nearly a week secured in Endeavour's payload bay. HST immediately acquired the sun and properly oriented its solar arrays. Communications through the Tracking and Data-Relay Satellite (TDRS) network was established, and HST controllers commanded the aperature door to reopen.

HST was released into an orbit 589 kilometers above the Earth's surface. Covey fired small thrusters twice to slowly separate away from HST without impinging thruster exhaust on the Telescope.

On December 12, the crew gathered in Endeavour's mid-deck, and talked with reporters for over an hour. The crew began a sleep period at 9:00 A.M. They were awakened for the final time of the mission at 5:00 P.M. Weather in the KSC area was expected to degrade during the early morning hours of December 13, so Mission Control decided to bring the shuttle home one orbit early.

A double sonic boom signaled Endeavour's return to the skies over Florida. Huge spotlights on the runway were turned off so that Covey could see just the landing lights. Endeavour flew behind the Vehicle Assembly Building on final approach and touched down on runway 33 at 12:26 A.M., December 13, 1993.

Knowledge Gained

Hubble Space Telescope scientific goals included measurement of Cepheid variable light curves for stars in galaxies at least 50 million light years away in an attempt to improve the determination of the expansion rate of the universe; examination of gravitational influences and signatures of massive black holes in both normal and active galactic cores; determination of the age of globular clusters; and imaging faint, distant objects.

Many of these lofty goals could not be accomplished using the telescope in the state in which it was launched. After STS-61, those goals were finally within reach. WFPC-2 was designed to detect objects over one hundred times fainter than those observable with ground-based telescopes, and with a ten-fold improvement in spatial resolution. The unit would restore HST to its prelaunch expectations for imaging of distant, faint objects.

Final HST science operations were concluded during the early morning hours of December 3, 1993. WFPC and HSP solar system observations were made before closing HST's aperature door. For HSP, these would be its last observations. HSP would be replaced by the COSTAR corrective optics package and be returned to Earth, its final disposition uncertain.

STS-61 astronaut activities were largely devoted to HST repairs, of course. However there were several science and technology payloads. The PILOT system, first flown on STS-58, designed to test shuttle piloting skills after a long exposure to weightlessness, was manifested on STS-61. Also, the Air Force Maui Optical System (AMOS) facility on Hawaii examined shuttle airglow, water dumps, and thruster firings when Endeavour passed over the island of Maui.

Five weeks after the first servicing mission, NASA held a news conference and a science briefing to discuss preliminary findings after testing the repaired Hubble. HST scientist Edward J. Weiler proclaimed that Hubble was fixed beyond the science team's wildest expectations.

Prior to the repair mission, only about 12 percent of a star's light converged in its central image. The original requirement for HST was to have 60 percent of the light in the central image. Thus, HST suffered from a serious spherical aberration. After the repair, it was determined that the amount of light was at least 70 percent and possibly more. The diffraction limit of HST's optics was 85 percent of the light in the central image. As a result of STS-61 repair work, HST was in even better shape than had been called for in the original design specifications. In terms of a more terrestrial analog, the diffraction capability of HST after the repair mission was equivalent to detecting the light of a firefly in Tokyo, Japan, from Washington, D.C., and being able to clearly resolve two fireflies at that distance if those fireflies were no less than three meters apart.

Context

During the January 13, 1994, conferences, NASA Administrator Daniel S. Goldin formally declared the Hubble Servicing Mission successful. "This is phase two of a fabulous, two-part success story. The world watched in wonder last month as the astronauts performed an unprecedented and incredibly smooth series of spacewalks," emphasized Goldin. "Now, we see the real fruits of their work and that of the entire NASA team. Men and women all across this agency committed themselves to this effort. They never wavered in their belief that the Hubble Space Telescope is a true international treasure."

The valiant actions of the well-trained and well-equipped STS-61 crew demonstrated the value of sending humans into space, by adapting to the unexpected and performing tasks not possible with machines. STS-61 partially restored the tarnished public image of NASA after a series of major problems such as the loss of the Mars Observer, failure to properly deploy the antenna on Galileo, the original flaws in HST's optics, and a series of nagging delays in launching space shuttle missions. With HST repaired, NASA had a world-class observatory uniquely qualified to address fundamental astrophysical issues.

Bibliography

National Aeronautics and Space Administration. *The Space Telescope*. NASA SP-392. Washington, D.C.: Superintendent of Documents, 1976. A collection of scientific works describing the capabilities of HST and the potential for advancing our understanding of the universe. Not for the casual reader.

National Aeronautics and Space Administration. *Hubble Space Telescope: Media Reference Guide*. Sunnyvale, Calif.: Lockheed Missiles & Space Company, Inc., 1990. A thorough reference guide to the Hubble Space Telescope prepared for NASA by one of the HST prime contractors. Filled with diagrams and charts describing all aspects of the HST program.

National Aeronautics and Space Administration. *Exploring the Universe with the Hubble Space Telescope*. NP-126. Washington, D.C.: Superintendent of Documents, 1990. A prelaunch NASA publication that illustrates the design, construction, and mission of the Hubble Telescope. Amply illustrated. Provides historical context for the Hubble Space Telescope.

Smith, Robert W. *The Space Telescope: A Study of NASA Science, Technology, and Politics*. Cambridge, England: Cambridge University Press, 1989. A prelaunch tour de force of the history of the development of HST. Includes internal NASA decision-making processes, reactions of the scientific community, the role of politics in bringing HST to realization, and the interation of contractors with government agencies. David G. Fisher

SPACE SHUTTLE FLIGHTS, 1994

Date: February 3 to November 14, 1994
Type of mission: Manned Earth-orbiting spaceflight

Seven space shuttle missions were flown in 1994. Unlike most years in the space shuttle program, the missions of this year were all dedicated science missions. Astronauts collected some of the best data ever on the Earth's atmosphere and conducted numerous experiments measuring the effects of microgravity on materials and living organisms.

PRINCIPAL PERSONAGES

CHARLES F. BOLDEN, JR., STS-60 Commander
KENNETH S. REIGHTLER, JR., STS-60 Pilot
N. JAN DAVIS,
RONALD M. SEGA,
FRANKLIN R. CHANG-DIAZ, and
SERGEI K. KRIKALEV, STS-60 Mission Specialists
JOHN H. CASPER, STS-62 Commander
ANDREW M. ALLEN, STS-62 Pilot
PIERRE J. THUOT,
CHARLES D. "SAM" GEMAR, and
MARSHA S. IVINS, STS-62 Mission Specialists
SIDNEY M. GUTIERREZ, STS-59 Commander
KEVIN P. CHILTON, STS-59 Pilot
JAY APT,
MICHAEL R. "RICH" CLIFFORD,
LINDA M. GODWIN, and
THOMAS D. "TOM" JONES, STS-59 Mission Specialists
ROBERT D. CABANA, STS-65 Commander
JAMES D. HALSELL, JR., STS-65 Pilot
RICHARD J. HIEB,
CARL E. WALZ,
LEROY CHIAO, and
DONALD A. THOMAS, STS-65 Mission Specialists
CHIAKI MUKAI, STS-65 Payload Specialist
RICHARD N. "DICK" RICHARDS, STS-64 Commander
L. BLAINE HAMMOND, JR., STS-64 Pilot
JERRY M. LINENGER,
SUSAN J. HELMS,
CARL J. MEADE, and
MARK C. LEE, STS-64 Mission Specialists
MICHAEL A. BAKER, STS-68 Commander
TERRENCE W. "TERRY" WILCUTT, STS-68 Pilot
STEVEN L. "STEVE" SMITH,
DANIEL W. "DAN" BURSCH,
PETER J.K. "JEFF" WISOFF, and
THOMAS D. "TOM" JONES, STS-68 Mission Specialists
DONALD R. "DON" MCMONAGLE,
STS-66 Commander
CURTIS L. "CURT" BROWN, JR., STS-66 Pilot
ELLEN OCHOA,
JOSEPH R. "JOE" TANNER,
JEAN-FRANCOIS CLERVOY, and
SCOTT E. PARAZYNSKI, STS-66 Mission Specialists

Summary of the Missions

Seven space shuttle missions were flown during the year, with all four active orbiters participating in missions. The orbiters *Columbia* (OV-102), *Discovery* (OV-103), and *Endeavour* (OV-105) each flew two missions. The orbiter *Atlantis* (OV-104) flew only one mission, near the end of year. All of the space shuttle missions of 1994 were dedicated science missions. Among the new records set during the year were the longest space shuttle mission and the first Russian cosmonaut to fly aboard an American spacecraft.

The first space shuttle mission of 1994 was the STS-60 mission, which began on February 3, 1994, with the launch of the *Discovery* from the Kennedy Space Center's Pad 39-A at 7:10 A.M., eastern standard time. The primary payloads were SPACEHAB-2 and the Wake Shield Facility (WSF). Landing was at on February 11, 1994, at 2:19 P.M., eastern standard time, on Runway 15 at the Kennedy Space Center.

The SPACEHAB-2 was a pressurized laboratory module which occupied part of the *Discovery*'s cargo bay. Astronauts working in the SPACEHAB module performed various experiments measuring the effects of microgravity on materials processing. While many of the experiments involved research that may one day lead to improved manufacturing of semiconductor crystals, several experiments were conducted that may lead to new techniques for pharmaceutical or biological processing.

The other major payload for STS-60 was the WSF. The WSF was designed to conduct additional materials science experiments. Many materials science investigations must be performed in a vacuum to avoid contamination of the samples. WSF was designed to take advantage of the fact that space offers an environment that is much more nearly a total vacuum than can be produced on Earth. To further minimize contamination, WSF was designed to operate autonomously from the shuttle at a distance of several miles. Unfortunately, numerous difficulties over several days prevented astronauts from being able to deploy the WSF. Many of the planned experiments were conducted in the cargo bay of the space shuttle, though this environment was less than ideal.

The second space shuttle mission of the year was STS-62. The mission began with the launch of the *Columbia* from Pad 39-B at the Kennedy Space Center on March 4, 1994, at 8:53 A.M., eastern standard time. A primary payload for STS-62 was the United States Microgravity Payload-2 (USMP-2). Landing occurred on March 18, 1994, at 8:10 A.M., eastern standard time, on Runway 33 at the Kennedy Space Center.

USMP-2 investigations studied materials and biological systems under the influence of microgravity. In orbit, the experiments are not subject to the same gravitational forces that they would be on the Earth's surface. This allows scientists to study systems as they would behave without the influence of gravity. Though the experiments in orbit can be performed with almost no gravitational influences, the term "microgravity" is used instead of "zero gravity" because tiny gravitational forces do still play a part and in fact can never be truly eliminated.

Among the material science investigations, astronauts studied crystal growth and fabrication techniques for semiconductors. Astronauts also performed experiments designed to observe the element xenon at its critical point. A fluid's critical point is the combination of temperature and pressure at which the fluid is simultaneously a liquid and a gas. Biological investigations included experiments designed to minimize the muscle atrophy that astronauts experience after extended periods in space.

The third space shuttle mission of 1994, STS-59, was a mission dedicated to studying the Earth from space. The mission began with the launch of the orbiter *Endeavour* from Kennedy Space Center Pad 39-A on April 9, 1994, at 7:05 A.M., eastern daylight time. The *Endeavour* landed eleven days later on April 20, 1994, at 12:55 P.M., eastern daylight time, on Runway 22 at the Kennedy Space Center.

One of the major payloads of STS-59 was the Space Radar Laboratory (SRL). A major component of SRL was an imaging radar system designed for detailed investigations of surface features on the Earth. The radar could image details of the surface of the Earth through clouds, vegetation, or even very dry soil. This ability provides scientists the ability to study parts of the Earth that would otherwise be inaccessible.

Another portion of the mission included the Measurement of Air Pollution from Satellite (MAPS) system. MAPS took data on carbon monoxide concentrations over the Earth's surface. Among the sites studied were areas of the Earth scarred by forest fires in an attempt to measure how well the Earth's atmosphere rids itself of greenhouse gases. Greenhouse gases are gases that hold in heat from the sun, thus raising the temperature of the Earth.

The fourth space shuttle mission of the year was an extended duration mission with the orbiter *Columbia*. The mission, STS-65, began with liftoff at 12:43 A.M., eastern daylight time, on July 8, 1994, from Pad 39-A of the Kennedy Space Center. *Columbia* carried the second International Microgravity Laboratory (IML-2) into orbit. Landing was at 6:38 A.M., eastern daylight time, on July 23, 1994, on Runway 33 at the Kennedy Space Center.

IML-2 was an international cooperative effort of scientists from six different nations. The scientific investigations centered on studying materials and biological systems under microgravity conditions. The material science investigations generally focused on crystal growth and semiconductor manufacturing experiments. The life science investigations were targeted at investigating how gravity affects living organisms.

Among the experiments performed on board *Columbia* during the STS-65 was an experiment designed to study crystal growth of proteins in an attempt to learn more about the structure of those proteins. Many of the experiments involved studies of human physiological changes in microgravity, with the astronauts themselves acting as test subjects. Critical point investigations similar to those of the STS-62 mission were also performed. *Columbia* also carried into orbit numerous aquatic animals, such as jellyfish, goldfish, and sea urchins, to see how these animals react to microgravity. Plant studies were performed in a centrifuge to determine the lowest levels of gravity that begin to affect plant roots.

The fifth space shuttle mission of 1994 was STS-64. The mission began at 6:23 P.M., eastern daylight time, on September 9, 1994, with the launch of the space shuttle *Discovery* from Pad 39-B at the Kennedy Space Center. The mission had a variety of basic science objectives. Landing was on September 20, 1994, at Edwards Air Force Base, in California, on Runway 04 at 2:13 P.M., Pacific daylight time.

A major component of the mission was the use of a laser system to study particles in the Earth's atmosphere. This information can be used to determine human impact on the atmosphere. These measurements of the atmosphere from above were done in conjunction with ground-based observations of the atmosphere.

A separate experiment was conducted using the SPARTAN-201 satellite. This satellite was designed to measure solar activity and to make measurements of the Sun's corona. The satellite was deployed on September 13, 1994, and it moved slowly away from *Discovery*, far enough to avoid an interference the space shuttle may have had with its measurements of the Sun. The satellite was retrieved on September 15, 1994.

Another component of the STS-64 mission was an untethered spacewalk. During this activity, astronauts tested equipment that was designed as a rescue aid for future astronauts who accidentally become untethered during a spacewalk and float away from their spacecraft or space station.

The sixth space shuttle mission of the year was STS-68. This mission was a continuation of the studies done by STS-59 earlier in the year, with the space shuttle *Endeavour* once again carrying Space Radar Laboratory (SRL-2) and the Measurement of Air Pollution by Satellite (MAPS) into orbit. Launch occurred at 7:16 A.M., eastern daylight time, on September 30, 1994, from Pad 39-A at the Kennedy Space Center. Landing was on October 11, 1994, at 10:02 A.M., Pacific daylight time, at Edwards Air Force Base, in California, on Runway 22.

STS-68 continued the radar observations of the Earth that the STS-59 mission had started in April, 1994. The system made radar images of volcanos that had erupted during the interval between the two missions. Radar images of the Sahara and the North Atlantic were made in an attempt to understand climate changes on Earth.

The MAPS system measured carbon monoxide levels around the world from above. This information is important in our understanding of human interactions with our atmosphere. Furthermore, MAPS measured carbon monoxide levels in the area of several deliberately set small forest fires. These fires had been planned prior to the space shuttle mission as part of forest management strategies and were not set solely for the benefit of the STS-68 mission.

The seventh, and final, space shuttle mission of 1994 emphasized atmospheric studies. This mission, STS-66, began on November 3, 1994, at seventeen seconds before noon with the launch of the space shuttle *Atlantis* from Pad 39-B at the Kennedy Space Center. The primary payload was the Atmospheric Laboratory for Applications and Sciences-3 (ATLAS-3). Landing was at Edwards Air Force Base, in California, on Runway 22 at 7:34 A.M., Pacific standard time, on November 14, 1994.

ATLAS-3 was the third in a series of ten planned ATLAS missions to study the Earth's atmosphere from above. Astronauts aboard *Atlantis* made numerous measurements of atmospheric composition during the mission. A major goal of the mission was to further the understanding of the chemical reactions involving ozone in the upper atmosphere.

Knowledge Gained

All of the space shuttle missions of 1994 were dedicated science missions. As a consequence, a wealth of scientific data was recorded over the course of the year. Much of this data is still being studied.

Numerous materials science experiments were performed under microgravity conditions. These experiments allowed scientists to study the formation and growth of crystals without the influence of gravity. Eliminating the influence of gravity allows crystals to grow larger and more perfectly than those grown on Earth. By studying these crystals, scientists can learn more about the atoms and molecules themselves that make up the materials being studied. The critical point experiments permitted scientists to study critical points of materials, such as xenon, which are difficult or impossible to study on Earth due to the convection effects that are augmented by the larger gravitational effects on the Earth's surface.

Biological investigations studied the growth and development of animals and plants under microgravity. Several of these investigations utilized the astronauts as test subjects to study how the human body adapts to microgravity. Among the findings was the fact that calcium loss from bones, a major problem in long duration spaceflights, can be lessened through physical exercise. Bone tissue replenishes calcium in bones under normal circumstances on Earth. However, under microgravity, bone tissue is much less efficient at replacing the calcium in bones. The experiments done during STS-65 indicated that compressing and stressing the bone tissues through physical exercise can minimize some of the detrimental effects of microgravity.

The two SRL missions, STS-59 and STS-68, provided valuable data on volcanic activity and fault motion. Furthermore, dry river beds under the sands of the Sahara Desert indicate that the northern portions of Africa were at one time much wetter than they are today. This information may be useful in understanding climate changes on Earth.

Atmospheric studies were an important aspect of the space shuttle missions of 1994. MAPS made measurements of carbon monoxide around the world. By flying on two missions during the year, MAPS was able to make some preliminary estimates of seasonal changes in carbon monoxide levels. This information is important in tracking greenhouse gas concentrations on Earth.

The ATLAS-3 was the third in a series of ten ATLAS missions. By comparing the findings of each of the ATLAS missions, scientists hope to understand how the Earth's atmosphere reacts to both human activities and solar activity. ATLAS-3 measurements were some of the best of the series. Excellent measurements were made of solar activity. Another major emphasis of the mission was measurements of concentrations of ozone in the upper atmosphere and measurements of other chemicals that interact with ozone. Scientists are using these measurements to gain a better understanding of ozone depletion in the upper atmosphere.

Context

Unlike most years, in which several space shuttle missions involved transporting satellites to or from orbit, the space shuttle missions of 1994 were all dedicated science missions. The science missions scheduled for the space shuttle could not have been done on the Earth, and thus had to be done in orbit. In accomplishing the objectives for the year, the space shuttle program set at least two new records. Sergei Krikalev was the first Russian cosmonaut to fly aboard a U.S. space shuttle. The space shuttle *Columbia* set a new record for the longest space shuttle mission with STS-65 with 14 days, 17 hours, and 55 minutes.

Materials science experiments, such as those done on STS-60, STS-62, and STS-65, helped scientists understand crystal growth. Crystal structure is fundamental to modern semiconductor electronics, and a better understanding of crystal structure is vital in the search for ways of building better semiconductors. Information learned about proteins on STS-62 and STS-65 may yield new drugs to help fight diseases on Earth.

The life science experiments performed during several missions throughout the year yielded information about the effects of microgravity on plants and animals. This information is essential to reduce the detrimental effects of microgravity on astronauts during long duration space flights, such as a space station or a mission to Mars. Furthermore, some of the techniques used by astronauts to prevent calcium loss in their bones may also aid people on Earth suffering from certain bone disorders, such as osteoporosis.

The two SRL missions provided valuable data for geologists to study. This method of using radar to study a planetary surface is very similar to that used by the Magellan Spacecraft at the planet Venus, where dense clouds prevent the spacecraft from visually studying the planet's surface.

The atmospheric investigations of the year were among the most sophisticated performed on Earth's atmosphere. The MAPS data provided scientists with a tool for measuring the effects of greenhouse gases in the atmosphere. ATLAS-3 studied ozone depletion, among other things. Both ozone depletion and the buildup of greenhouse gases have been postulated to be the result of interactions between humans and the atmosphere of our planet. These missions help to understand the effects of pollution on the atmosphere.

Bibliography

Branley, Franklyn M. *From Sputnik to Space Shuttles.* New York: Thomas Y. Crowell, 1986. This is a very good juvenile book on the development of space exploration, from unmanned satellites to the modern space shuttle. The book emphasizes the uses of the space shuttle more than the shuttle itself.

DeWaard, E. John, and Nancy DeWaard. *History of NASA: America's Voyage to the Stars.* New York: Exeter Books, 1988. This book is intended for the general audience. The last chapter of the book covers the beginnings of the space shuttle program. The need for a system such as the space shuttle is discussed. The book ends with the explosion that destroyed the *Challenger* in 1986.

Hawkes, Nigel. *Space Shuttle.* New York: Gloucester Press,1983. This juvenile book includes a very nice description of the design and layout of the space shuttle. It also includes a very brief discussion of living in space and the sorts of payloads and missions available to the space shuttle.

Joels, Kerry M., Gregory P. Kennedy, and David Larkin. *The Space Shuttle Operator's Manual.* New York: Ballantine Books,1982. This is the definitive description of the space shuttle itself, available for general audiences. It includes diagrams and charts illustrating key parts of the space shuttle, and even includes fold-out sections showing instrument panels. It includes a very good description of living facilities on board the shuttle. There are even detailed timelines for both the launch and landing phases of the missions.

Rycroft, Michael, Ed. *The Cambridge Encyclopedia of Space.* New York: Cambridge, 1990. This general audience level work contains a section on space launch facilities, including the Kennedy Space Center, where space shuttle missions are launched. Further sections discuss living in space, with a very thorough description of the various types of microgravity experiments that can be conducted in space. The space shuttle system itself is given considerable coverage.

Shapland, David, and Michael Rycroft. *Spacelab: Research in Earth Orbit.* New York: Cambridge, 1984. Written in the early days of the space shuttle program, this general audience level book describes the types of scientific missions anticipated for the space shuttle. The book describes a Spacelab, which is a living/working module that is flown in the space shuttle's cargo bay during science missions.

Torres, George J. *Space Shuttle: A Quantum Leap.* Novato, Calif.: Presidio Press, 1986. This general audience level book traces the development of manned spaceflight, culminating with the space shuttle. Included are brief descriptions on many different types of shuttle missions. The book was written at the time of the *Challenger* explosion, and thus does not include any of the missions that occurred after the resumption of the shuttle program.

Raymond Benge

Space Shuttle Flights, 1995

Date: February 3, 1995, to November 20, 1995
Type of mission: Manned Earth-orbiting spaceflight

The space shuttle flights of 1995 included seven missions, three of which involved a rendezvous with the Russian space station Mir. Ultraviolet image data of the universe was collected, along with experiments performed under reduced gravity. Astronauts engaged in extended spacewalks to test the thermal performance of spacesuits.

Principal personages

JAMES D. WETHERBEE, STS-63 Commander
EILEEN M. COLLINS, STS-63 Pilot
BERNARD A. HARRIS, JR.,
C. MICHAEL FOALE,
JANICE E. VOSS, and
VLADIMIR G. TITOV, STS-63 Mission Specialists
STEPHEN S. OSWALD, STS-67 cCommander
WILLIAM G. GREGORY, STS-67 Pilot
JOHN M. GRUNSFELD,
WENDY B. LAWRENCE, and
TAMARA E. JERNIGAN, STS-67 Mission Specialists
RONALD A. PARISE and
SAMUEL T. DURRANCE, STS-67 Payload Specialists
ROBERT L. "HOOT" GIBSON, STS-71 Commander
CHARLES PRECOURT, STS-71 Pilot
ELLEN L. BAKER,
GREGORY J. HARBAUGH, and
BONNIE J. DUNBAR, STS-71 Mission Specialists
ANATOLY Y. SOLOVYOV and
NIKOLAI BUDARIN, STS-71 Payload Specialists
TERENCE T. "TOM" HENRICKS, STS-70 Commander
KEVIN R. KREGEL, STS-70 Pilot
DONALD A. THOMAS,
NANCY J. CURRIE, and
MARY ELLEN WEBER, STS-70 Mission Specialists
DAVID M. WALKER, STS-69 Commander
KENNETH D. COCKRELL, STS-69 Pilot
JAMES S. VOSS,
JAMES H. NEWMAN, and
MICHAEL L. GERNHARDT, STS-69 Mission Specialists
KENNETH D. BOWERSOX, STS-73 Commander
KENT W. ROMINGER, STS-73 Pilot
CATHERINE G. "CADY" COLEMAN,
MICHAEL E. LOPEZ-ALEGRIA, and
KATHRYN C. THORNTON, STS-73 Mission Specialists
FRED W. LESLIE and

ALBERT SACCO, JR., STS-73 Payload Specialists
KENNETH D. CAMERON, STS-74 Commander
JAMES D. HALSELL, JR., STS-74 Pilot
CHRIS A. HADFIELD,
JERRY L. ROSS, and
WILLIAM S. "BILL" MCARTHUR,
STS-74 Mission Specialists

Summary of the Missions

Mission 63 with the *Discovery* space shuttle entered an orbit to catch up with the Russian space station Mir after launch on February 3, 1995 at 12:22 P.M., eastern standard time (EST). On February 6, after careful maneuvering by Navy Commander James D. Wetherbee and Lt. Colonel Eileen M. Collins, *Discovery* approached Mir to within 11.2 meters, the two largest spacecraft ever to rendezvous in orbit.

Training for the historic meeting was aided with a laptop computer program that simulated the actual flight paths of both spacecraft. When an actual orbit of 340 kilometers was reached, *Discovery* approached Mir from below, closing in from ahead but along Mir's flight path. From 700 meters away, Wetherbee and Collins used manual control of the shuttle's thrusters to ease into a target located on Mir's module docking port. Live television cameras from both spacecraft documented the flight. The entire fly-around lasted one half of an orbit or forty-five minutes.

Throughout the flight, the crew of *Discovery* performed a variety of experiments. Vladimir G. Titov, a Russian cosmonaut on *Discovery*, used the orbiter's manipulator arm to lift and position the imaging spectrograph to measure the glow produced by atomic oxygen striking the spacecraft. The spectrograph was also used to image the interstellar medium.

A major objective of the mission was to gather data on spacesuit comfort in extreme cold. Two astronauts performed extended spacewalks for 4.5 hours to provide spacesuit temperature data. Activities involved moving a 1,300 kilogram satellite with a special tool manipulated by hand. STS-63 ended with the landing of *Discovery* at the Kennedy Space Center Shuttle Landing Facility on February 11, 1995, at 6:50 P.M., EST.

STS-67 with the *Endeavour* orbiter was launched on March 2, 1995, at 1:38 A.M., EST. The primary objective of the mission was to operate the ultraviolet astronomy instruments, mapping out major uncharted regions of space that are usually blocked by ozone layers in the atmosphere. Scientists expected to detect the diffuse hydrogen and helium gas composing the intergalactic medium believed to be uniformly distributed after the "Big Bang" episode.

A key to the success of that task was ensuring that the Instrument Pointing System (IPS) functioned properly. The IPS used improved software that allowed astronauts and scientists on the ground to lock on to certain target stars closer to the Earth's limb, permitting longer observations of those stars and sources with three major instruments. The crew of *Endeavour* carried out detailed round-the-clock observations for the first days of the mission in two teams. One team, including Mission Commander Stephen S. Oswald, flew the orbiter while the others were involved with the experiments.

During the mission, the astronauts conducted more than forty hours of experiments with the middeck active control experiment. This package is designed to detect and reduce uncontrolled vibrations that can occur on a free-floating surface. Programs abased on preflight configurations provided the information to control vibrations and maintain platform stability. Data from this experiment was expected to result in the development of control systems that would eliminate vibrations that interfere with the operations of structures on satellites and other space vehicles. *Endeavour* touched down on Edwards Air Force Base Runway 22 on March 18, 1995, at 1:47 P.M., Pacific standard time (PST), setting an orbiter endurance record of 16 days, 15 hours, 8 minutes, and 48 seconds. The launch of STS-71 with the *Atlantis* orbiter occurred on June 27, 1995 at 3:32 P.M., eastern standard time (EDT), and marked the one hundredth American manned space flight. The primary objective of the mission was to rendezvous and dock with the Mir 18 spacestation, exchanging crew members. Techniques for the close approach were pioneered on STS-63 except for one major difference. On the earlier mission, the *Discovery* orbiter approached Mir from the front along Mir's flight path. This time the *Atlantis* closed toward the space station from underneath directly along a line termed the "R-bar." This maneuver was easier to accomplish since gravity would naturally act as a brake, conserving fuel from *Atlantis'* thrusters.

The new docking procedure was initiated by the concerns that *Atlantis* might cause damage to Mir's solar panels and the use of gravity would reduce the need for *Atlantis'* crew to use the thrusters to lessen the approach speed. The actual docking was achieved on June 29 at an altitude of 340 kilometers and was assisted by television cameras, including one placed on the centerline of the docking system. The docking was the result of more than two years of preparation by flight controllers and resulted in new communication and procedures. *Atlantis* remained docked to the Mir Complex for five days of joint experiments. On July 6, 1995, at 10:55 A.M., EDT, the orbiter glided to a landing at the Kennedy Space Center.

The launch of STS-70, commanded by Tom Henricks, originally scheduled for lift-off on June 8, was not a reality until July 13th, six days after Mission 71 had already landed. The delay in the launch was due to the discovery of 205 holes in the urethane foam insulation on the vehicle's external fuel tank. The holes were apparently caused by several woodpeckers seeking a nest. The delay cost the agency more than two million dollars, one hundred thousand of this alone just to roll back the shuttle and return it to the pad.

The eight-day mission of the *Discovery* orbiter began at 9:42 A.M., EDT, July 13, 1995 and featured the new Block 1 shuttle main engine. The new design, which included a dual duct hot-gas manifold and a redesigned main injector, was expected to provide greater reliability and performance. A second major objective was the orbital launching of the Tracking and Data Relay Satellite-G (TDRS-G). TDRS-G was positioned in a geosynchronous orbit complementing five similar satellites already in orbit.

Discovery's crew also tested a camera system designed to take high resolution images of ship tracks for the purpose of analysis of wave formations. The crew also performed a number of biological and materials experiments including one involving the determination of the effects of microgravity on physiological changes in rats. The flight concluded at 8:02 A.M., EDT, on July 22, 1995, on Runway 33 at KSC's Shuttle Landing Facility.

The eleven-day flight of STS-69 with the *Endeavour* orbiter, launched on September 7, 1995, at 11:09 A.M., EDT, and commanded by David M. Walker, carried a variety of experiments including the Wake Shield Facility (WSF), a saucer shaped satellite designed to fly free from the shuttle. The WSF allowed growth of thin films in the almost perfect vacuum produced by the wake of the satellite. This has an application to the semiconductor industry and the chemical growth of films for electrical instruments. To ensure a clean environment, the WSF is allowed to trail the *Endeavour* by as much as sixty-five kilometers.

On-board payloads investigated the effects of microgravity or reduced gravity on cell changes and on the repeated melting and freezing of fluoride salts. The National Institute of Health's experiment investigated the loss of bone mass during spaceflight. Other experiments determined the gravity sensing mechanisms within mammalian cells and the effect of microgravity on neuromuscular disorders. The thermal energy storage experiment was designed to understand the long-term behavior of the spaces of a lithium fluoride salt produced in a repeated environment of melting and refreezing in microgravity. This salt is used to store thermal energy in solar powered designs. *Endeavour* landed on September 18, 1995, at 7:38 A.M., EDT, at Kennedy Space Center.

STS-73 with the *Columbia* orbiter was launched on October 20, 1995, at 9:53 A.M., EDT, with Kenneth D. Bowersox as Command Pilot. The fifteen-day-mission objectives included deployment of the second microgravity laboratory, providing insights into the role of gravity in theoretical models of fluid physics, crystal growth, plant growth, and drop encapsulation. A particle dispersion experiment was placed on board to confirm the behavior of dust and particles in space.

The Geophysical Fluid Flow Cell (GFFC) experiment modeled planetary atmospheres and simulated fluid flow within the Earth's mantle. The planet Jupiter's atmosphere was modeled on several of the runs. On the final session, a lower rotation similar to that found in the fluids of the Earth's mantle was simulated. All these runs were performed by the crew members through the payload general support computer, with the status reported to ground investigators.

With a device called the glovebox, mission specialists monitored protein crystal growth exposed to a constantly changing gravity vector. The changing gravity was achieved by periodic changes in the vernier thrusters on board the spacecraft and was expected to cause unusual structural changes. The glovebox was used to study how the behavior and movement of fluids was affected by the attitude and shape of the container. In another container, five small potatoes were placed to study the feasibility of growing edible plants in space.

Drop encapsulation was successfully accomplished in the drop physics module, a device that uses sound waves to levitate and maneuver liquid drops for close study. The coalescence of two drops into one was achieved as was the formation of a chemical membrane between the two drops. Drop fissioning, the spinning of one drop until it splits into two, was accomplished on this mission.

The particle dispersion experiment was designed to confirm a theory about the behavior of dust and particle clouds. The experiment tested the theory that attraction occurs in the dust clouds of space due to static electrical charges. Variables such as particle size, cloud density, and type of material (volcanic material, quartz, and copper) were all tested. *Columbia*'s mission came to a successful conclusion at the Kennedy Space Center on November 5, 1995, at 6:45 A.M., EST.

The final shuttle mission of 1995 occurred with the launch of STS-74 and the *Atlantis* orbiter on November 12 at 7:31 A.M., EST. The crew, with Colonel Kenneth D. Cameron as commander, utilized the experience of STS-71, which achieved the first shuttle docking with the Mir space station. The mission objectives included transporting food, water, and experiments up to the space stations, retrieving test samples and hardware, and deploying and installing a new docking module and solar arrays. Both Russian and U.S. crew members as well as other officials were able to work closely together on the demanding task of assembling a new space station.

The new forty-one kilogram docking module was installed on Mir on November 14. The docking module was deployed from the cargo bay of *Atlantis* by means of a fifteen-meter remote manipulator arm. The docking module, which would link the *Atlantis* to Mir, replaced the older Kristal module, enabling future shuttles to dock with adequate clearance. Two sets of solar arrays were attached to the docking module and were deployed by the cosmonauts on Mir by means of Extravehicular Activities (EVA) or spacewalks. *Atlantis* ended its historic mission with a landing at the Kennedy Space Center on November 20, 1995, ar 12:01 P.M., EST.

Knowledge Gained

The joint missions of the Russian Mir spacecraft and STS-63, STS-71, and STS-74 led to the development of new techniques for rendezvous and ultimately docking in orbit. STS-71 pioneered the "R-bar" approach and docking method of closing in on the space station from below and using gravity to act as a brake, conserving fuel from the thruster engines. These missions developed the ODS or Orbital Docking System, which was an automated system with a capture ring that facilitated docking by removing residual motion upon contact of the spacecraft.

STS-74 delivered a new docking module to Mir that was installed in space with a remote manipulator arm. The remote manipulator arm was operated by Mission Specialist Chris Hadfield. When docking was completed, the shuttle and Mir formed the largest combined spacecraft ever attempted, with a total mass of more than 226,000 kilograms. This was considered a major step in the assembly of an international space station.

The handshake between Robert "Hoot" Gibson and Vladimir N. Dezhurov across the *Atlantis* and Mir docking tunnel on STS-71 was the first international spacecraft linkup since the U.S. Apollo and Soviet Soyuz flight on July 17, 1975. Three of the Mir crew, including Dezhurov, returned to Earth on *Atlantis* July 7 after almost four months in space. While in orbit the *Atlantis* crew transferred seventy-six kilograms of water to Mir and used *Atlantis* to raise Mir's cabin pressure by 11 percent. STS-74 off-loaded 970 kilograms of food, water, and equipment with supplies of nitrogen and oxygen to Mir. In turn, 371 kilograms of equipment and samples were taken on board *Atlantis* for the return trip. All of these achievements demonstrated the feasibility of servicing an operating space station.

The extended spacewalks performed by STS-63 specialists Michael Foale and Dr. Bernard Harris provided valuable data on suit comfort under extreme cold. It was discovered that the astronauts' fingertips became too cold for them to work in space. As a result, spacesuits were modified with heater coils placed in each fingertip of their gloves. Temperature sensors were also added to the astronaut's boots. Spacesuits were redesigned, allowing the coolant to flow around the body without interrupting flow to the suit electronics and other components.

The second flight of the Wake Shield Facility on STS-69

to test materials in the near vacuum created by the satellite's wake demonstrated the ability to grow exotic and ultrapure thin films in space. The wake created on the top side of the satellite produced a vacuum several orders of magnitude greater than could be reached on Earth. The satellite also carried an experiment that measured how much electrical fields caused by ionized particles around the spacecraft interfered with communications.

The reduced gravity or microgravity of the shuttle's environment led to successful results in a number of experiments. Scientists observed various heat-driven flow patterns in the GFFC experiment when the initial conditions were changed. These findings appear to substantiate theoretical mathematical models of planetary and solar fluid flow. The Glovebox experiment showed differences in the way fluids adhered to container walls, indicating that there are additional physical factors involved not predicted by current theory.

The ultraviolet measurements taken by the crew of STS-67 included the first observations of an active volcano on Jupiter's moon Io. The first measurements of radiation and polarization were made on Nova Aquilae, a binary star system. The largest telescope locked on to distant quasars and measured the abundance of hydrogen and helium in the intergalactic medium.

Context

The space shuttle flights of 1995 demonstrated the importance of international cooperation in achieving the major goals of space exploration, including building an international space station in the late 1990's. Russia and the United States were able to coordinate their space programs precisely not only to communicate and rendezvous in space but jointly to build a docking module that will serve future missions. Both of these nations were able to use their separate facilities to train astronauts and cosmonauts for common missions.

Crew members on these flights were able to deploy and retrieve a remote satellite with capabilities for growing exotic films in the nearly perfect vacuum created in a wake. The high quality chemical growth produced under these conditions will have application for the development of semiconductors and other high-tech electrical instruments.

Observations of fluid flow in microgravity is expected to help scientists better understand the behavior of fluid flows on the Earth, as well as movements of water in the oceans and circulation within the atmosphere. The glovebox experiment showed how surfaces form in low gravity, and these insights will help in the design of fluid systems used in space, including those utilizing fuel.

Results from ultraviolet experiments indicated the advantages of using orbital telescopes for observing wavelengths that would otherwise be filtered out by the Earth's atmosphere. A nova star was imaged for the first time with the 3.7-meter-long telescope. A clearer image of the white hot gas pulled off a white dwarf star by a normal star in this binary star system was obtained than is possible from ground-based telescopes. Research based on this data may provide clues about the triggering mechanism for the resulting thermonuclear fusion process.

The relative abundance of hydrogen to helium in the intergalactic medium was also detected with ultraviolet telescopes, and data gathered will be used to confirm or rule out the quantities of helium predicted by the Big Bang Theory. The medium of hydrogen and helium is thought to have formed in the first few minutes after the Big Bang, the epic that astronomers generally believe created the universe. Data collected from ultraviolet measurements are being compiled into a galactic atlas that may prove an important key in unlocking some of the mysteries of the universe.

Bibliography

Asker, James R. "Mission 63 To Test U.S.-Russian Teaming." *Aviation Week and Space Technology* (January 30, 1995). An excellent source of information with detailed coverage on all of the current space flights. The shuttle flights of this manuscript are covered in considerable depth in subsequent issues. James Asker has done a commendable job in keeping the reader informed of developments in all phases of the space program. Available at most public libraries.

Gore, Rick. "When the Space Shuttle Finally Flies." *National Geographic* (March, 1981). The development of the space shuttle is documented in this thirty-page article. There are abundant color photographs and diagrams featuring the cockpit interior, an exploded view of the shuttle's interior, and the power plants. Photographs show astronauts practicing at the underwater facility in Huntsville, Alabama. The first test flights in California are also described.

Kerrod, Robin. *The Illustrated History of NASA.* New York: Gallery Books, 1987. A superb reference of color plates depicting advancements in the space age. Chapter 5 documents the development and first missions of the shuttles *Challenger* and *Columbia*. Chapter 6 is profusely illustrated with the flights of *Atlantis* and *Discovery*. A chronology of NASA's first twenty-five years covers significant events by year from 1960 to 1983. A fitting epilog describes the last flight of *Challenger* on January 28, 1986.

Powers, Robert M. *Shuttle: The World's First Spaceship.* Harrisburg, Pa.: Stackpole Books, 1981. An history of the development of the shuttle detailing how the design evolved from earlier rocket configurations. Excellent diagrams and photographs include cockpit detail and component dimensions. Specifications are included in the appendices. A glossary of NASA abbreviations and jargon is provided for the reader.

Time-Life Books. *Life In Space.* Alexandria, Va.: Author, 1983. Three chapters are devoted entirely to the shuttle with the third chapter containing color photographs of six shuttle missions. A description is given of the various mission specialists and their functions. The arrangement and composition of the more than 30,000 ceramic tiles covering the shuttle's outer skin is discussed. Cutaway diagrams reveal internal structure and a full-page diagram depicts twenty one positions of the shuttle's trajectory.

———— *Outbound.* Richmond, Va.: Author, 1989. A photographic history of man's exploration in space. Photographs show controllers at the Johnson Space Center monitoring the fourth flight of the Shuttle *Challenger* in February, 1984. An good reference for information on the suits worn by the shuttle crew. Color diagrams reveal the interior details of the suit life support and power systems.

————. *Spacefarers.* Richmond, Va.: Author, 1989. Depicts the shuttle as a delivery vehicle for the construction of the space station. Diagrams show how the elements of the space station will fill the shuttle's cargo bay. Additional color illustrations display the space station under construction and the use of the mobile transporter that will carry the astronauts and their supplies to the work area.

Michael L. Broyles

SPACE SHUTTLE MISSION
STS-63

Dates: February 3, 1995, to February 11, 1995
Type of mission: Manned Earth-orbiting spaceflight

STS-63 began an important phase of cooperation between the National Aeronautics and Space Administration (NASA) and the Russian Space Agency (RSA), in which techniques for joint operations between the NASA space shuttle and the Russian Mir space station would demonstrate the feasibility of construction of an International Space Station.

PRINCIPAL PERSONAGES

JAMES D. WETHERBEE, STS-63 Commander
EILEEN M. COLLINS, STS-63 Pilot
BERNARD A. HARRIS, JR.,
C. MICHAEL FOALE, and
JANICE E. VOSS, STS-63 Mission Specialists
VLADIMIR G. TITOV, Russian cosmonaut
    and STS-63 Mission Specialist
ALEXANDER VIKTORENKO, Soyuz TM-20/Mir 17
    Commander
ELENA KONDAKOVA, Soyuz TM-20/Mir 17
    Flight Engineer
VALERY POLYAKOV, Soyuz TM-18/Mir 15 physician-
    cosmonaut, who spent over a year on Mir

Summary of the Mission

NASA announced the crew for STS-63 and its primary focus on September 9, 1993. That crew included one important and historic distinction — the first female shuttle pilot. Selected as STS-63 Commander was United States Navy commander James D. Weatherbee. His Pilot was United States Air Force Major Eileen M. Collins. Mission Specialists included Dr. C. Michael Foale, Dr. Janice E. Voss, Dr. Bernard A. Harris, Jr., M.D., and Russian Air Force Colonel cosmonaut Vladimir G. Titov. STS-63's primary payloads were Spacehab 3, and the Spartan-204 astronomical research platform for studying the solar wind.

Weather conditions for launch were perfect. *Discovery* lifted off at 12:22 A.M. on February 3, 1995, into clear starlit skies. The vehicle appeared somewhat sluggish at Solid Rocket Booster (SRB) ignition, but performed nominally. SRB separation was clean. The second-stage performance was nominal and steered *Discovery* into the proper insertion point at main engine cutoff. *Discovery* had returned to space on its twentieth mission.

Once in orbit, it was soon noticed that *Discovery* had a leaking Reaction Control System (RCS) thruster on its right Orbital Maneuvering System (OMS) pod. Flight rules for STS-63 required that *Discovery* have all of its aft-firing thrusters up and running nominally before it could approach to within the final 300 meters of Mir. Mission Control ordered Weatherbee to change *Discovery's* attitude so that the offending thruster was facing the sun in order to allow it to warm up for several hours. The thruster was losing about one kilogram of propellant per hour. This was certainly a manageable loss, one with no impact on the ability of the orbiter to remain in space for the duration of the scheduled mission. But the gas could contaminate structures on Mir if it was to impinge upon them, structures such as the solar panels and high-optical-quality glass portholes.

Shortly after midnight on February 4, the astronauts grappled the Spartan-204 satellite with the Remote Manipulator System (RMS) arm's end effector, and lifted the payload out of *Discovery's* cargo bay for a shuttle glow measurement. The payload was held about eleven meters above the bay for several hours and was then returned to its berthing position. Titov was primarily responsible for RMS operation.

The crew transmitted to Mission Control televised pictures of the leaking RCS thruster. Streams of fuel coming out of the thruster cone were plainly visible. Weatherbee and Collins performed some work to stop the leak, but their attempts failed. Russian engineers were reluctant to allow *Discovery* to get closer than 121 meters of Mir. Flight directors in Houston were decidedly reluctant not to fly to within the 11 meters previously agreed upon.

On Feburary 6 the crew received its wake-up call at 12:21 A.M. For the first time in nearly twenty years, an American and Russian manned spacecraft were about to fly in formation. When the crew arose, they were still "no go" for the close approach. There were still six hours to go until completion of the rendezvous, but the astronauts already saw Mir as a bright dot in the pre-dawn skies several hundred kilometers away. The astronauts transmitted televised images of Mir, the space station appearing only as a bright dot.

Titov first made contact with his comrades on Mir, calling from *Discovery* at 10:00 A.M. The two spacecraft were still about 160 kilometers apart.

The Russian Space Agengy (RSA) eventually approved

the close approach to Mir at 10:25 A.M. *Discovery* would keep the leaking thruster and its companion on the same manifold shut down during the close approach, and a new combination of thrusters would be used instead.

The orbiter's nose pointed forward, its payload bay to Mir. Mir Commander Viktorenko reported seeing thruster firings from his vantage point, noting that Mir's solar arrays were not affected by the thruster plumes. Dr. Polyakov reported he could see Weatherbee through the aft overhead window of *Discovery*. Weatherbee responded with a wave back to the cosmonaut.

Weatherbee assumed manual control of the remainder of the approach when *Discovery* was only 606 meters away from Mir. He switched *Discovery* to the low-Z mode when 310 meters from Mir. This prevented thrusters from firing up against Mir structures. *Discovery* moved down a 16 degree cone eventually approaching close to the Kristall module of Mir at less than three centimeters per second.

The astronaut crew waited for a final "go" to get within 11.5 meters of the Mir docking port as the two vehicles passed over the South Atlantic Ocean. Foale continued to work as a flight engineer at the aft station, and *Discovery* remained 121 meters in front of Mir while final preparations were made prior to getting a "go" to proceed in further.

Dr. Polyakov, the current world space flight endurance holder— going for eighteen months, well past a year already — could see flames from thruster firings as Weatherbee moved in on Mir. None of the thrusters he used fired up.

The cosmonauts on Mir asked Titov to describe the condition of Mir from his vantage point on *Discovery*, specifically looking to note changes since the time he spent a year on board the space station. The shuttle crew saw no motion of Mir's solar arrays in response to *Discovery*'s thruster firings. The Mir cosmonauts agreed with that assessment.

Weatherbee began to brake *Discovery*'s approach and held the two vehicles in separation at a distance of ten meters. The two vehicles were flying over the mid-Pacific Ocean in conditions of daylight on *Discovery*'s fifty-seventh orbit of the mission. Viktorenko noted that the cosmonauts inside Mir never felt any movement of Mir from the activities of *Discovery* during the close approach phase. After data refinement, it was announced that the closest distance between the top of Spacehab and the Kristall module on Mir was 11.2 meters, a separation achieved at 1:20 P.M.

Discovery backed away to a 121-meter separation distance at 3:26 P.M. and began a fly-around of the space station. Live television was downlinked from inside Mir showing Dr. Polyakov as *Discovery* crossed directly above the space station at a distance of 144 meters. To complete the fly-around, *Discovery* flew to a position directly behind Mir and then flew underneath the space station and returned to the point from which it started the fly-around.

Mir was currently in a so-called T-configuration. It would not be in this particular arrangement when *Atlantis* would attempt a docking during STS-71. Plans were in place to launch another module, Spekter, from Kazakhstan prior to the STS-71 docking mission. Kristall would be moved to another axial port on Mir to make room for Spekter. STS-71 would still make its docking to the Kristall module, regardless of its location on Mir.

The final separation burn was performed by *Discovery* at 4:13 P.M. *Discovery* was directly above Mir at the time. The burn pushed *Discovery* behind, away, and down from Mir, entering a lower orbit that would speed up the orbiter and eventually put it ahead of Mir.

The flight was hardly over, although this aspect was by and large the most exciting part, something of which flight director Phil Engelauf reminded the press. He called the joint flight a very invigorating exercise that had required a great deal of work.

On February 7, Titov grappled the Spartan-204 satellite again and raised it high above the payload bay as *Discovery* passed over Brazil at an altitude of 384 kilometers. This time, he let the satellite go, and *Discovery* backed away from the payload so that it could begin to make its far ultraviolet measurements autonomously over the next several days. By the end of the day, Mir and *Discovery* were about 260 kilometers apart, continuing to separate further.

EVA preparations filled much of February 8, as did experiment work with Spacehab investigations. On February 9, Voss used the RMS arm to grapple the Spartan-204 satellite at 6:53 A.M., as *Discovery* flew over the Aleutian Islands. For most of its autonomous flight, the free-flier had been about 68 kilometers ahead of *Discovery*.

By 7:00 A.M., the Spacehab hatch was open and the EVA was about to start. The two spacewalkers were both making their first excursion outside a spacecraft in orbit. When he exited the airlock, Harris became the first African American to perform an EVA.

The EVA astronauts wore warm thermal undergarments, sporting modifications made for the STS-61 Hubble Space Telescope (HST) First Servicing Mission EVA work. Cooling water was stopped short of the suit arms and thermal mittens were available. When cold, each man could pull his fingers back out of the glove and into the palm to make a fist and restore circulation and warmth.

When Foale and Harris reported just how cold they eventually got (the temperature in Foale's glove went well under 4.5° C), Collins joked with them, suggesting that they were in a deep freeze. Because of the cold, the orbiter's attitude was eventually changed to point its payload bay to Earth for reflected light to warm up the spacewalkers. But that was not done until most thermal testing and mass-handling exercises were accomplished. Further, because of the cold, the spacewalk was terminated about twenty-five minutes early.

With Titov at the RMS controls, Harris and Foale were maneuvered to a position high above the payload bay. This test lasted twenty minutes, with both men doing very little except serving as guinea pigs for the temperature exercise. Both men got uncomfortably cold in the fingers.

Special handholds were attached to the Spartan-204 payload to facilitate its manipulation. By 9:10 A.M., the astronauts were working on Spartan, evaluating how easily it could be maneuvered manually about the payload bay. Harris held the 1,273-kilogram satellite while on the RMS arm, judging how well he could handle such a bulk. He was moving something almost three times as massive as had been moved about on the HST First Servicing Mission (STS-61).

A thermal cube had been attached to the RMS arm to record temperatures encountered during the EVA. Before the EVA concluded at 11:30 A.M., the astronauts demated that thermal cube and stowed it for return to Earth.

On February 10, Weatherbee and Collins thoroughly checked out *Discovery's* flight control system. Thrusters were test fired and the computer software used on entry was verified. The GLO experiment was allowed to observe the effect of the steering jets' firings on the orbiter's airglow phenomenon.

The astronauts had one final look at Mir. At 1:35 P.M., the maneuver was performed that allowed the crew to see Mir near the Earth's horizon at a distance of over 1,360 kilometers. To the crew, and the payload bay television cameras, Mir appeared only as a small flashing light.

On February 11, *Discovery* was given a "go" for a nominal deorbit. As the vehicle entered the upper atmosphere, however, crosswinds at the Kennedy Space Center (KSC) Shuttle Landing Facility (SLF) picked up in intensity. Weatherbee would have to steer on runway 15 during touchdown and rollout as a result.

About 250 visitors gathered at a viewing site to watch *Discovery* glide out of the post-dawn skies at KSC to what could only be described as a picture-perfect touchdown at 6:50 A.M.

Knowledge Gained

The STS-63 flight plan was packed with diverse mission activities and objectives, the most exciting of which was undoubtedly the flyaround of the Russian space station Mir. In addition, *Discovery* carried the third Spacehab module with twenty individual experiments, a Hitchhiker payload holding low-cost shuttle experiments, and the Spartan-204 free-flier to study stellar emissions in the far ultraviolet. Also, as part of preparations for the upcoming shuttle/Mir missions and space station assembly flights, a five-hour spacewalk by Foale and Harris tested modifications to the shuttle EVA suit. They also practiced handling large objects in weightlessness.

The primary instrument on Spartan-204 was the Far Ultraviolet Imaging Spectrograph (FUVIS). The objective of FUVIS was to observe astronomical (and shuttle-created) sources of far ultraviolet (UV) radiation, looking for information on the composition, physical and chemical properties, and distribution of far UV-emitting materials in the interstellar medium. Some of the objects that Spartan observed included the Barnard Loop, North America nebula, Cygnus Loop, comets, and galactic structures outside the Milky Way. The principal investigator for FUVIS was Dr. George Carruthers at the Naval Research Laboratory.

The Spacehab module on STS-63 contained twenty experiments sponsored by the NASA Office of Space Access and Technology, NASA Office of Life and Microgravity Sciences and Applications, and the Department of Defense (DOD). Many of these experiments had flown before on the shuttle, several having flown on numerous occasions.

One of the experiments that the crew performed on February 5 in the Spacehab module was the Solid Surface Combustion Experiment. Pieces of Plexiglas were burned in a weightless environment, a part of a continuing experiment program that had flown on several previous shuttle missions. Another experiment conducted in Spacehab was the Charlotte robotic device. This experiment was set up by the astronauts, then left to run by its own devices. Cables were strung for the robot to maneuver across.

Context

Discovery touched down at KSC, having completed 129 orbits of the Earth and traveled a total of 3,348,480 kilometers over the course of an 8-day, 6-hour, and 28-minute flight, one of the most exciting and activity-packed shuttle missions, one that laid the groundwork for yet an even more exciting shuttle mission: STS-71, the first shuttle/Mir Complex docking.

Collins was notified during flight that she was now in line for a shuttle commander position. Often a shuttle pilot flies twice in the pilot seat before being given a command assignment. There were no details released about any specific flight assignment for the first female shuttle pilot.

Down on Earth, the events in space between NASA and the Russians sparked debate concerning the merits and even wisdom of attempting joint operations, based largely on two concerns: that the Russians were once a bitter rival, and that the Russian economy and government are both unpredictably volatile and unstable.

With threats of further cuts in the NASA budget from Congress, NASA administrator Daniel S. Goldin affirmed his commitment to reduce the $3.2 billion annual shuttle operating budget. Goldin stated that the budget could not hide behind issues of safety but stressed also that safety would not be compromised, either.

The Republican-controlled House of Representatives raised an old issue: privatization of the shuttle fleet. Consideration was given to plans to turn the shuttle maintenance and operational aspects of flight over to a commercial concern for profit and release NASA to perform purely research and development tasks.

Bibliography

"'Ballet' at Mir Opens New Space Era." *Aviation Week and Space Technology* (Feburary 13, 1995): 68-69. Provides details of the in-flight experiences of STS-63. Shows a photograph of the docking apparatus on Mir, and an image of *Discovery* taken from inside Mir as the shuttle approached the Russian space station.

"Houston, Moscow To Rely On New Com Links." *Aviation Week and Space Technology* (January 30, 1995): 36. Discusses how NASA Mission Control would coordinate with the Russian equivalent in Kaliningrad during the flight of STS-63.

"Mission 63 To Test U.S.-Russian Teaming." *Aviation Week and Space Technology* (January 30, 1995): 36-38. Describes the interaction between NASA and RSA, astronauts and cosmonauts, necessary to carry out STS-63.

National Aeronautics and Space Administration. *STS-63 Press Kit.* January, 1995. Fully describes the Mir space station, shuttle Orbiter *Discovery*, and the flight plan for this first shuttle flyaround of the Russian space station.

"U.S., Russia Plot Shuttle/Mir Flights." *Aviation Week and Space Technology* (January 16, 1995): 20-21. A description of the agreement between NASA and RSA that led up to joint Mir/shuttle activities. Includes details about the STS-63 flight in particular.

"U.S., Russian Suits Serve Diverse EVA Goals." *Aviation Week and Space Technology* (January 16, 1995): 40-41. Describes the differences between NASA shuttle spacesuits and Russian EVA suits. Relates to joint Mir/shuttle operations, not specifically STS-63.

David G. Fisher

SPACE SHUTTLE MISSION STS-71
MIR PRIMARY EXPEDITION 18

Date: June 27 to July 7, 1995
Type of mission: Manned Earth-orbiting spaceflight

STS-71, the National Aeronautic and Space Administration's (NASA) one-hundredth human spaceflight, conducted the United States' first docking with Russia's space station Mir, bringing it supplies and a new crew from Earth, returning other materials and the old crew, and gaining valuable experience for both countries in complex, joint operations in space. Already on board the Mir Complex were two Russians and one American participating in the Mir Primary Expedition 18 mission. The Russian abbreviation for Primary Expedition (Ekspeditsya Osnovnaya) is EO. This flight was the eighteenth long-duration mission aboard the Mir Complex.

PRINCIPAL PERSONAGES

ROBERT L. "HOOT" GIBSON, STS-71 Commander
CHARLES PRECOURT, STS-71 Pilot
ELLEN L. BAKER, STS-71 Mission Specialist
GREGORY J. HARBAUGH, STS-71 Mission Specialist
BONNIE J. DUNBAR, STS-71 Mission Specialist
  and backup to Norman Thagard on EO-18
ANATOLY Y. SOLOVYEV and
NIKOLAI M. BUDARIN, STS-71 Payload Specialists
VLADIMIR N. DEZHUROV, Soyuz TM-21/EO-18
  Commander
GENNADY M. STREKALOV, Soyuz TM-21/EO-18
  Flight Engineer
NORMAN E. THAGARD, Soyuz EO-18 Cosmonaut
  Researcher

Summary of the Mission

After decades of rivalry in the exploration and utilization of space, interrupted by brief collaborations, the United States and Russia agreed in the early 1990's to conduct joint operations in human spaceflight. Following STS-60, with the first flight of a cosmonaut on a space shuttle, and STS-63, which demonstrated the ability to rendezvous and orbit as close as about 11 meters of Russia's space station Mir ("peace"), STS-71 would be the first of a series of space shuttle missions to dock with the station.

When the space shuttle lifted off from the Kennedy Space Center in Florida at 3:32 P.M., eastern daylight time, on June 27, 1995, Mir was in orbit over Iraq. When *Atlantis* reached orbit eight minutes later, it was 13,000 kilometers behind Mir

but in an orbit that carried it around Earth faster than the space station. Initially, with each orbit the distance closed by 1,630 kilometers. Over the course of the next two days, the astronauts would modify their orbit to match that of Mir.

STS-71 was commanded by Robert Gibson, and his pilot was Charles Precourt. Bonnie Dunbar, Ellen Baker, and Greg Harbaugh were the mission specialists. The two other crew members, cosmonauts Anatoly Solovyev and Nikolai Budarin, would take up residence aboard Mir after the two ships had docked.

Already in space were Vladimir Dezhurov, Gennady Strekalov, and NASA astronaut Norman Thagard, conducting the eighteenth long duration mission on Mir, denoted EO-18 by Russia. They had been living there since two days after they left Earth on March 14 and would exchange places with Solovyev and Budarin during STS-71. In addition to performing experiments and maintaining the station, they had finalized the reconfiguration of its movable modules and solar arrays in preparation for *Atlantis'* arrival.

On the second day of STS-71, as *Atlantis* continued to close the distance to Mir, Gibson, Precourt, and Dunbar activated the Spacelab module carried in the orbiter's payload bay. This laboratory's extensive complement of scientific instruments would be used to measure the effects of the Mir crew's long exposure to the microgravity of orbit. Meanwhile, Baker verified that the Russian-built docking system, also in the payload bay, was in good condition. By the end of the day, *Atlantis* was within 2,500 kilometers of Mir, and the crew was ready to focus on performing the complicated rendezvous and docking the following day.

On June 29, as Gibson flew the orbiter to Mir, Precourt monitored *Atlantis'* position and orientation on a computer, Harbaugh measured the range and speed between the two spacecraft, and Dunbar transmitted these data to the crew on Mir. When they were about seventy-five meters from Mir, Gibson halted the approach to allow time for Russian controllers to verify that Mir was in the correct orientation for docking. In addition, the network of solar arrays on the station had to be adjusted so that they would be edge-on to *Atlantis*. This would ensure that exhaust from the orbiter's maneuvering thrusters would cause minimal disturbance to these sensitive appendages.

Continuing to close slowly, Gibson used a camera mounted in the docking apparatus to provide a view of the docking target attached to the Kristall ("crystal") module of Mir. When he had maneuvered *Atlantis* to about nine meters from the station, he halted his approach again for five minutes. This would allow the actual link-up to occur within range of a Russian tracking station and permitted a final check on all critical systems before joining the two massive spacecraft. Finally Gibson nudged *Atlantis* up to Mir at 3.3 centimeters per second. Precourt immediately fired *Atlantis'* thrusters to engage the docking mechanism and make a firm connection.

Besides providing a strong coupling, the docking system included a tunnel that allowed the astronauts and cosmonauts to move freely between the linked spacecraft. After pressurizing the passageway and confirming that both spacecraft were in good condition, the two commanders, Gibson and Dezhurov, opened the hatches and greeted each other.

Following a short welcoming ceremony, the resident crew on Mir gave a safety briefing, including emergency evacuation procedures, to their guests. *Atlantis* carried custom-made contoured seat liners that Solovyev and Budarin would use when they returned to Earth the following October aboard the Soyuz ("union") spacecraft the EO-18 crew had used to reach the station. Before the EO-19 crew could take up residence on Mir, these liners were installed in the return craft, and those used by the EO-18 crew on their ascent were transferred to *Atlantis*. With the completion of the safety briefing and the seat-liner exchange, Dezhurov, Strekalov, and Thagard officially became members of the *Atlantis* crew, and Solovyev and Budarin took over as the newest Mir crew. That night, after 105 days on board Mir, the EO-18 crew slept in *Atlantis*.

Apart from some ceremonies commemorating the first American-Russian docking in twenty years, most of the next four days was devoted to transferring supplies between the two craft and conducting experiments designed to help understand the effects of the long stay in space on the physiology of the EO-18 crew members. The crews also took measurements to assess the hygienic and radiation conditions on Mir and conducted joint engineering experiments on the most massive and complex spacecraft ever assembled.

A substantial effort went into the time-consuming transfer of equipment and supplies. The shuttle orbiter would return the bounty from the EO-18 crew's nearly four months in space. They had accumulated many samples of blood, saliva, and urine from their human physiology investigations, and these were carefully stowed in *Atlantis* for the ride back to Earth. Cassettes and disks with scientific and engineering data were removed from Mir as was some equipment no longer needed aboard the station.

Atlantis resupplied Mir with many of the items it needed to support humans in space for extended periods. Water, produced in excess as a by-product of the orbiter's generation of electricity, was a precious resource on Mir. Transferring it was slow, as it required the astronauts to fill containers in *Atlantis'*

galley and carry them to Mir, but they managed to deliver nearly 500 kilograms. By raising the air pressure in *Atlantis* (and allowing the air to mix through the docking tunnel), the astronauts increased the oxygen and nitrogen supply on board the station.

During their stay on Mir, Solovyev and Budarin planned to try to free a jammed solar array on Mir's newest module, Spekter ("spectrum"). Special tools for this task had been constructed in the United States and Russia and were ferried to Mir by *Atlantis*.

Some items delivered during STS-71 were specifically intended to make life on board an orbiting station more pleasant. Fresh fruits, chocolates, flowers, and even replacement strings for the guitar on board Mir would boost the spirits of people spending long times away from Earth.

Engineering tests of the operation of the large structure were conducted to aid in the design of future space stations. By firing the thrusters on *Atlantis*, the crew could measure how firmly the docking system gripped the two spacecraft, and the stability of the complex could be inferred by measurements of vibrations induced in the solar arrays.

On July 3, after a farewell ceremony in Mir, the final transfers were completed and the crews left each other's spacecraft for the last time. The next day, at 6:55 A.M., eastern daylight time, Solovyev and Budarin separated their Soyuz craft from Mir so they could film *Atlantis'* undocking. Fifteen minutes later, springs gently pushed *Atlantis* away from the complex, and Gibson and Precourt began a flight around the station to make detailed assessments of its condition after more than nine years in space. A computer problem on Mir forced the cosmonauts to redock sooner than planned, but they managed to capture some spectacular views first.

As *Atlantis* receded from Mir, the crew conducted more medical examinations on the EO-18 crew members, who also exercised to prepare their bodies for the return to gravity. By the next day, Mir appeared to them as only a distant point of light.

The day before landing, the astronauts performed a routine prelanding evaluation of *Atlantis*, deactivated Spacelab, and installed special seats that would allow the EO-18 crew to be recumbent during the landing. This position minimizes the physiological stress of returning to gravity, a necessity after a long absence from the effects of that constant force. A communications test from NASA's new flight control room showed that it operated as intended, so subsequent flights could be controlled from this modernized facility.

On July 7, Gibson guided *Atlantis* to a smooth landing at 10:55 A.M., eastern daylight time, at Kennedy Space Center, completing two missions at once. The five STS-71 astronauts added more than 9 days, 19 hours to their spaceflight experience. At the same time, the EO-18 crew completed nearly 115 days, 9 hours of spaceflight. Yet their work was not complete. Several weeks of medical tests to monitor their bodies' readaptations to gravity awaited them. More immediate, however, were some rewards, including Thagard's being reunited with his family and hot-fudge sundaes for the men who had been deprived of many Earthly pleasures during their extended stay in space.

Knowledge Gained

More than two years of engineering analysis and planning preceded the flight of STS-71. The success of the mission not only was a testament to the skill and dedication of the Americans and Russians who made the preparations, but it served as a valuable validation of the techniques they developed and used to prepare for such a complex operation. This new knowledge would serve well for future, even more ambitious flights with the space shuttle and Mir and, it was hoped, with the International Space Station, which was in advanced design and construction during STS-71.

Activities that would be important for the International Space Station and future Mir docking missions were evaluated for the first time on STS-71. For example, the transfer of several hundred items from each spacecraft to the other proved to be more time consuming than expected. This experience led to methods for improving efficiency, thus helping to ensure that subsequent flights would be as productive as possible.

Over many years, NASA had honed its skills in the complex tasks of planning before and replanning during space missions to make optimal use of the crews and the systems. STS-71 illuminated an important difference between what NASA had become accustomed to, with space shuttle missions generally lasting one to two weeks, and the longer missions that would be typical for people on the International Space Station. When circumstances necessitated a change in plans on a short space shuttle flight, ground controllers frequently developed a new plan overnight so the crew could incorporate it into their activities as soon as possible. Such a hectic pace could not be sustained during extended stays on Mir, and a delay of a few days was not so important, so cosmonauts were more accustomed to receiving new plans several days in advance of implementing them. This difference in style contributed to some difficulty in working with the EO-18 crew during STS-71 and provided a valuable lesson for NASA.

After their long stay in the station, the EO-18 crew members were occasionally irritable and not as cooperative as ground personnel expected them to be. The men were not always willing to participate, complaining that some experiments were too strenuous or uncomfortable. To accommodate this reticence, ground controllers were forced to make quick changes in the plans and schedules. Nevertheless, Baker and Dunbar were able to perform extensive cardiovascular, pulmonary, metabolic, neurological, and behavioral experiments on the EO-18 crew, providing scientists in the United States their first opportunity since Skylab, more than twenty-one years earlier, to investigate the effects of prolonged stays in space on the human body. Experiments performed during and after the flight added to a small but growing database on humans' adaptation to space and readaptation to Earth. The EO-18 crew's extensive program of exercising served the dual purpose of aiding in the evaluation of their physiological changes during spaceflight and acting as a countermeasure to those changes to ease the return to Earth.

Context

Although NASA had gained extensive experience in orbital docking during the Gemini and Apollo programs, STS-71 was its first docking since the Apollo-Soyuz Test Project, twenty years earlier. Because of the tremendous mass of the space shuttle orbiter and Mir, and their distinct lack of symmetry, their docking was much more complicated than that of the spacecraft in earlier programs. Nevertheless, Gibson and his crew executed a flawless docking, and their structural experiments showed that the docking system formed a very secure attachment.

The second shuttle/Mir docking mission, STS-74, incorporated several changes based on the results of its predecessor. For example, after the orbiter's propellant usage, when controlling the docked assembly, was discovered to be higher than predicted on STS-71, computer control systems were improved, and this led to reduced expenditure of precious propellant during STS-74. Mir had experienced power shortages after it assumed the orientation required for docking with *Atlantis*, still about seventy-five meters away. To reduce the time the station would spend without pointing its solar arrays at the sun, on STS-74 *Atlantis* closed to about fifty meters before halting to allow Mir to orient itself. In addition, the geometry of the approach of the two spacecraft was modified to improve the lighting on Mir's solar arrays rather than to guarantee coverage by Russian ground stations.

The excellent cooperation demonstrated during STS-71 was a stark contrast to most of the history of space exploration. During the Cold War, the space programs of the United States and the Soviet Union were strong political weapons, helping each country demonstrate to the world its technological strengths. By the beginning of the 1990's, however, priorities had shifted, and economic aspects of the space programs grew in importance compared with the political rivalries. The United States Congress became increasingly reluctant to fund NASA's planned International Space Station and other expensive programs, and a collapsing economy in Russia made it harder and harder for the new government to sustain the former Soviet Union's costly space program.

As both sides recognized that to accomplish their goals they could benefit more from cooperation than competition, they signed a set of agreements in 1992 and 1993 to conduct three phases of joint operations in human spaceflight. The principal objectives of Phase One, of which STS-71 was a key element, were to gain experience in working together on complex space missions; to resupply Mir with water, air, food, and other essentials; and to test hardware, software, and techniques to be used on the International Space Station, which already included the United States, Canada, the European Space Agency, and Japan. Phases Two and Three, extending from 1997 to 2002, would cover the joint construction and initial operation of the vast orbiting complex.

Phase One would provide the opportunity for each country to gain experience with the complex rules, procedures, and organizations that the others had developed during more than thirty years of human spaceflight. This would help greatly as they entered into the challenging task of building and operating the International Space Station. Russia had a well-developed capability to resupply Mir and return equipment to Earth, but the payload capacity was limited. The space

shuttle, as demonstrated on STS-71 and subsequent missions, could provide a very large capacity. At the same time, both countries were interested in developing countermeasures to the physiological and psychological effects of extended spaceflight. Russia had greater experience in long-duration flights on the Soviet Salyut ("salute") stations, operated between 1971 and 1986, and Mir, inhabited nearly continuously since 1986. By combining this experience with the comprehensive space-borne scientific facilities the United States could apply to this research in Spacelab, both countries expected to garner better data. This was expected to serve the future well, as people from those countries and others worked together on Earth and in space.

Bibliography

Allen, Joseph P., and Russell Martin. *Entering Space: An Astronaut's Odyssey.* New York: Stewart, Tabori and Chang, 1984. This exquisite book, coauthored by a space shuttle astronaut, describes the experiences of spaceflight, from the routine chores of living and working to the excitement and drama of being in space. Suitable for all audiences, it presents more than 200 color photographs displaying the beautiful views available in space as well as the activities performed by astronauts.

Clark, Phillip. *The Soviet Manned Space Program.* New York: Orion Books, 1988. In addition to descriptions of every Soviet human spaceflight, this book contains useful information on related robotic missions, thus helping the reader understand the plans, goals, and significance of much of the Soviet space program. Speculation on intended objectives are carefully explained and make fascinating reading. Includes many drawings and photographs.

Cooper, Henry S. F., Jr. *Before Lift-off: The Making of a Space Shuttle Crew.* Baltimore: The Johns Hopkins University Press, 1987. The long process of training astronauts for a flight on the space shuttle is described by following the crew of STS 41-G. Although they train for a specific mission, the book contains a great deal of interesting insight into the difficulty and importance of the preflight training for any mission. All readers will gain an appreciation of the challenges that precede a space shuttle flight and how hard the astronauts have to work in order to make their mission proceed so smoothly.

Furniss, Tim. *Manned Spaceflight Log.* Rev. ed. London: Jane's Publishing Company, 1986. With a description of every human mission into space through Soyuz T-15 in March, 1986, this book is entertainingly written and should be enjoyed by general audiences. It provides the essential facts from each flight and allows the reader to understand any spaceflight in the context of humankind's efforts to explore and work in space.

Joels, Kerry M., and Gregory P. Kennedy. *The Space Shuttle Operator's Manual.* Rev. ed. New York: Ballantine Books, 1987. This book contains a wealth of information on space shuttle systems and flight procedures. It is written as a manual for imaginary crew members on a generic mission and will be appreciated by anyone interested in how the astronauts fly the orbiter, conduct experiments in Spacelab, and live in space. The book contains many drawings and some photographs of equipment.

Marc D. Rayman

THE SPACE TASK GROUP

Date: November 5, 1958, to November 1, 1961
Type of organization: Space agency

The Space Task Group was the United States' first civilian agency for manned spaceflight. It was the core team responsible for the Mercury, Gemini, and Apollo projects, and it was the seed from which grew the Manned Spacecraft Center, now the Johnson Space Center, near Houston.

PRINCIPAL PERSONAGES

ROBERT R. GILRUTH, Manager and Director, Space
Task Group
CHARLES J. DONLAN, Assistant Manager, STG
MAXIME A. FAGET, Chief Designer, STG

Summary of the Organization

Unofficially established by the brand-new National Aeronautics and Space Administration (NASA) on October 8, 1958, the Space Task Group (STG), created by a memorandum bearing thirty-five names and dated November 5, 1958, was destined to place the first humans on the Moon.

The thirty-five scientists from the old aeronautical laboratory at Langley Field, Virginia, and the ten additional professionals soon to join them from Lewis Research Center in Cleveland, Ohio, foresaw the possibilities that might arise from combining aviation with rocket and missile technologies. Their initial charge was simply to create a team of humans and machines for manned space exploration, and their early projects involved one-man ballistic and orbital spaceflight. Soon, however, they were responsible for a two-man maneuverable spacecraft and for three-man circumlunar and lunar-landing vehicles.

In the mid-1950's, well before Soviet Sputniks 1 and 2 spurred the creation of NASA from the National Advisory Committee for Aeronautics (NACA), certain farsighted engineers within and outside the government were studying rockets' potential to send humans beyond Earth's atmosphere. The U.S. Air Force had long been interested in expanding its flight regime into space, and NACA's X-15 rocket research airplane was another example of future-oriented work. Perhaps the most vigorous group of aerospace visionaries, however, was gathered around Robert R. Gilruth and his Pilotless Aircraft Research Division (PARD) at Langley Field and Wallops Island in Virginia. Another group at the NACA Lewis Propulsion Laboratory, led by Abe Silverstein, also vied for attention, but when NACA became NASA in October, 1958, President Dwight D. Eisenhower appointed T. Keith Glennan as the first NASA administrator, and Glennan needed Silverstein's assistance. They delegated authority to Gilruth's group to proceed with a manned satellite program. Officially designated Project Mercury on November 26, 1958, the manned satellite program began the first American series of flights into space.

During the spring and summer of 1958, a series of competitive planning conferences around the country gradually led NACA engineers to a consensus that the best proposal for a method of manned spaceflight was the one championed by Maxime A. Faget and his colleagues at Langley, near Norfolk. After years of experience with Gilruth's PARD testing drones and guided missiles, Faget and his associates advocated a wingless, nonlifting, nose cone configuration for the first manned satellite. Rather than follow the pattern of the X-15 rocket research airplane, they wanted to adapt a small, inhabitable cockpit to the first operationally tested Intercontinental Ballistic Missile (ICBM). This idea was at first received without enthusiasm by the Air Force and by General Dynamics/Astronautics, whose Atlas ICBM was the only viable candidate at that time for the job of launching a person into orbit. While the orbital flight plan rapidly took shape at the field centers, NASA headquarters expanded and helped STG to complete preliminary designs, to issue specifications, to choose the prime contractors, and to manage the entire project. Criticism abated as creative engineering activities moved ahead rapidly.

At the beginning of 1959, McDonnell Aircraft Corporation in St. Louis was chosen, out of a competitive bidding group of a dozen companies, to manufacture a dozen manned satellite capsules according to the Faget concept. John F. Yardley of McDonnell quickly assumed leadership in the development of the Mercury hardware. He and his corporation, together with Faget and his STG colleagues, became the core of the Mercury team. By midyear, when seven military test pilots, to be called astronauts, joined the project, most of the basic decisions as to how NASA would try to put a human in space were firm.

Three central principles guided the Mercury program: Use the simplest and most reliable approach, attempt a minimum of new developments, and conduct a progressive series of tests. In the hope of saving time and money and ensuring safety, NASA's policymakers tried to minimize trial and error. Five approaches to major aspects of the project were determined as soon as the government-industry team began to cooperate: The manned satellite capsule would be launched into orbit by the Atlas ICBM; it would be equipped with a tractor escape system, in case the booster malfunctioned; it would be a frustum-shaped vehicle with an attitude control system; it would be braked in orbit by retro-rockets; and it would be slowed on descent by parachutes. Although these plans and the mission profiles were remarkably well laid, nearly all the details of their implementation were yet to be incorporated and verified.

Patents for inventions made in the course of work on Project Mercury were conferred only after the designs were proved in practice, so that official awards tended to obscure the actual process of innovation. Seven men were credited by NASA with designing the Mercury spacecraft: Faget, Andre J. Meyer, Jr., Robert G. Chilton, Willard S. Blanchard, Jr., Alan B. Kehlet, Jerome B. Hammack, and Caldwell C. Johnson. For their conceptual designs and preliminary tests of components, these members of Faget's team were recognized some eight years later in the issuance of U.S. Patent 3,270,908. In addition, Faget and Meyer were credited with the tractor-pylon emergency rocket escape system, and Meyer was credited with the parachute and jettison system design; along with Faget, William M. Bland, Jr., and Jack C. Heberlig were recognized for the pilot's contour couch. Later still, R. Bryan Erb and Kenneth C. Weston shared honors with Meyer for the ablation heatshield, and Matthew I. Radnofsky and Glenn A. Shewmake were recognized for their inflatable life rafts and radar reflectors.

McDonnell employees, led by Raymond A. Pepping, Edward M. Flesh, Logan T. McMillan, John F. Yardley, and, later, Walter F. Burke, took an active and at times initiating role in the creation of the Mercury spacecraft. NASA's policy of retaining ownership of inventions was highly controversial at first, but it did not stanch industrial initiative; STG grew from thirty-five to more than 350 members within its first nine months of existence, but the industrial team grew even faster.

Many subcontractors and third-tier vendors, as well as the prime contractor working with STG, suggested and completed systems engineering studies and components for the Mercury project. Especially noteworthy examples were the McDonnell "pig-drop" impact studies of the aluminum honeycomb shock absorber, the research work of Brush Beryllium Company and of Cincinnati Testing Laboratories on the heat sink and ablation heatshield, and the extraordinarily careful design and development of the environmental control system by AiResearch Manufacturing. The contractors were not limited to hardware development; new techniques and procedures, notably human factors engineering led by Edward R. Jones of McDonnell, originated as often from contractors as from NASA workers.

Because of concerns over weightlessness and its effects on humans and mechanical parts in orbit, the automation experts held sway over the development of Mercury during 1959. By the end of 1960, however, the automatons had failed so often and the astronauts had been trained so well that Mercury's managers were beginning to place more reliance on men than on machines for mission success. At all critical points, redundant, automatic safety features were built in, but the pilots were given manual control over their vehicle wherever feasible. Missile and aircraft technology were rapidly converging.

Meanwhile, STG was continuously testing each part and the whole Mercury configuration in the laboratory and in flight. Three levels of testing had originally been specified: development, qualification, and performance. To these were added, in mid-1960, reliability tests of many varieties to ascertain the life and limits of all the systems. Most dramatic was the extensive flight testing program, which used the unique Little Joe boosters for several tests and the Atlas booster for a single Big Joe shot that demonstrated reentry capability. The Big Joe mission was accomplished successfully on September 9, 1959, and so paved the way for a series of seven more Little Joe missions during the next two years.

By the beginning of 1960, a presidential election year, Gilruth's STG was in high gear and accelerating. Military liaisons had been established, a worldwide tracking and data network was being arranged, an industrial priority rating for Mercury was obtained, a class of seven military test pilots had been chosen and were undergoing astronaut training, and intensive studies and renovations were under way to "man-rate" the booster rockets (that is, to make them safe enough for humans). Politically, however, 1960 was to be a rough year. STG's personnel roster contained about five hundred names, and its prime contract with McDonnell, already modified in more than 120 particulars, was nearing $70 million and rising. Gilruth's group, still housed and hosted by Langley, was supposed to be moving to the new Goddard Space Flight Center being built at Beltsville, Maryland, between the capital and Baltimore. It was unclear, however, whether construction would go forward on the Marshall Space Flight Center at Huntsville, Alabama, which would be occupied by Wernher von Braun's team of rocket experts, at work on the Saturn series of engines and boosters. Political rhetoric about the so-called missile gap, the U-2 incident in May, continuing Soviet launches of dogs and robots in orbital spacecraft, and several widely publicized failures of NASA flight tests in October and November helped make 1960 a most suspenseful year.

The appointments of a number of senior engineers, who distinguished themselves further as Gilruth's group evolved, added to STG's strength during this critical year. Walter C. Williams, Kenneth S. Kleinknecht, Robert O. Piland, James A. Chamberlin, and G. Merritt Preston were a few of the managers. George M. Low and others at NASA headquarters decided between administrations, in January, 1961, to make STG separate from the Goddard center. By then, STG employed 680 persons.

After monkeys had survived flights in boilerplate spacecraft propelled by Little Joe solid rockets, the McDonnell-built spacecraft were mated to Atlas and Redstone liquid-fueled rockets for their combination qualification flight tests. The first two attempts at mated flight failed, because of the boosters more than the capsules. By February, 1961, however, successful flights of both the Mercury-Redstone (MR) and the Mercury-Atlas (MA) combinations had gone far toward man-rating the machines. The performance and recovery of the chimpanzee Ham in MR-2 seemed to indicate that a human could make a similar suborbital hop. On April 12, 1961, however, Yuri Gagarin orbited Earth in 108 minutes aboard the Soviet Union's Vostok 1. Thus the parabolic test flights in May of Alan B. Shepard, Jr., in *Freedom* 7 (MR-3), and in July of Virgil I. "Gus" Grissom, in *Liberty Bell* 7 (MR-4), set no world records, but merely tested the ability of the Mercury men and machines to work in space for a few minutes. Shepard and Grissom did prove, however, that STG had designed and developed a primitive spacecraft and not merely a manned bullet.

By mid-1961, the tiny Mercury spacecraft, encasing forty thousand components and eleven kilometers of wiring, was widely publicized around the world. Designed for a reference mission of three orbits, the basic systems in Mercury were advertised openly and often described as falling into ten categories: heat protection, mechanical, pyrotechnical, control, communication, instrumentation, life support, electrical, sequential, and network. Some sixteen major subsystems were novel and critical enough to worry reliability experts and STG managers. STG was upstaged again when, on August 6 and 7, 1961, the Soviet cosmonaut Gherman S. Titov made a seventeen-orbit, day-long circumnavigation of Earth in Vostok 2. In contrast to Mercury, the details of the Vostok spacecraft were shrouded in secrecy.

Difficulties and delays in manufacturing the Mercury capsule and in man-rating its boosters had afforded the seven American astronauts more than an extra year of training. Because they were active as consulting engineers as well as test pilots, the Mercury astronauts contributed to quality control, mission planning, and operational procedures before they ventured into space. Their specialty assignments indicated another way of categorizing the most critical features of the Mercury program. Shepard became the expert on tracking and recovery operations, Grissom studied the complicated electromechanical spacecraft control systems, and John H. Glenn, Jr., worked on the cockpit layout. M. Scott Carpenter specialized in the communications and navigation systems, Walter M. Schirra, Jr., handled the life-support systems and spacesuits, and L. Gordon Cooper, Jr., and Donald K. Slayton analyzed the Redstone and Atlas boosters.

For all the exotic training and trips undergone by the astronauts, only three activities proved to have been indispensable: weightlessness conditioning, accomplished through flights of Keplerian parabolas; acceleration endurance tests in human centrifuges; and, most important, the overlearning of

mission tasks in McDonnell-built capsule procedures trainers. Many other training aids were helpful in bolstering the astronauts' confidence that they could endure and overcome any eventuality, but they were confident men; learning to live and work within the pressure suits and within the sealed pressure vessels was an exceedingly difficult job in itself.

At times in 1960 and 1961, all members of the Mercury teams were stymied by some recalcitrant system, process, or device. The recurrent balkiness of the smaller thrusters in the reaction control system, the overassigning of pilot tasks in flight planning, and difficulties with the Department of Defense in scheduling support operations typified tendencies that threatened to become permanent. Both STG and McDonnell underwent several reorganizations of personnel and divisions of labor to meet changing program situations. Moving to the Cape Canaveral launch site, establishing an operations team, responding to new hardware integration needs, and riding the tide of a new political administration all caused confusion and elicited new organization.

Nevertheless, the flight test series began to experience success. By late 1961, it was obvious that STG was to become institutionalized as a permanent, separate NASA installation devoted to the long-term development of manned spaceflight and space exploration. John Glenn was ready to fulfill the Mercury mission, the capsule had evolved from a container into a spacecraft, and the boosters had been refined to the point of deserving to be called man-launching vehicles.

On May 25, 1961, President John F. Kennedy called upon Congress to approve a decade-long lunar landing-and-return program. Already funds had been approved for a site selection process. STG itself proposed a manned spacecraft development center. On September 19, NASA announced that a site near Houston, Texas, had been selected, and by October 13, NASA headquarters had approved construction plans for at least eighteen buildings. More important, STG's responsibility for Project Mercury had escalated into responsibility for the Apollo spacecraft Mercury Mark 2, soon to be renamed Gemini. Thus it was no surprise when on November 1, 1961, STG personnel, now numbering about one thousand, learned that "the Space Task Group is officially redesignated the Manned Spacecraft Center."

Context

The Space Task Group, headquartered in Virginia from 1958 to 1961, fulfilled its initial mission with Glenn's three-orbit flight in *Freedom* 7 (MA-6) on February 20, 1962. By that time, the STG that had become virtually synonymous with Project Mercury was anticipating relocating under its new name, the Manned Spacecraft Center (MSC), to southeast Texas around Houston and Galveston Bay. There, its members would design, develop, manage, and control the missions for several new generations of spacecraft. The influence of Faget's flight systems design team and Gilruth's directorship pervaded the next decade of U.S. spacecraft developments, as attested by the similarities in the Mercury, Gemini, and Apollo command modules. In addition to twelve men brought back to Earth safely after six lunar landings, fifteen astronauts had circumnavigated the Moon and returned in Apollo command modules by the end of 1973. On February 17 of that year, the MSC was officially renamed the Lyndon Baines Johnson Space Center (JSC).

A space task group of a different sort passed quickly into obscurity during this period. In January, 1969, President Richard M. Nixon appointed his vice president, Spiro Agnew, to chair a special advisory committee on future directions for manned spaceflight. This commission was formed in the wake of celebrations of mankind's first circumnavigation of the Moon, in Apollo 8, and met amid the excitement of the Apollo 9, 10, and 11 achievements. In September of 1969, it published a report titled *Post-Apollo Space Program: Directions for the Future*. The group advocated manned missions to Mars, but it was so marred by the political scandals that soon enveloped its chairman that its recommendations were quickly forgotten.

Bibliography

Brooks, Courtney G., James M. Grimwood, and Loyd S. Swenson, Jr. *Chariots for Apollo: A History of Manned Lunar Spacecraft*. NASA SP-4205. Washington, D.C.: Government Printing Office, 1979. This is the semiofficial history of the initial achievements of the Apollo spacecraft as seen from Houston. Part of the NASA Historical Series, the work is a sequel to two earlier books that cover the Mercury and Gemini programs. Its stops short of considering the Apollo 12 through 17 missions. Several more volumes in the series deal with other aspects of the Moon-landing program.

Ertel, Ivan D., and Mary Louise Morse. *The Apollo Spacecraft: A Chronology*. Vol. 1, *Through November 7, 1962*. NASA SP-4009. Washington, D.C.: Government Printing Office, 1969. This first of four volumes covering the Apollo program chronicles key events from the 1920's through the lunar orbital rendezvous decision of November 7, 1962. Includes a foreword by Robert O. Piland as well as forty-four illustrations, abstracts of key events, seven appendices, and an index.

Ezell, Linda Neuman. *Programs and Projects*. Vol. 2 in *NASA Historical Data Book, 1958 – 1968*. NASA SP-4012. Washington, D.C.: Government Printing Office, 1988. This reference work complements the volume by Van Nimmen, with five chapters documenting launch vehicles, manned spaceflight, space science and applications, advanced research and technology, and tracking and data acquisition. Charts, tables, maps, diagrams, and drawings abound, but there are no photographs.

Grimwood, James M. *Project Mercury: A Chronology*. NASA SP-4001. Washington, D.C.: Government Printing Office, 1963. This is the first of a series of historical chronologies and programmatic accounts of U.S. manned spaceflight projects. Features a preface by K. S. Kleinknecht and a foreword by Hugh L. Dryden. Includes sixty-eight illustrations, ten appendices, and a good index.

Grimwood, James M., Barton C. Hacker, and Peter J. Vorzimmer. *Project Gemini, Technology and Operations: A Chronology*. NASA SP-4002. Washington, D.C.: Government Printing Office, 1969. Focuses on the technology and operations of Gemini, from its concept and design in April, 1959, to its abolition and the summary conference in February, 1967. This book would serve as a good introduction to *On the Shoulders of Titans* (see below). The foreword is by Charles W. Mathews. Includes 131 illustrations, eight appendices, and a thorough index.

Hacker, Barton C., and James M. Grimwood. *On the Shoulders of Titans: A History of Project Gemini*. NASA SP-4203. Washington, D.C.: Government Printing Office, 1977. A volume in the NASA Historical Series, this work is a history of the Gemini program. It describes how the Mercury Mark 2 became a first-class maneuverable spacecraft, suitable for rendezvous and docking in orbit.

Rosholt, Robert L. *An Administrative History of NASA, 1958 – 1963*. NASA SP-4101. Washington, D.C.: Government Printing Office, 1966. With an interesting foreword by James E. Webb, this book presents a political scientist's analysis of the first five years of NASA administration. It is heavily documented but poorly illustrated, and it focuses almost exclusively on NASA headquarters.

Swenson, Loyd S., Jr., James M. Grimwood, and Charles C. Alexander. *This New Ocean: A History of Project Mercury*. NASA SP-4201. Washington, D.C.: Government Printing Office, 1966. This 681-page narrative is the first program history to be published in the NASA Historical Series. Organized in three parts —"Research," "Development," and "Operations"— this book is the semiofficial account of the Mercury program. It emphasizes the history of Mercury's technology and field management. Profusely illustrated and fully documented, the work was designed as a model for NASA spaceflight histories and is aimed at the intelligent layperson.

Van Nimmen, Jane, Leonard C. Bruno, and Robert L. Rosholt. *NASA Resources*. Vol. 1 in *NASA Historical Data Book, 1958 – 1968*. NASA SP-4012. Washington, D.C.: Government Printing Office, 1976. This reference work traces the growth of NASA over its first decade, with six topical chapters and two appendices. With a brief foreword by George M. Low, the book presents tabular and graphical data on NASA's facilities, personnel, finances, procurement, installations, awards, and organization. The largest section is chapter 6, which details basic facts about NASA's fourteen-largest field installations.

Loyd S. Swenson, Jr.

THE SPACEFLIGHT TRACKING AND DATA NETWORK

Date: Beginning in 1972
Type of organization: Tracking network

The Spaceflight Tracking and Data Network was a network of fifteen ground communications and tracking stations, located in countries around the world, which provided data relay, data processing, communications, and command support to the U.S. space shuttle program and to other orbital and suborbital spaceflights.

PRINCIPAL PERSONAGES
    DANIEL A. SPINTMAN, Division Chief, Ground and
        Space Networks, Goddard Space Flight Center
    VAUGHN E. TURNER, Chief, NASA Communications
        Division, GSFC
    ROBERT T. GROVES, Chief, Flight Dynamics Facility,
        GSFC
    JOHN T. DALTON, Chief, Data Systems Technology
        Division, GSFC

Summary of the Organization

The Spaceflight Tracking and Data Network (STDN) is part of a complex and rapidly changing group of programs designed to provide a two-way communications and command link between flight control centers on the ground and manned and unmanned space missions. In the 1980's, STDN also provided primary support for U.S. space shuttle missions.

As of the end of 1995, STDN operated three tracking stations, located at Merritt Island, Florida; Bermuda; and Dakar, Senegal. Merritt Island provides preflight and launch support to the Shuttle, expendable launch vehicles, and the integrated payloads. Bermuda provides launch and range safety support. Dakar provides back-up voice communications for shuttle launches. The stations are being replaced by the constellation of satellites that form the Tracking and Data-Relay Satellite System (TDRSS).

STDN ground stations are equipped with ultrahigh-frequency and television hardware, with 4.3-, 9-, and 26-meter S-band antenna systems, used for radio transmissions, and with C-band radar systems, used for tracking objects by radar. Department of Defense facilities are frequently used to supplement existing STDN hardware with additional S-band and C-band equipment.

STDN operates in cooperation with the National Aeronautics and Space Administration (NASA) — specifically, with NASA's Communications Division (NASCOM) and the Flight Dynamics Facility at GSFC. NASCOM is the communications link for launch and landing sites, for mission and network control centers, and for all U.S. spacecraft. It provides voice, low- and high-speed telemetry, and television transmissions to more than a hundred NASA facilities. The Flight Dynamics Facility receives the tracking data relayed by STDN and calculates the information necessary to orient the spacecraft being tracked.

By 1988, STDN was providing tracking, communications, and command services to a total of nineteen scientific, weather, communications, and environmental U.S. satellites in Earth orbit. STDN also has the capability to support European, Soviet, and Chinese satellites with similar services.

Part of STDN and other NASA tracking networks, the NASA Ground Terminal at White Sands, New Mexico, would serve as a backup space shuttle mission control facility if Johnson Space Center, in Houston, were rendered inoperative for any length of time. GSFC would serve as an interim mission control center while the flight control personnel transferred from Houston to White Sands.

Each STDN station was able to track and communicate with a spacecraft only during the period when the spacecraft's orbit brought it into the station's "line of sight," or when Earth's curvature did not block direct radio and radar contact. Each station could track or remain in contact with a spacecraft for a maximum of approximately 15 percent of its orbit. When one station lost contact, responsibility for tracking and communications passed to the next ground station in the network.

Knowledge Gained

On both manned and unmanned missions, the data received by STDN gave mission managers and technicians a complete picture of the health and reliability of the spacecraft in orbit, something that the astronauts on manned missions often did not have the time or opportunity to do. The information let the mission managers on Earth serve as "extra crew members" who could help prevent or overcome problems with the spacecraft.

The space shuttle *Challenger*'s launch of the first Tracking and Data-Relay Satellite (TDRS) in April, 1983, demonstrated the interaction between ground control and spacecraft made possible by STDN. TDRS-A was successfully released from the space shuttle on April 5, 1983, but the booster rocket attached to the satellite failed to fire, leaving it in a uselessly low Earth orbit. STDN allowed TDRS mission managers to assess TDRS's situation and devise an alternate way for it to reach a geosynchronous orbit (an orbit wherein a satellite travels once around Earth every twenty-four hours) 35,900 kilometers above Earth. Sending commands via STDN, ground control workers used the satellite's tiny reaction control thrusters to move it slowly to the proper altitude.

STDN and other NASA-operated tracking networks have allowed the United States to participate in the growth of the international space community by providing launch and data tracking support for the French Ariane rocket program. STDN is also capable of providing support to other foreign satellites.

Context

STDN works in conjunction with the Deep Space Network, which is controlled by the Jet Propulsion Laboratory, in Pasadena, California, and the Tracking and Data-Relay Satellite System (TDRSS), made up of satellites in geosynchronous orbit.

STDN is part of an effort to develop U.S. communication and tracking capabilities that began in the earliest days of the nation's space program. In 1958, as part of the country's plan to launch an artificial satellite into orbit as the United States' contribution to the eighteen-month International Geophysical Year (July 1, 1957, to December 31, 1958), NASA took over the U.S. Naval Research Laboratory's Minitrack network of ground stations. These facilities were designed only to track satellites and receive data and did not have the capacity to transmit commands to spacecraft from the ground.

There were only ten stations in the Minitrack system when NASA first began using the network; by 1963, however, eighteen ground facilities were in use. Their locations were San Diego; Goldstone, California; Blossom Point, Maryland; Fort Meyers, Florida; East Grand Forks, Minnesota; Fairbanks, Alaska; Rosman, North Carolina; Antigua, West Indies; Quito, Ecuador; Lima, Peru; Antofagasta and Santiago, in Chile; Canberra and Woomera, Australia; Saint John's, Newfoundland; Winkfield, England; and Eselen Park and Johannesburg, in South Africa.

During the years that the Minitrack network was in operation, NASA began expanding the technological capabilities of its ground stations, adding new and more powerful antennae and better data retrieval and processing systems. With additions in 1963 of 12- and 26-meter antennae to several

Minitrack stations, the system was renamed the Satellite Network. By 1964, NASA had brought into use the Satellite Telemetry Automatic Reduction (STAR) system, which provided not only better tracking and data processing but enabled ground stations to issue commands to unmanned satellites. The improved network, which operated from 1964 to 1972, was known as the Space Tracking and Data Acquisition Network (STADAN). STADAN operated ten ground stations at former Minitrack locations, with an additional station at Tananarive, Madagascar.

In 1962, NASA had separated tracking and communications functions into a satellite division and a manned division, creating the Manned Space Flight Network (MSFN), which operated concurrently with the STADAN satellite-tracking system. In addition to land-based stations, MSFN used eight aircraft and five ships to provide a comprehensive network of facilities that could communicate with Mercury, Gemini, and Apollo astronauts, receive telemetry signals, and command both manned spacecraft and unmanned target vehicles such as those used during Gemini flights VIII through XII.

A total of twenty-two MSFN ground stations were located in White Sands, New Mexico; Corpus Christi, Texas; Eglin Air Force Base and Merritt Island, in Florida; Point Arguello and Goldstone, in California; Kauai, Hawaii; Antigua; Ascension Island; the Canary Islands; Bermuda; Canton Island; Grand Bahama Island; Grand Turk Island; Guam; Canberra, Carnarvon, and Muchea, in Australia; Guaymas, Mexico; Kano, Nigeria; Madrid; and Tananarive, Madagascar. In 1972, the STADAN and MSFN systems were unified to create the STDN system.

Because of the complexities of receiving data from manned and unmanned spacecraft and relaying data among the several STDN facilities, NASA inaugurated the TDRSS system with the 1983 launch of TDRS-A, which became TDRS 1 when it was successfully placed in orbit. TDRS-B was on board the space shuttle *Challenger* when it exploded shortly after launch on January 28, 1986. TDRS-C was the payload on the space shuttle *Discovery*, launched in September, 1988. TDRS-D was deployed in March, 1989, and TDRS-E in August, 1991. In January, 1993, the fifth TDRS (TDRS-F) was deployed from the cargo bay of *Endeavour*. TDRS-G (TDRS-7) was deployed from *Discovery* in July, 1995.

The TDRSS network was designed to replace STDN as NASA's primary tracking system, using satellites in geostationary orbits above the equator to receive data from other spacecraft and relay them to the White Sands Ground Terminal. Only three STDN stations remain now that TDRSS is fully operational.

Bibliography

Elliott, James C. *Goddard's Worldwide Communications Network Set to Provide Support for STS-26 Mission.* NASA News Release. Washington, D.C.: National Aeronautics and Space Administration, 1988. This news release details the role of STDN and TDRSS in the 1988 flight of the space shuttle *Discovery*. It includes a list of the times during the mission when *Discovery* was in contact with Goddard Space Flight Center through STDN.

————. *NASA/Goddard Space Flight Center*. NASA Release 88-43. Washington, D.C.: National Aeronautics and Space Administration, 1988. This brochure gives an overview of the activities and functions performed by Goddard Space Flight Center. It was written to help reporters and broadcasters better communicate information about Goddard, STDN, and other NASA organizations.

————. *Questions and Answers on the Space Flight Tracking Data Network (STDN)*. NASA Release 88-47. Washington, D.C.: National Aeronautics and Space Administration, 1988. This news release, written in easy-to-understand language, looks at the changing nature of STDN and its future after TDRSS.

Furniss, Tim. *Manned Spaceflight Log*. Rev. ed. London: Jane's Publishing Co., 1986. This is a concise, fact-filled listing of the primary mission objectives and results of all manned spaceflights up to Soyuz T-15. The book provides a broad overview of the progress made in space exploration. One of the best books for the beginning space enthusiast.

National Aeronautics and Space Administration. *Entering the Era of the Tracking and Data Relay Satellite System: NASA Facts/Goddard Space Flight Center*. Washington, D.C.: Author, 1987. This brochure introduces STDN and TDRSS to the layperson. It also discusses the importance of ground stations to the success of both manned and unmanned space missions.

Rosenthal, Alfred. *The Early Years, Goddard Space Flight Center: Historical Origins and Activities Through December 1962*. Washington, D.C.: Government Printing Office, 1964. This commemorative manual provides a precise and comprehensive look at the founding of Goddard Space Flight Center, the Minitrack Network, and the beginnings of the Satellite Network.

Eric Christensen

SPACEHAB

Date: Beginning 1983
Type of program: Manned space shuttle experiment facility

Spacehab is a commercial mini-laboratory that fits into the cargo bay of the space shuttle. Designed to be leased to private corporate interests, it represents the expansion of outer space from a governmental research arena to a locale for business and commerce.

PRINCIPAL PERSONAGES

ROBERT CITRON, Founder of Spacehab, Inc.
JAMES M. BEGGS, Spacehab, Inc., Chairman
RICHARD K. JACOBSON, President and
    Chief Executive of Spacehab from 1987 to 1991
DR. JANICE VOSS, Mission Specialist
    aboard first Spacehab flight
DAVID LOW, Payload Commander
    aboard first Spacehab flight

Summary of the Program

Spacehab (Space Habitat Module) is a privately produced, pressurized, cylindrical research module designed to fit in the space shuttle's cargo bay. It reflects the National Aeronautics and Space Administration's (NASA) increased emphasis on cost savings through use of private money.

The intent of Spacehab is to promote the use of the microgravity environment of space to produce products such as drugs, crystals, and fine machine parts more efficiently, perfectly, and economically than Earth manufacturing. For example, some materials can be produced at high temperatures in space without the use of containers, which on Earth can add contaminants to the product.

Spacehab was founded in 1983 by Robert Citron, a former Smithsonian Institution scientist, who at first intended to build a module that would carry tourists into orbit. This idea was soon abandoned in favor of a module for scientific experiments. Spacehab, Inc., contracted the actual building of the modules out to McDonnell Douglas Space Systems Company, under the leadership of Richard K. Jacobson, Spacehab president and chief executive during Spacehab's design period. Under an agreement with NASA, Spacehab buys launch services from the space agency and in turn leases its capacity to users, such as corporations and universities.

A major hurdle in initiating for-profit laboratory space on the shuttle was the reluctance of private industry to invest in such a new and untested venture. To address this reality, the

U.S. government, in 1985, promoted the establishment of the Centers for Commercial Development of Space (CCDS), a non-profit consortium for conducting space-based, high technology research and development. The intent was to encourage U.S. industry leadership in commercial space-related activities. The seventeen CCDS centers operate with government grants to provide American companies and universities with opportunities to carry out low-cost, space-based commercial research and development. Often, a company interested in Spacehab works with one of the CCDS centers.

A Spacehab Middeck Augumentation module consists of up to sixty-one separate compartments or "lockers," each capable of housing a separate experiment. An individual locker has a volume of .06 cubic meters (2.2 cubic feet). There is also space for accommodating larger experiments in one or more single or double "racks" (volume limit .63 cubic meters, 22.5 cubic feet and 1.3 cubic meters, 45 cubic feet, respectively. The module is 2.7 meters long and 4.1 meters in diameter and weighs 4,220 kilograms. It has a truncated top and flat end caps and sits in front of the shuttle's cargo bay, occupying about a fourth of the bay's space. Astronauts are able to enter and leave the module through a special hatch. A single Spacehab module doubles the available living and working space on the otherwise cramped shuttle and quadruples the experimentation space. It can be flown on any of the four shuttle orbiters, which are modified with special attachments to accommodate the unit. Each Spacehab mission has one full-time mission specialist to tend the module's experiments, as well as one mission specialist required part time for this task. Before a launch, the experiments are processed, integrated into lockers, and installed in the modules at Spacehab, Inc.'s payload processing facility in Port Canaveral, Florida.

In 1988, NASA signed an agreement with Spacehab, Inc., allowing the company to load Spacehab on six shuttle flights. NASA itself purchased back two hundred of the three hundred experiment lockers that would be available. The first of these flights, Spacehab One (SH-01), took place on June 21, 1993, aboard the shuttle *Endeavour* mission STS-57.

Spacehab One was the maiden flight of the program and module Flight Unit One. A six-person crew operated Spacehab experiments, with the majority being carried out by Mission Specialist Dr. Janice Voss and Payload Commander David Low. Twenty-two experiments were flown in the module and the shuttle mid-deck. These experiments were designed by both American and European scientists and involved investigations in the biomedical and materials sciences. Most were of a scale small enough to fit into Spacehab's bulkhead-mounted lockers.

As it has long been known that gravity causes the growth of imperfect crystals on Earth, several Spacehab experiments took advantage of the shuttle's microgravity environment to study the growth of crystals in space. Five other experiments measured the environment within Spacehab, including the light, sound, high energy particle, and acceleration levels. Four payloads investigated the growth and separation of living cells. One experiment used rodents to test the effects of drugs on adaptation to space. Two others examined the effects of weightlessness on the astronauts themselves, including a study of the so-called "zero-g-crouch," a postural change that affects humans in space. To document this phenomenon over the duration of a space mission, still and video images of crew members in a relaxed position were recorded at both early and late stages in the mission. Three hardware experiments were directly related to space station concepts: One dealt with water filtration; another involved the in-orbit repair of electronics, which included the first soldering ever conducted on a spaceflight; and a third examined the lighting and nutrient needs of plants.

In addition, experiments were carried out dealing with the transfer of fluids in weightlessness without creating bubbles. This experiment, called the Fluid Acquisition and Resupply Experiment (FARE), was designed to study filters and processes connected with the refueling of orbiting spacecraft.

The majority of the first Spacehab experiments achieved their scientific objectives, although nine required resources and support from the crew that had not been foreseen at the time of initial request. All of the support systems cooling, AC and DC power, computer monitoring, and video functioned without problem.

Spacehab Two (SH-02) was the maiden flight of module Flight Unit Two, carried aboard the space shuttle *Discovery* launched on February 3, 1994, for an eight-day mission of STS-60. Twelve experiments were housed in the Spacehab module and the mid-deck of the shuttle itself.

Spacehab Two contained several experiments that were reflown from Spacehab One: two acceleration measurement setups, an upgraded test of a plant growth system, protein crystal growth hardware, and an investigation of the organic separation of cells. New experiments examined space adaptation in rodents as well as cell and crystal growth. A hallmark for Spacehab Two was the first externally mounted experiment, which collected cosmic dust from the top of the module. As with Spacehab One, the support systems of Spacehab Two functioned flawlessly, supplying electricity, cooling, computer data, and video.

The third Spacehab flight (SH-03), involving the module Flight Unit Three, was also aboard the space shuttle *Discovery*, launched February 3, 1995. This mission was unique in that its primary objective was to perform a rendezvous and fly around of the Russian space station Mir. NASA had already signed a contract leasing Spacehab to resupply Mir in 1996 and 1997.

The experiences with Spacehab One and Two, although by and large successful, reemphasized the precious nature of the astronauts' time. Due to the demands of operating the shuttle itself and tending to mission priorities, a plan for reducing human interface time with the Spacehab module was developed prior to the launch of Spacehab Three. Spacehab, Inc.'s response was to develop improved equipment to automate a number of manual tasks. The first of these new features was a video switch to reduce the time the crew had to dedicate to video operations. Another was the installation of a system to relieve the astronauts of some of the responsibilities for the downlink of data. By far the most creative, interesting, and efficient of these labor-saving innovations was the development of a robot, named Charlotte, to carry out many of the experiment-tending tasks previously done by the astronauts themselves. The robot moves along cables and can perform many routine procedures, such as changing experiment samples. Charlotte was controlled by a mission specialist using a laptop computer while scientists on Earth observed their experiments on television transmitted through cameras on the front of the robot. Charlotte was also able to digitize, compress, and downlink still images taken from the video system. The experience with Charlotte marked the first time that a robot has worked together with astronauts in the same area in a space vehicle.

During Spacehab Three's flight, the *Discovery* crew carried out some twenty experiments, mostly associated with the research and development of commercial products, including experiments for new pharmaceuticals and advanced materials for improved contact lenses. The *Discovery* crew also conducted an experiment that examined how materials burn in weightlessness, using Plexiglas in this instance.

As result of Spacehab's superior performance on three shuttle flights through 1995, its near-term future looks promising. Spacehab, Inc., is developing a Double Module and a connecting tunnel to the shuttle. The Double Module is slated to fly on three of the Mir resupply missions in 1996 and 1997, carrying more than 6,000 pounds of U.S. and Russian food, water, supplies, and experiments on each flight. This is viewed as an opportunity for U.S. industry to play a role in the build-up phase of the space-station era. It is also indicative of the feasibility of transforming space from an exclusively government concern to an area of private and commercial interest.

Knowledge Gained

The intent of the Spacehab program is to determine whether outer space can serve as a part of the national economy, a place where resources are produced rather than consumed.

Spacehab performed flawlessly on its first three flights, the research experiments it carried generating revolutionary advances in biotechnology, materials sciences, and other technologies. This type of product-oriented research is essential to preparing U.S. industry for the era of the space station. The successes achieved aboard Spacehab have demonstrated the viability of a partnership in space between government and the private sector.

Although many of Spacehab's experimental results, such as those from the external cosmic dust collector, are still undergoing evaluation, many findings became clear almost immediately. In the area of advanced materials science, Paragon Vision Sciences Corporation of Mesa, Arizona, a leader in the production of oxygen-permeable contact lenses, placed a polymerization experiment on board the first Spacehab flight. The experiment mixed raw materials currently in use on Earth with a new type of material that demonstrated a high permeability. The resulting product proved to be more permeable than what was produced on Earth, allowing almost four times as many lenses to be sliced from the same amount of material. Paragon forecasts the introduction of the new lenses onto the commercial market as early as 1998.

Spacehab flight two carried an experiment dealing with the growing of insulin crystals in microgravity. The results revealed new information about insulin's molecular structure, prompting the development of a time-release substance that can be combined with insulin, reducing the frequency of injections for diabetics.

Spacehab flight three carried an experimental unit called Astroculturex, sponsored by the Wisconsin Center for Space Automation and Robotics, a NASA Center for the Commercial Development of Space. This setup demonstrated the successful flowering of plants in space, a powerful indication of the degree of environmental control attainable in a compact and reliable flight package.

The small robot, Charlotte, which flew aboard the third Spacehab mission, was unique in that it was an experiment that interacted with the operations of the shuttle itself. Charlotte was suspended from eight cables emanating from the corners of Spacehab, creating a "web" along which the robot moved. This type of suspension configuration eliminated heavy gantries, reducing the robot's weight and increasing its flexibility. This represented a revolutionary advance in robotics technology. The flawless performance of Charlotte demonstrated the ability of robots to assist astronauts in space with routine tasks. The robot's camera equipment also allowed scientists on Earth to monitor their experiments during the astronauts' sleep period.

The performance of Spacehab is providing guidance for NASA as it develops plans for the approaching age of the space station. Spacehab's agreement with NASA, allowing it to ferry supplies to the Russian space station Mir in 1996 and

1997, will give U.S. industry opportunities to participate in the build-up phase of the space-station era, furthering the development of space as an area of private and commercial interest.

Context

When Spacehab One blasted off aboard the space shuttle *Endeavour* on June 21, 1993, hopes were high that the microgravity found only in space would allow experiments to be conducted that would lead to improved manufacturing processes, better electronic components, and life-saving drugs.

There were also tremendous doubts about the time being right for the successful commercialization of space. Cost was the main concern that caused hesitation on the part of corporations to invest in Spacehab's promise: The price of the transportation is still too exorbitant. As long as products can be manufactured more cheaply on Earth, space will not be an option. At present it costs about $100,000 per pound to ship materials into space for processing and return a commercial product. Therefore, this product would have to be worth much more than one hundred thousand dollars per pound.

Another complication is that no one at present has a clear idea of which space-made products will be commercially feasible. Spacehab, then, cannot be considered at this time to be a manufacturing plant in the commercial sense. Rather, it is a host for corporate experiments to determine whether space can, in fact, manufacture products more efficiently, economically, and perfectly than on Earth. Assuming positive answers to these questions, the next phase would be full-scale manufacturing in space.

Because of these doubts, Spacehab, Inc., initially had great difficulty securing financing from banks. It had no customers and therefore no immediate way to repay its loans. The breakthrough came when NASA bought most of the locker space on the first three Spacehab flights, which inspired confidence in a number of private entities to invest in space experimentation. This enabled Spacehab, Inc., to secure the funding it needed to proceed with the program.

There is still one fundamental disadvantage, however, that Spacehab has not overcome: it is dependent on buying government launch services to get its modules into space. NASA, in turn, is unable to guarantee Spacehab's launch agreements, dependent as it is on fluctuating government support. This means that Spacehab cannot promise its corporate customers that its modules will be launched on schedule. For this reason many potential clients remain leery, and Spacehab's continued success cannot be realized without ongoing support from NASA.

In spite of these realities, Spacehab's first three flights proved the feasibility of space-based manufacturing and increased enthusiasm for such ventures. Research in Spacehab's laboratories has generated revolutionary advances in biotechnology, advanced materials, and other technologies. NASA's purchase of locker space on board the next three flights is further guarantee that continued opportunities for such experimentation will be there. Spacehab's gamut of successes has created a sense of urgency about creating space facilities for manufacturing on a larger scale, specifically the International Space Station, in which Spacehab is envisioned to play a role. The European Space Agency, the National Space Agency of Japan, and corporations around the world are developing payloads for future Spacehab missions.

Bibliography

Banham, Russ. "Insuring the Next Frontier." *Risk Management* 40 (November 11, 1993): 29 (6). The only comprehensive article available on Spacehab. It offers detailed information about some of the difficulties involved in getting Spacehab off the ground, as well as a discussion of the feasibility of space-based manufacturing. Also examines the private sector's view of Spacehab's potential. Non-technical, accessible reading.

Gerondakis, George C. *Get Away Special (GAS) Educational Applications of Space Flight.* Washington, D.C.: National Aeronautics and Space Administration, 1989. Describes how anyone can place a small, self-contained, scientific payload on the space shuttle at low cost. The goal of the GAS program is to promote interest in the space program among young people. Although this is one of NASA's "technical" papers, the language is clear, non-technical, and appropriate for high school level and above.

Joels, Kerry M., and Gregory P. Kennedy. *The Space Shuttle Operator's Manual.* New York: Ballantine Books, 1982. An engagingly written atlas to space shuttle operations and architecture. This book is an annotated collection of line drawings and photographs detailing all aspects of the space shuttle for the interested layperson. Includes a section on conducting experiments in space, with a "walk-through" of a daily routine in a space shuttle-based laboratory.

McCurdy, Howard E. *The Space Station Decision.* Baltimore and London: The Johns Hopkins University Press, 1990. A historical account of the growth of thought from the space program's early days regarding the construction of a space station. This is a thoroughgoing study of the politics, people, and plans for an orbiting space station, written almost like a novel, full of strongly defined personalities and moments of high drama.

Summers, Carolyn. *Toys in Space: Exploring Science with the Astronauts.* Blue Ridge Summit, Pa.: Tab Books, 1994. Directed at upper elementary grades through high school, this book combines concise, well-written doses of theory with experiments young people can carry out to emulate the experiments conducted aboard the space shuttle as part of the Toys in Space project, which explored how common toys behave in the zero-gravity conditions of space.

Space Station Task Force, National Aeronautics and Space Administration. *Space Station Program Description, Applications, and Opportunities.* Park Ridge, N.J.: Noyes Publications, 1985. This formidable volume, written in a matter-of-fact style, is for the informed layperson with a serious interest in the space station program. Pages 160-190 are relevant to the Spacehab program, as they detail the commercial possibilities for space-based manufacturing. The book suffers, however, from lack of an alphabetized index.

Robert Klose

The Spacelab Program

Date: Beginning August, 1973
Type of program: Manned space shuttle experiment facility

Spacelab is a major space shuttle payload designed to provide scientists with facilities approximating those of a terrestrial laboratory.

PRINCIPAL PERSONAGES
JAMES FLETCHER, NASA Administrator
ALEXANDER HOCKER, Director General of ESRO
ROY GIBSON, Director General of ESA
DOUGLAS R. LORD, Spacelab Program Manager
T. J. LEE and
JOHN THOMAS, Spacelab project managers
JAMES DOWNEY and
JESSE MOORE, Spacelab missions managers

Summary of the Program

Because Spacelab was designed to operate within the payload bay of the space shuttle orbiter, configuration interface between it and the concurrently designed shuttle was sometimes problematic. Components of Spacelab often had to be redesigned in order to meet changing shuttle requirements. In particular, a major redesign of Spacelab's instrument pointing system was required. Starting in 1974, Spacelab passed through many tests and design reviews as hardware was planned and built. These led to final acceptance reviews in 1981 and 1982, when the elements of what was termed Flight Unit I were delivered to the National Aeronautics and Space Administration (NASA).

Meanwhile, NASA had organized the management of Spacelab within its network of facilities. Marshall Space Flight Center (MSFC) in Huntsville, Alabama, was to oversee the work of the European Space Agency (ESA) on Spacelab and assure that agency's compliance with shuttle standards. (Later, MSFC was given responsibility for developing additional missions and for providing the hardware to other NASA centers that also prepare and conduct Spacelab missions.) NASA issued an "announcement of opportunity" to space scientists, asking them to propose experiments that might be performed aboard the first two Spacelab missions. Since these were verification flights, NASA tried to accommodate as many scientific disciplines as possible. The payload mass was allocated equally between NASA and the ESA for Spacelab 1, while Spacelab 2 was primarily an American mission (but European scientists were invited to propose experiments). Researching a

path for the complete Spacelab, NASA flew engineering models of Spacelab pallets on the STS-2 and STS-3 shuttle missions in 1981 and 1982. As part of the exercise, the pallets carried science instruments that gathered useful data.

The final configuration for Spacelab comprised pressure modules and open pallets in addition to equipment designed to join these components and provide supports for the experimental gear they would carry. Spacelab is controlled by crew members operating a computer either in the module or in the aft flight deck of the shuttle.

The module was designed with core and experiment components. Each segment is 2.70 meters wide and 2.88 meters long. With end cones, a short module measures 4.27 meters in length and a long module 6.96 meters in length. The interior arrangement includes a floor to cover the support systems, equipment racks placed on each side of the module, overhead storage areas, and a small access science port. Designed as "singles" and "doubles," the racks were 1.48 centimeters wide and capable of holding up to 290 and 580 kilograms of experimental gear, respectively.

In the Spacelab core module, the two forwardmost double racks were dedicated as the control station (starboard) and the workbench (port). That left two double racks and two single racks (one each, port and starboard) for use by experimenters. The experiment segment added another four double racks and two single racks.

Additional experiments can be accommodated by an optical quality viewport mounted over the core segment and a small science air lock in a similar position in the experiment segment. The viewport has a removable exterior cover so that it is protected from the space environment and shuttle contamination, except when used by medium-sized cameras mounted by the crew. The air lock allows the payload crew to expose equipment up to 1 meter long and 0.98 meter wide into space.

Linking the module to the shuttle cabin is the transfer tunnel, 1.02 meters in diameter. It is assembled from a set of cylindrical sections to match different module lengths and locations. It also has flexible sections to allow for slight bending in the airframe during ascent and entry.

The other major element of the Spacelab system is the pallet, a U-shaped platform that provides an interface between experiment hardware and the shuttle itself. Each pallet is 4 meters wide and 2.9 meters long and, like the modules, can be joined with other similar components. Each pallet is made up of five angular U-shaped frames joined by longitudinal members and covered with metal plates. The inner plates have a pattern of bolt holes in a 14-by-14-centimeter grid for mounting lightweight hardware; twenty-four hard points are provided for heavy equipment. The pallets also provide routing for cooling equipment, electrical cables, and other support services.

Both the pallets and the modules are held in the payload bay by sill and keel trunnions, 8.25-centimeter pins which are locked down by special clamps bolted to the orbiter structure. The modules and pallets can be grouped in almost a dozen configurations depending on mission needs. The module can be flown "long" or "short," with or without one to three pallets. Up to three pallets may be joined in a train, and up to five may be flown at once.

Spacelab is totally dependent upon the space shuttle for electrical power, environmental control, and life support. Power and environmental gear provided in Spacelab's subfloor area is designed to assist the shuttle in that respect. Spacelab does have its own Command and Data Management System (CDMS), through which the crew may control experiments. The CDMS commands experiment apparatus and collects data from them by way of Remote Acquisition Units (RAUs), which function somewhat like sophisticated telephone exchanges. The rapid advance of micro-electronics in the late 1970's, however, has relegated the CDMS to the role of traffic controller for the various experiments, which often have their own microprocessors. The Spacelab CDMS actually comprises three central processing units: one to operate Spacelab proper, one to operate the experiments, and a third held as a manually selected backup. In missions using the module, the CDMS is housed in the starboard forward double rack. For pallet-only missions, it is housed in a pressurized container, the "igloo," mounted on the forwardmost pallet. The igloo provides a sea-level environment for the CDMS, thus eliminating the need to prepare the computer for the environment of raw space.

For pointing large telescopes or telescope clusters at targets, an Instrument Pointing System (IPS) is provided. The IPS has three electrically driven gimbals that can point the IPS payload within extremely close range of a target. The IPS is mounted on a support framework on a pallet, and, in turn, provides a large, circular equipment platform for the payload. Payloads can weigh up to 7,000 kilograms and can be several meters long. Not all payloads requiring pointing can justify use of the IPS, so experimenters have developed smaller pointers tailored to their investigations.

Assembly of a Spacelab mission is a long, complex process involving several levels of effort. After the science community has identified important investigations, NASA performs a preliminary study of the kinds of instruments that might satisfy this need. Instruments generally fall into two classes, the principal investigator and the facility. In the first, an individual scientist or science team develops an instrument for a narrow investigation. In the second, NASA and a contractor develop an instrument that can serve a number of scientists on many missions. Experiments on Spacelabs 1, 2, and 3 were developed from announcements asking specifically for them. In 1978, NASA issued a broader announcement soliciting instruments in physics and astronomy. Forty instruments were selected, some of which were grouped for Spacelab or other missions and some of which were later canceled. Other announcements were issued for life-science missions and facility-class instruments.

After the science investigations are selected, an Investigators' Working Group (IWG) is formed from the lead scientists. NASA appoints a mission manager and mission scientist from its own ranks. The IWG and NASA engineers work closely together to develop the flightplan and details of how and when each investigation is to be conducted during the mission. It is not unusual to discover that some experiments will not fit in or will be late. This normally results in an instrument's being moved to a later mission rather than it being canceled.

As it becomes ready, experiment hardware is delivered to Kennedy Space Center, Florida, for integration into the complete Spacelab. The first step in the process is to install the experiment elements in racks or on pallets. The racks and floor are then fitted into the module, the module is closed, and the module and pallets are physically and electrically joined. The complete assembly is placed inside the Cargo Integrated Test Equipment (CITE) stand, where all the components are exercised as they would be in flight. Finally, the complete Spacelab is installed in the space shuttle, and an "end-to-end" test is conducted to validate all links from the experiment to the control center.

Typically, a Spacelab mission includes three types of crew members: Pilot Astronauts, the Mission Commander and Pilot, who fly the shuttle itself; Mission Specialists, career NASA scientist astronauts who have overall responsibility for the payload; and Payload Specialists, members of the IWG selected to fly on the mission and to conduct the experiments. Two Payload Specialists, prime and alternate, are selected for each flight opening.

The inclusion of Payload Specialists on the Spacelab missions was a major point used by NASA in selling Spacelab and the shuttle to the science community. Previously, scientists could only listen or watch from the ground while their experiments were conducted by career NASA astronauts. With the routine operations to be provided by the shuttle, scientists could fly, almost passenger-like, with the experiments that they had developed. The process has turned out to be slightly more complex, but the basic philosophy holds.

Spacelab missions start a few hours after the shuttle achieves orbit and last until about four hours before reentry. When the shuttle's in-orbit time is shorter than originally planned — for example, ten days instead of thirty — mission

View of the Challenger's payload bay and the SOUP experiment. (NASA)

activities are intense and go around the clock. Typically, there will be a six- or seven-man crew, which operates in three-man, twelve-hour shifts.

Spacelab missions are directed from two control centers. The first, Mission Control at Johnson Space Center, retains overall control of and responsibility for the completion of the flight. For the most part, Mission Control defers to the Payload Operations Control Center (POCC), where the science phase of the mission is directed. Thus, Spacelab is heavily dependent on the Tracking and Data-Relay Satellite System (TDRSS) to relay telemetry from the experiments to the POCC and commands back from the POCC.

Knowledge Gained

Spacelab has proved a versatile and useful facility for conducting space science research. Spacelabs 1, 2, 3, and D-1 (this last set of experiments were sponsored by West Germany) — and the various single-pallet payloads flown on STS-2, STS-3, and STS-41G — were all successful. Unfortunately, in the view of many scientists, NASA has made use of the facility too difficult. In fact, in the era preceding the 1986 *Challenger* tragedy, the agency replaced its own science payloads with commercial and military equipment. Thus, scientists soon found themselves in a sort of inflationary spiral where the cost of a mission required extreme efforts to guarantee success, which, in turn, raised the cost of the mission.

To combat this problem, NASA conducted a Spacelab mission integration cost analysis and developed a concept known as the Dedicated Discipline Laboratories (DDL). Each DDL would comprise a group of experiments with similar or complementary mission requirements. For example, it would be logical to carry astronomical instruments and solar instruments on one mission, since they would have similar pointing requirements during a mission. One would not, perhaps, think to carry materials and life-science experiments together until one compared their needs: heavy electrical power demands and intermittent tuning for materials experiments, and intense manpower and low-power demands for life sciences. Yet procedures often required of biomedical experiments can be disruptive to crystal-growth and other fluid experiments. Thus, carrying them together would require innovative scheduling to avoid conflicts.

At the very least, the DDL concept would reduce the integration cost of Spacelab missions by reducing the analysis and paperwork required for each mission. At the best, much work could be avoided by allowing clusters of instruments to remain intact until their next flight. Even requests to upgrade instruments were disregarded in order to cut costs.

The Spacelab and shuttle experiences have also contributed to a better understanding of what is required to support a vigorous experiment program. This knowledge led to the development of intermediate payload carriers between Spacelab and the Get-Away Specials, an innovative payload system. The effort required to replace even a single rack inside the module affected the design of the U.S. space station, so its racks are now better designed for easy replacement in orbit.

Context

Spacelab has proved difficult for scientists from some disciplines to use. Materials scientists need as smooth a ride as possible so that samples are not jostled (excessive motion disrupts the formation of crystals and the study of fluid flows). These required conditions are at odds with necessary crew exercise periods and even with pumps and fans that cool Spacelab. Early shuttle missions discovered a phenomenon known as "shuttle glow," an eerie luminescence that peaks in the infrared spectrum. The cause remains under debate but appears to be some chemical reaction between the shuttle itself and rare molecular species in the upper atmosphere. The shuttle glow hampers observations in the infrared and low-light levels under certain conditions.

Spacelab grew out of a 1969 invitation by NASA for the European Space Research Organization (ESRO) to become involved in the post-Apollo space program. European involvement in U.S. space activities had been commonplace since the origins of NASA but rarely had been larger than limited partnerships on small satellite projects. The European Space Conference in 1970 authorized studies with the United States in the post-Apollo area. In 1972, NASA selected the space shuttle program as its major effort for the 1970's.

As conceived, the space shuttle was to have a reusable third stage, called the Space Tug, to carry satellites to and from geostationary orbit and other destinations. ESRO was very interested in developing this vehicle, which it saw as having potential uses aboard European launch vehicles then under study and possibly providing more jobs for the European aerospace industry. Yet because the shuttle also was to serve a number of U.S. military payloads, the Department of Defense opposed any foreign role in the Space Tug, especially since ESRO might try to veto launches of defense satellites it found objectionable. In 1972, both the Department of Defense and the State Department formally denied ESRO a role in the Space Tug, and an alternate was sought by NASA and ESRO. The two possibilities were structural elements of the shuttle orbiter and a science lab that would fit in the payload bay. Of the two, ESRO found the latter more attractive because it would provide research opportunities for European scientists and provide the community with direct experience in manned spaceflight.

NASA had for some time been studying a Research and Applications Module (RAM), which would function as a lab facility and turn the shuttle into a temporary space station. Since a permanent space station was on indefinite hold, that was seen as necessary to continue manned space research.

Between December, 1972, at the ministerial meeting of the European Space Conference, and August, 1973, NASA and ESRO officials conducted concept and definition studies of the laboratory facility, soon called Spacelab. An intergovernmental agreement was reached in August, 1973, and a memorandum of understanding was signed by NASA Administrator James Fletcher that month and the ESRO director general, Alexander Hocker, in September.

Under the terms of the memorandum, ESRO would design and build a complete Spacelab flight unit for use by NASA and ESRO aboard the space shuttle "for peaceful purposes," and NASA agreed to buy a second flight unit at a price to be negotiated later. Although the term "peaceful purposes" is subject to debate, it has been interpreted by NASA and ESRO (later ESA) as permitting Department of Defense research missions but not weapons missions. In 1974, a West German consortium was selected as the prime contractor for Spacelab. In keeping with ESRO's international nature, contracts were awarded to ESRO member nations in proportion to their contributions to Spacelab. In this manner, each nation recouped most of the money that it had invested in Spacelab. Finally, in 1975, ESRO merged with the European Launcher Development Organization (ELDO) to become the European Space Agency.

ESA's experience in developing Spacelab, and in flying it less often than expected, led that agency to assume a tougher negotiating stance on participation in the space station missions and to demand treatment as an equal partner in the space community. It has also provided the basis for ESA's own Columbus program to develop a human-tended station.

Bibliography

Dooling, Dave. "Future Spacelab Missions." *Space World* T-10-238 (October, 1983): 33 - 37. Describes efforts by NASA to reduce the cost of future Spacelab missions and plans for dedicated discipline laboratories. Written for the general reader. Illustrated.

————. "Spacelab 1." *Space World* T-8-9-236/237 (August/September, 1983): 8-14. This article provides an overview of how a Spacelab mission is developed and traces the plans for Spacelab 1. Describes preliminary results from the Spacelab pallets carried on STS-2 and STS-3.

Froelich, Walter. *Spacelab: An International Short-Stay Orbiting Laboratory*. NASA EP-165. Washington, D.C.: Government Printing Office, 1983. A booklet designed for teachers and students. Describes the development of Spacelab and the work required to assemble a mission. Includes color illustrations.

National Aeronautics and Space Administration. *Spacelab 1*. NASA MR-009. Washington, D.C.: National Aeronautics and Space Administration, 1984. A NASA publication written for teachers and reporters, with color illustrations and capsule summaries of experiments. It includes a discussion of how an IWG functions and how the mission was conducted.

Shapland, David, and Michael Rycroft. *Spacelab: Research in Earth Orbit*. New York: Cambridge University Press, 1984. A broad description of the development of Spacelab through the first mission, with descriptions of various scientific disciplines it can serve. Written for a general audience.

Dave Dooling

THE DEVELOPMENT OF SPACESUITS

Date: Beginning May 5, 1961
Type of technology: Humans in space

Spacesuits initially provided astronauts with an emergency intravehicular backup system in case of the loss of spacecraft cabin pressure. Early spacesuit designs were based on the technology developed for high-altitude aircraft pressure suits. More complex spacesuit mobility systems now allow astronauts to venture beyond the protective limits of the spacecraft.

PRINCIPAL PERSONAGES

> JOHN SCOTT HALDANE, a British physiologist
> MARK EDWARD RIDGE, an American balloonist
> WILEY POST, an American aviator
> RUSSELL COLLEY, a pressure suit designer
> YURI A. GAGARIN, a Soviet cosmonaut
> ALAN B. SHEPARD, an American astronaut
> ALEXEI LEONOV, a Soviet cosmonaut
> EDWARD H. WHITE, an American astronaut
> NEIL A. ARMSTRONG, an American astronaut
> JOSEPH KOSMO, Subsystem Manager, spacesuit development, NASA
> HUBERT VYKUKAL, Senior Research Scientist, NASA

Summary of the Technology

It has long been recognized that humans cannot survive the conditions of space without special protection. The development of spacesuits contributed significantly toward making space exploration possible.

In 1920, John Scott Haldane, a British physiologist, first proposed the use of a pressure suit to provide protection for flight crew members against the lack of oxygen at altitudes above 12,200 meters. This idea, however, was not tested until November 30, 1933, when American balloonist Mark Ridge donned a modified deep-sea diver's suit and was exposed to 25,600-meter altitude conditions in a low-pressure chamber.

In the same year that Ridge was testing the British-built "high-altitude" suit, Wiley Post, an American aviator, initiated the development of a full pressure suit. Post needed the pressure suit to help him break the existing world aircraft altitude record. Wearing a suit designed by Russell Colley of the B.F. Goodrich Company, Post made a number of high-altitude flights during 1934 and 1935. Unfortunately, because of problems with the recording barographs, no official altitude record could be verified. Unofficially, however, Post had reached altitudes in excess of 12,200 meters with the help of the full pressure suit. More important, the efforts made and risks taken by Ridge and Post proved that pressure suits were practical systems that would safely enable humans to fly at high altitudes and, in time, into space.

Before and during World War II, full pressure suits were developed by various European countries and the United States. All the early suits were very cumbersome to wear and, when inflated, caused serious mobility restrictions. These suits were primarily developed to protect high-altitude flight crew members from lack of oxygen, and, as such, they were regarded as precursors of aircraft cabin pressurization systems. With the onset of World War II, the pressure cabin became standard in most aircraft, and from then on, pressure suit research focused on emergency situations in military and experimental high-altitude aircraft.

In 1961, the first men to wear full pressure suits in space were Russian cosmonaut Yuri Gagarin and American astronaut Alan Shepard. The pressure suits they used were worn uninflated in a pressurized cabin and would only have been inflated in the event of a failure of the Vostok or Mercury capsule pressurization system. The spacesuit configurations developed by the National Aeronautics and Space Administration (NASA) for the Mercury project and Gemini program originated from the earlier high-altitude aircraft full pressure suit.

The Mercury project and Gemini program spacesuits were essentially modified versions of existing military full pressure suits. The Mercury project utilized the Navy's Mark 4 full pressure suit, built by B.F. Goodrich Company, and the Gemini program spacesuit was derived from the Air Force's AP/22 full pressure suit, manufactured by David Clark Company. The early Soviet spacesuits also originated from military aircraft pressure suit technology. These early spacesuits lacked sophisticated mobility systems; because the suits served primarily as backup systems against the loss of cabin pressure, only limited pressurized intravehicular mobility was required.

The development of the mobile spacesuit was spurred by the requirement for astronauts to perform tasks outside the spacecraft. In 1965, cosmonaut Alexei Leonov, of Voskhod 2, and astronaut Edward White, of Gemini IV, performed the world's first "spacewalks." During this phase of the Gemini program, U.S. scientists recognized that astronauts needed improved mobility systems and protection from extravehicular environmental hazards.

The Apollo spacesuit was designed to function as an emergency intravehicular suit and as an extravehicular suit that would enable lunar surface exploration. A variety of prototype Apollo spacesuit configurations evolved between 1965 and 1970. The International Latex Corporation was responsible for the design, development, and fabrication of the A7L and A7LB Apollo spacesuits, which were selected for use. On July 20, 1969, Neil Armstrong, wearing an A7L spacesuit, was the first human to set foot on an extraterrestrial surface and collect data while being sustained in and protected from a hostile environment. Later Apollo astronauts wore the improved A7LB spacesuit when they explored the lunar surface. The Skylab program adopted the basic Apollo A7L spacesuit with minor modifications for use in various planned orbital extravehicular activities.

During the early 1960's, as spacesuits were being developed for the Gemini program and the planned Apollo and Skylab programs, Joseph Kosmo of the Johnson Space Center (JSC) and Hubert Vykukal of the Ames Research Center embarked on the development of advanced spacesuit mobility systems for potential future application. NASA long-range program planners had been studying the feasibility of establishing lunar surface bases and conducting manned Mars planetary exploration. In support of these envisioned post-Apollo operations, NASA initiated a series of spacesuit technology development programs.

The first of the JSC-sponsored advanced technology suit concepts was the rigid experimental suit assembly, or RX-1, developed by Litton Industries' Space Sciences Laboratory. The suit was a radical departure from the basic, soft-fabric spacesuits of early 1962. The RX-1 was developed to demonstrate the feasibility of low-force mobility joint systems in the arms and legs. Additional suit features included hard torso structure and an easily fastened single-plane body seal closure. (The Mercury, Gemini, and Apollo spacesuits used pressure-sealing zippers.) Between 1962 and 1968, numerous RX models were developed. Each version incorporated mobility joint and structural improvements made possible by evaluation and testing of the previous model. The RX-5A was the final configuration of the RX series.

In conjunction with the development of JSC's RX, Ames Research Center initiated investigations into a hard spacesuit that would use a combination of bearings and metal bellows for mobility joint systems. Between 1964 and 1968, two "Ames experimental" hard suit assemblies, identified as AX-1 and AX-2, were developed by Ames.

As the Apollo program matured, it became apparent that spacecraft payload weight and stowage volume limitations were constraints on the various hard suit concepts. This realization resulted in the initiation of new approaches to spacesuit design, and JSC produced a family of advanced spacesuit configurations representing a hybridization of hard suit and soft suit technologies. With the completion of the Apollo program, much of the advanced mobility system technology that had been developed earlier was shelved. In the 1970's and 1980's, various elements of the advanced spacesuit technology base were incorporated in the spacesuits designed for the U.S. space shuttle and space station programs.

Unlike previous flight program spacesuits, the shuttle spacesuit was designed for extravehicular use only. Emphasis was therefore placed on providing astronauts with a high degree of extravehicular operational capability, uncompromised by other requirements. The shuttle spacesuit incorporated a hard upper-torso shell of fiberglass, a horizontal single-plane body seal closure ring, pressure-sealed bearings in the shoulder, upper arm, and lower torso areas, and flat, all-fabric arm, waist, and leg joints. All these elements had evolved from earlier advanced spacesuit technology.

The space station program focused on expanding Extravehicular Activity (EVA) capabilities beyond those of previous space programs. The space station suit was to have a higher operating pressure — 8.3 pounds per square inch — so that astronauts would not need to spend costly time prebreathing pure oxygen before performing an EVA. Previously, prebreathing operations served to wash nitrogen gas from an astronaut's bloodstream so that nitrogen bubbles would not form when the astronaut moved from the shuttle cabin, pressurized at 14.7 pounds per square inch, to the spacesuit, pressurized at 4.3 pounds per square inch. If, however, the spacesuit could operate at higher pressure, prebreathing could be eliminated and EVA operations would be able to be conducted more routinely. The core technology for advanced mobility systems established over the previous years enabled the development of higher operating pressure spacesuits. JSC's Mark 3 and Ames's AX-5, designs that eliminated the need for prebreathing, were developed in response to need of the space station program.

Minimizing energy expenditure by the astronaut, made difficult by the tendency of the volume of gas in joint elements to change during pressurized operations, continues to be the primary impetus behind research in improved mobility systems. The systems developed between 1960 and the 1980's have demonstrated the technical viability of certain design features and have served as a base for the development of future lunar and planetary exploration spacesuits.

Knowledge Gained

As space missions became more complex, so did spacesuits. The experience gained from a variety of space missions has influenced spacesuit development.

From the manned EVAs of the Gemini program, it was learned that improved cooling techniques to remove astronaut-produced metabolic heat would be required if longer and more involved EVAs were to be conducted. As a result, NASA scientists developed an undergarment, to be worn inside the spacesuit, that contained small tubes through which water could be pumped. The liquid-cooled garment became a standard design feature of the Apollo spacesuit and all subsequent suits.

In the Apollo and Skylab programs, the spacesuit fulfilled two roles. It was worn at launch and during critical spacecraft docking operations, and its function during these phases was to provide a backup pressurized environment in case of cabin pressure failure. It also acted as a pressurized mobility system and a portable life-support system during orbital and lunar surface EVAs. The environmental hazards and hostile conditions of extended orbital operations and lunar surface exploration meant that scientists had to develop improved materials and designs to protect the spacesuited astronaut. The requirements of intravehicular versus extravehicular operations posed a number of design problems, including limitations on bulk, operational complexity, and mobility.

For the shuttle program, the spacesuit was designed for the single purpose of supporting manned extravehicular operations, and features were optimized for enhanced EVA performance and reduced cost. The shuttle spacesuit, with its corresponding life-support system, represented the first completely integrated extravehicular assembly. No cumbersome external life-support hoses, connections, or harness straps were required to allow an astronaut to leave the spacecraft.

Previously, each spacesuit was custom built to fit the astronaut and was used on only one mission. The shuttle spacesuit design featured modular components that could be combined in various ways, enabling both male and female astronauts to be fitted with a minimum number of spacesuits and reducing overall program cost.

Elimination of prebreathing operations through the development of spacesuit mobility systems that operate at higher pressure levels will make EVAs more routine and easier to perform. The modular design of the candidate space station pressure suits, being similar to that of the shuttle spacesuit, will enable the suits to be reused numerous times. In addition, in-orbit maintenance and replacement of various fabric components and the reuse of hardware components will reduce overall spacesuit and flight program costs.

Context

Throughout the Mercury project and until the feasibility of EVAs was established on the Gemini IV mission, spacesuits were simply backups for cabin pressure systems. The development of the true spacesuit occurred almost simultaneously in two separate parts of the world. In 1965, Soviet cosmonaut Leonov and American astronaut White performed independent spacewalks outside the confines of their respective spacecraft. The Gemini program provided the first EVA in the U.S. manned space effort.

The Gemini program's accomplishments were significant for spacesuit development. For the first time, spacesuits were used to allow humans to work in space. It was recognized that manned EVAs would increase spacecraft's capabilities and enable the development of new operational techniques. EVA technology from the Gemini program was incorporated wherever possible in the Apollo spacesuit. Improved body-cooling and mobility systems were direct results of the Gemini experiences.

Spacesuits make EVAs possible in three ways. First, in combination with a portable life-support system, the spacesuit maintains the physiological well-being of the astronaut, which includes supplying oxygen for breathing and ventilation and removing carbon dioxide and metabolic heat. Second, the spacesuit incorporates various mobility joint system features that enable the astronaut to perform tasks in the extravehicular environment. Finally, the spacesuit provides protection against the hazards of space, which include thermal extremes, meteoroid and debris particles, radiation, and, on the lunar surface, sand and dust. The pressure retention layer of the spacesuit is both a protective barrier and a structural foundation for various mobility systems. A separate outer garment comprising layers of various materials protects the astronaut from hostile environments.

None of the materials used in the early spacesuits were originally developed with space exploration in mind. As more complex spacesuit systems evolved, special needs were identified that required the development of new materials or combinations of materials to provide structural integrity and increased protection.

The space station created a need for a spacesuit that would operate at higher pressure, which in turn stimulated the development of improved mobility system technology. Extensive EVA experience has been accumulated in space environments ranging from near-Earth orbit to the lunar surface. In all cases, a spacesuit has been designed to accommodate the unique conditions encountered. As scientists plan future space missions that will encompass longer and more diversified EVAs, established spacesuit technology will form the foundation for the development of more advanced spacesuits.

Bibliography

Cortright, Edgar M., ed. *Apollo Expeditions to the Moon*. NASA SP-350. Washington, D.C.: Government Printing Office, 1975. The personal accounts of eighteen men, including NASA managers, scientists, engineers, and astronauts, who directed, developed, and conducted the Apollo missions. Suitable for general audiences, it describes the various political events and engineering projects that influenced the Apollo program. Includes numerous illustrations and photographs covering the historical period of the Apollo program, along with pictures transmitted from Apollo spacecraft showing the use of spacesuits.

Faget, Maxime Allen. *Manned Space Flight*. New York: Holt, Rinehart and Winston, 1965. Describes some of the technical problems engineers are faced with in the building of manned spacecraft systems. Offers insights into the various scientific principles that were used to provide engineering solutions to those problems. Contains numerous charts and illustrations. Recommended for high school and college-level readers.

Kozloski, Lillian D. U.S. *Space Gear: Outfitting the Astronauts*. Washington, D.C.: Smithsonian Institution Press, 1994. This book is a comprehensive look at the space suits developed for each of the U.S. manned spaceflight projects. It begins with the pressure suits developed for high-altitude flying and shows how these led to the Mercury pressure suit used in America's first journeys to space. The evolution of the space suit through its current use in the Space Shuttle Program is chronicled. There are more than 150 illustrations. Appendices detail the pressure suits in the Preservation/Study Collections and summarize the U.S. manned spaceflights from Project Mercury through Space Shuttle mission STS-53 in December, 1992.

Machell, Reginald M., ed. *Summary of Gemini Extravehicular Activity*. NASA SP-149. Springfield, Va.: Clearinghouse for Federal Scientific Information, 1967. An official summary of the Gemini program extravehicular operations described from the developmental viewpoint. Discusses the systems used, the testing and qualification of those systems, the preparation of the flight crews, and operational and medical aspects of the missions. Contains numerous illustrations, charts, and photographs relating to the Gemini program. Suitable for advanced high school and college-level readers.

Mallan, Lloyd. *Suiting Up for Space: The Evolution of the Space Suit*. New York: John Day, 1971. Presents a historical perspective of the development of early pressure suits for high-altitude flight and the evolution of spacesuits from these beginnings. Contains numerous photographs of early prototype and flight model pressure suits and spacesuits. A highly accessible work.

Oberg, James E. *Mission to Mars: Plans and Concepts for the First Manned Landing*. Harrisburg, Pa.: Stackpole Books, 1982. Describes the feasibility of a manned Mars mission and discusses topics such as spaceship design, propulsion, life-support systems, spacesuits, Martian surface exploration, cost factors, and political and social issues relating to future plans for colonization. Contains photographs and illustrations. Suitable for general audiences.

Paine, Thomas O. *Pioneering the Space Frontier: The Report of the National Commission on Space*. New York: Bantam Books, 1986. Presents a programmatic view of the steps the United States must take to remain competitive with the Soviet Union for the next fifty years of space exploration. Discusses NASA's long-term goals. Contains photographs, charts, graphs, and illustrations related to future missions. Includes a glossary of technical terms and an extensive bibliography identifying a wide variety of space-related reference sources. Accessible to the layman.

Swenson, Loyd S., Jr., et al. *This New Ocean: A History of Project Mercury*. NASA SP-4201. Washington, D.C.: Government Printing Office, 1966. The official history of the Mercury project, this book details the elements of research, development, and operations that made up the program. Identifies and describes the importance of exploring the human factor in regard to spaceflight and records the beginning of the space age. Includes extensive footnotes and a thorough bibliography.

Joseph J. Kosmo

SPY SATELLITES

Date: Beginning February 28, 1959
Type of satellite: Military

Spy satellites provide countries with an accurate and fast means of gathering sophisticated information where other means of reconnaissance are less effective and more dangerous. The United States and the Soviet Union developed the most highly developed spy satellite programs.

Summary of the Satellites

Although reconnaissance airplanes had been in use since the outbreak of World War I, the development of antiaircraft weapons made the airplane increasingly vulnerable. It was not until the late 1950's that technology was advanced enough to permit an alternative method by which intelligence could be gathered.

In 1946, the Research and Development (RAND) Corporation published a report in which the feasibility of launching a reconnaissance satellite into orbit was discussed. Additional reports were published in 1956 and 1957. These reports played a large part in the eventual development and launching of spy satellites.

In 1958, President Dwight D. Eisenhower approved a reconnaissance program which was to be operational by 1959. Under this program, the Missile Defense Alarm System (MIDAS), Discoverer, and the Satellite and Missile Observation System (SAMOS) were developed.

MIDAS, later renamed Program 239A, was first launched February 26, 1960. Relying on an infrared scanner that was sensitive to the heat emitted by a rocket, it was to provide warnings of any Intercontinental Ballistic Missile (ICBM) attack. The advantage of MIDAS over earlier warning systems, such as the Canadian Distant Early Warning System, was that it was capable of detecting ICBMs more quickly.

Discoverer 1 was launched on February 28, 1959, and Discoverer 38 on February 27, 1962. Of the thirty-eight launches attempted, twenty-six were successful. During the time that this program was in effect, emphasis was placed on the biomedical experiments conducted, such as the one involving the orbiting of the chimpanzee Pale Face.

The first United States photoreconnaissance satellite to be launched was Discoverer 14. The first twelve Discoverer enterprises had all ended in failure; finally, the successful launch of Discoverer 13 on August 10, 1960, convinced scientists that it was feasible to include a camera in Discoverer 14.

The average perigee of the Discoverer satellites was 220.3 kilometers. The perigee is that point at which a satellite makes its closest approach to Earth and, owing to increased gravitation, the point at which the speed is greatest. Photographs are therefore usually taken before or after the perigee. The apogee, or farthest point from Earth, was 706.3 kilometers. At a later stage, the resolution of the film was, according to the director of the program, on the order of 30.48 centimeters; in other words, the satellite was able to detect objects that were 30.48 centimeters or larger. This capability enabled the detection and identification of Soviet ICBMs. The lifetime of these satellites averaged 108 days.

Until the Kennedy Administration, the Discoverer program had had two sides, the public and the official. The public name for the program was Discoverer, and the official CORONA. President John F. Kennedy phased out the Discoverer program by simply removing the public name, after which essentially the same program continued for six more launches, but now under the name CORONA. CORONA was designated Keyhole (KH) 4, but commonly called Close Look. Six KH-4's were launched between March 7 and November 11, 1962. The perigee was reduced slightly, while the apogee now was only 337.63 kilometers. The lifetime was shortened to a mere 2.8 days.

SAMOS began as Weapon System-117L (WS-117L), with the code name Pied Piper, which was changed to Sentry and finally, during the Kennedy Administration, to KH-1. KH-1, which was operated by the United States Air Force, carried a conventional camera to photograph the target. The film was then developed and scanned by a fine beam of light, after which the signal was transmitted to a station on the ground, where it was used to construct a picture. Although this system was intended to pioneer spy activities, owing to unforeseen delays it was the last of the three to be launched. SAMOS was in operation from October 11, 1960, until November 27, 1963.

A second generation of U.S. spy satellites had been initiated with the launch of KH-5 on February 28, 1963, to replace SAMOS. On July 12, 1963, KH-6 replaced CORONA. The success rate of these new satellites was considerably greater than that of their predecessors, with forty-six of fifty KH-5's succeeding between February 28, 1963, and March 30, 1967, and thirty-six of thirty-eight KH-6's between July 12, 1963, and June 4, 1967.

In the summer of 1966, a new, third generation of satellites was introduced. On July 29, KH-8 was launched, followed almost immediately on August 9 by KH-7. The third-generation satellites supplemented their ordinary cameras with infrared scanners and with a new antenna that allowed a faster transmission rate. KH-7 satellites were phased out in 1972, but KH-8, with its excellent resolution, continued functioning until the early 1980's. It eventually became known that KH-8 was 7.3 meters long and 1.73 meters in diameter, with a weight of approximately 2,990 kilograms.

The more recent fourth generation of satellites consists of KH-9, KH-10, and KH-11. Of these, KH-9 is commonly dubbed Big Bird. KH-10 was to be a manned orbiting laboratory, but the success of KH-11, coupled with the growing cost of KH-10, resulted in the cancellation of the latter.

Big Bird was first launched on June 15, 1971. It weighed about 11,340 kilograms, was 15.25 meters in length, and had a diameter of 3.048 meters. The satellite was equipped with an ordinary camera and an infrared scanner. It was also believed to contain a multispectral camera, for use in detecting camouflage, and sensitive listening devices that would allow the Pentagon to intercept radio and microwave telephone signals as well as transmissions from Soviet satellites. On board there were four returnable film canisters. The average lifetime of KH-9 was 130 days. Its mean perigee was 166 kilometers and its apogee 269 kilometers.

KH-11 was the first satellite to report events in real time (as they occurred). This feat was achieved by using a Charge Coupled Device (CCD), first developed in the New Jersey Bell Telephone Laboratory in 1970. A CCD is activated when light strikes a silicon sheet divided into millions of small pixels. The silicon sheet converts photons into electrons; the electrons in each pixel are counted, and the information is then transmitted to the ground. The greater the number of electrons, the greater the light intensity. Since the electrons are captured for only milliseconds before they drain away, the thousands of pictures transmitted each second can be put together in much the same way as films are.

The resolution of KH-11 was inferior to that of KH-8 but exceeded that of KH-9, in the range of 6.6 to 8.8 centimeters. KH-11 transmitted its information directly to a communications satellite and from there either to Fort Meade, Florida, or to Fort Belvoir, Virginia.

The KH-11 program received a severe setback when a technical manual was lost to Soviet intelligence in 1977. According to some, the Soviets had been completely unaware that KH-11 was a reconnaissance spacecraft (because of its indirect transmission methods), imagining instead that it was a "ferret" satellite (a term used to describe a spacecraft used to probe foreign radar and to detect microwave signals). Others disagree, arguing instead that the Soviets were aware of its nature but had simply underestimated its capabilities.

To replace KH-11, researchers designed KH-12. It was projected that this fifth-generation reconnaissance satellite would carry advanced versions of the KH-11 sensors, as well as extra fuel that would allow it to move from a low to a high orbit when not in use. Operating in fours, the new satellites would provide instant coverage of any locale within twenty minutes. When the fuel supply was exhausted, the satellite was to be refueled by the space shuttle.

Apart from the KH series, the United States placed three other types of reconnaissance satellites in orbit: electronic, ocean surveillance, and early warning. In the first, electronic devices pick up radio and microwave signals, which are used to determine what type of radar waves are sent out, so that correct methods can be used to penetrate them in the event of war. Ocean surveillance satellites play the same role as those that gather information over land. These satellites are used in detecting ships and submarines. Early warning satellites depend on infrared sensors to detect missile and rocket launches. Sophisticated radar then tracks each object until it becomes clear that it does not pose a threat to the United States.

Of the more than two thousand military satellites that had been launched as of the late 1980's, more than half were reconnaissance spacecraft. These reconnaissance satellites were usually launched by modified Titan rockets. It was projected that satellites would also be sent into orbit by the space shuttle. The loss of the Titan 34D rockets in August, 1985, and April, 1986, left the United States with only one KH-11 in orbit at that time.

The United States is not alone in launching spy satellites. On April 26, 1962, the Soviets launched Kosmos 4, their first photoreconnaissance satellite. Even in the 1980's, Soviet reconnaissance spacecraft were, by American standards, still fairly crude. While the United States preferred to employ a small number of handmade, state-of-the-art satellites, the Soviet Union attempted to offset its lack of superior technology with sheer numbers. In 1985, the United States launched two KH-11 satellites, while the Soviets launched thirty-four camera-carrying spacecraft. Another major difference between Soviet and American satellites was the preparation time. Soviet satellites could be launched within a matter of hours, while an American launch followed months of preparation.

Several other nations have developed reconnaissance satellites that can be used for intelligence gathering, though that purpose is generally not overtly stated. Among these countries are China, India, Israel, France, and West Germany. France's SPOT, for Satellite Probatoire de l'Observation de la Terre (Satellite Probe for Earth Observation), despite its noncommittal title, is an example of an orbiting craft that can obtain information of military importance as well as data regarding weather and Earth resources.

Knowledge Gained

Reconnaissance satellites regularly provide information on new airplanes, ships, and submarines. Such spying often encounters difficulties. For example, in order to stop satellites from monitoring the construction of their submarines, the Soviets placed netting over them. The Americans temporarily overcame this obstacle — until they inadvertently divulged their tactics during the Strategic Arms Limitation Treaty (SALT) II negotiations — by using slant photography to view the submarines from the side.

Reconnaissance satellites continually monitor troubled areas of the world. Photoreconnaissance is believed to have been used in 1967 during the Arab-Israeli conflict, for the attempted rescue in 1979 of the American hostages in Iran, and, in 1986, in advance of the Libyan bombing and at the time of the nuclear accident at Chernobyl. The Persian Gulf War of the 1980's was presumably scrutinized by American satellites. When ships and submarines are at sea, infrared pictures allow their paths to be plotted.

Technology produced for spy satellites has had important ramifications for other space programs. For example, it is believed that military photographic technology was made available to the National Aeronautics and Space Administration (NASA) for its missions to Mars and the Moon.

Advanced radar has made it possible to plot the ground of the former Soviet Union and other countries accurately. Shots of the same area taken at different times are subsequently fed into a computer, which removes similar objects and highlights the discrepancies by using a process known as electro-optical subtraction. The remaining objects either are identified and noted or, if they are blurred, are enhanced by digital restoration.

The CCD technology developed in Bell Laboratories for use in spy spacecraft is now used in many other fields, including astronomy, medical imaging, and plasma physics — even in video cameras. The Galileo spacecraft and the space telescope also rely on the CCD.

The Soviet Union made use of its spy satellites not only in observing the Sino-Soviet border during the period of hostilities but also in attempting to trace the movements of the Afghan resistance. This last effort met with mixed results.

An obvious use of reconnaissance photographs is to study cloud formations and weather around the world. It is possible that pictures of clouds are used by meteorologists at the Central Intelligence Agency (CIA) to determine the nature of the weather, and this information is then passed on to the armed forces.

Context

In the years preceding the launch of the first U.S. reconnaissance satellite, it was widely believed that the Soviet Union had as many as two hundred ICBMs. As a result, millions of dollars were spent by the United States in an effort both to develop ICBMs and to modernize the airplanes that carried nuclear bombs. After the launch of the Discoverer spacecraft, the number of Soviet first-generation ICBMs was found to have been only around twenty. By exposing the true number of first-generation Soviet ICBMs, Discoverer saved millions of dollars on further U.S. military expenditures. It is possible that if Premier Nikita Khrushchev had not pushed the United States into the development of ICBMs with his constant hints that the Soviets would without hesitation use them on unfriendly countries, the United States might not have manufactured its own weapons for ten years, and the launching of reconnaissance spacecraft could well have been delayed by at least five years.

The significance of spy satellites to the United States during the Cold War can be gauged by the amount of money that was spent on the program. In March, 1967, President Lyndon B. Johnson told a small group of people that the total space-program expenditure up to that time was between $35 and $40 billion. At that time, the space program was less than twenty years old. He concluded that if nothing else had been achieved, the photographs from the spy satellites would have made the program worthwhile.

Since the beginning of the U.S. reconnaissance program, great advances have been made in photography. This photographic technology has been used in other programs, most notably in the space telescope, the Galileo spacecraft, and the probes sent to view the Moon and Mars. The excellent resolution of KH spacecraft cameras permits the identification of missiles and hence provides a means of verifying compliance with arms control treaties. In a political climate of mutual mistrust between the United States and the Soviet Union, spy satellites were relied on heavily to ensure that treaties are honored.

During the early years of the space program, the Soviet Union made strong condemnations of the intrusion into their air space of American satellites and threatened to shoot them down. The Soviets' rhetoric softened, however, after they began regular launches of their own reconnaissance spacecraft.

By the late 1980's, the United States had had one accident involving the reentry of a space satellite, while the Soviet Union had had two major mishaps in connection with spy satellites. In 1978, an area around Great Slave Lake in the Northwest Territories was littered with highly radioactive waste. Five years later, a nuclear reactor fell into the Indian Ocean.

Bibliography

Burrows, William E. *Deep Black: Space Espionage and National Security*. New York: Random House, 1986. This carefully researched and well-written book documents the progression of the American reconnaissance satellite from the early years until the mid-1980's, presenting its strengths and weaknesses. Contains photographs and a list of references.

Jasani, Bhupendra. *Space Weapons: The Arms Control Dilemma*. London: Taylor and Francis, 1984. Features a chapter on reconnaissance satellites and antisatellite (ASAT) weapons. A bibliography is included.

Klass, Philip J. *Secret Sentries in Space*. New York: Random House, 1971. Klass has written one of the best books available on early spy satellites. He discusses their significance and use, covering reconnaissance spacecraft from their inception up to the ill-fated Manned Orbiting Laboratory. Klass also includes a chapter on the Soviet effort. Contains photographs.

Richelson, Jeffrey. *American Espionage and the Soviet Target*. New York: William Morrow and Co., 1987. The book gives a detailed view of how American espionage functions. There is a chapter on the Keyhole program, showing how the different methods of gathering information fit together. A large bibliography, a section on acronyms, and photographs are included.

————. "The Keyhole Satellite Program." *Journal of Strategic Studies* 7 (June, 1984): 121-153. This essay is probably the most complete analysis of the Keyhole program. Includes, among other things, sections on sensors and resolution. All satellites from CORONA to KH-11 are mentioned, and their contributions to the reconnaissance program are listed. Bibliography included.

Taylor, John W. R., ed. *Jane's All the World's Aircraft, 1978 – 1979*. London: Jane's Publishing Co., 1978. Contains a small section on the Big Bird satellite, giving dimensions and orbit specification. Directly below are notes on the Titan rocket, which was used to launch KH-9.

John Newman

THE STRATEGIC DEFENSE INITIATIVE

Date: Beginning March 23, 1983
Type of program: Unmanned military defense

The United States' Strategic Defense Initiative and similar efforts in the former Soviet Union were multiphased programs designed to counter nuclear missile attacks. Both nations' projects involved the use of high-technology, space- and Earth-based, directed energy weapons to defeat intercontinental ballistic missiles.

PRINCIPAL PERSONAGES

RONALD REAGAN, fortieth President of the United
States

EDWARD TELLER, Associate Director Emeritus,
Lawrence Livermore National Laboratory

WILLIAM R. GRAHAM, JR., Director, Office of Science
and Technology Policy

JAMES A. ABRAHAMSON, Director, Strategic Defense
Initiative Organization

GEORGE A. KEYWORTH, former Science Adviser to
the president

BILL CLINTON, forty-second President
of the United States

Summary of the Program

The Strategic Defense Initiative (SDI), as first defined by U.S. president Ronald Reagan in 1983, was a research program designed to create an effective defense against nuclear missile attacks. SDI, or "Star Wars," as it was popularly termed, has been the subject of immense controversy in scientific and political circles.

On March 23, 1983, President Reagan used a nationally televised speech to announce a major research effort to discover ways to protect the United States from a strategic nuclear missile attack by the Soviet Union. The President stated his hope that technology developed through SDI could be used to make missile-delivered nuclear weapons obsolete in the twenty-first century. Achieving this end would require several advancements in existing technology and breakthroughs in hardware applications.

At the time of the president's announcement, both the superpowers were depending on the principle of Mutual Assured Destruction (MAD) to prevent nuclear war. MAD is predicated on the belief that, since both the United States and Soviet Union possessed enough nuclear weapons (six thousand for the United States, twelve thousand for the Soviet

Union), to destroy each other several times over, and since neither nation could reasonably expect to survive a nuclear exchange, neither would be willing to risk its own destruction to defeat the other.

Strategic defense is based on the premise that high technology will allow the building of a defensive system to bear the brunt of a first-strike nuclear attack so that the defender nation would then be able to use its own weapons to counterattack. The threat of such a counterattack would, according to the theory, be an adequate deterrent to a first strike.

For more than thirty years prior to the strategic defense program, both the United States and the Soviet Union had worked to find ways to defend against a nuclear missile attack. These efforts, known as BMD, for Ballistic Missile Defense, provided impetus for the Anti-Ballistic Missile Treaty signed by President Richard Nixon and Soviet premier Leonid Brezhnev in 1972. The ABM Treaty allowed the two nations to build two anti-ballistic missile sites of no more than one hundred missiles each and to continue research into BMD. That research led to some of the major developments in SDI.

The Strategic Defense Initiative program was to be deployed in three stages beginning in the 1990's and continuing through about 2115. It consisted largely of two types of technology: kinetic energy devices and lasers. When fully deployed, SDI was designed to use these weapons systems and a complex array of sensors, relay satellites, and battle management computers to construct ground-based and Earth-orbiting "screens" to "filter out" Intercontinental Ballistic Missiles (ICBMs) and their multiple nuclear warheads, known as reentry vehicles, before detonation.

Kinetic energy devices are nonexplosive projectiles that, when launched at an object, rely on their momentum, or kinetic energy, to destroy the target. Two types of kinetic energy weapon, Space-Based Interceptors (SBIs) and the ground-launched Exoatmospheric Reentry Interceptor System (ERIS), were to be used in SDI. SBIs are weapons platforms to be placed in Earth orbit that would fire kinetic energy projectiles at just-launched ICBMs, destroying the missiles before the release of their multiple reentry vehicle payloads. ERIS interceptors launched from the ground would destroy reentry vehicles outside Earth's atmosphere after they have been released from an ICBM. High endoatmospheric interceptors would be employed at later stages to destroy warheads after they entered the atmosphere but before they reached their targets and could inflict significant damage.

These kinetic energy weapons would be used in SDI's first phase of deployment, with refinements being put in place as development of the technology permitted. In 1984, the United States successfully conducted the first test of a ground-launched interceptor designed to destroy an ICBM, and 1995 was a target date for the first phases of a ground-based kinetic energy defense system.

The second weapon to be used in SDI was the ground-based or space-based laser. Any or all of several types of laser, possibly including free electron lasers, nuclear powered X-ray lasers, chemical lasers, and laser guided particle beams, would be fired from bases on Earth at reflecting mirrors in Earth orbit that would reflect and enhance the beams, directing them toward incoming ICBMs or reentry vehicles. An experiment in which a harmless laser beam was aimed at the space shuttle *Discovery* was successfully completed in 1985.

Research was also being conducted into the possible use of particle beams as defensive weapons against nuclear missiles. Although this plan had numerous technological drawbacks, some scientists believed it would be possible to use beams of highly charged hydrogen protons to destroy an ICBM in flight.

It was also thought that orbiting platforms equipped with particle beam or laser weaponry for an advanced BMD program would be feasible at some point, perhaps as late as 2115.

In addition to the weapons systems, SDI required a highly sophisticated command and communications system that would use massive satellites equipped with Earth-orbiting heat and radar sensors, ground-based sensors, aircraft, and specially launched sensor rockets. This complex network would detect, analyze, and relay information on incoming ICBMs to battle command centers and then relay and execute the commands of military and civilian leaders. So that those leaders could effectively recognize and react to a potential nuclear threat, command of the weapons systems and the sensor network would rely on a blend of computerized analysis and human judgment.

As originally conceived by the Reagan Administration, SDI served at least two major military and diplomatic purposes. First, SDI was a defensive system designed to make the cost of a nuclear attack by either nation against the other prohibitive and its success uncertain, and second, from the United States' viewpoint, SDI gave the Soviet Union an added impetus to negotiate further arms control and disarmament agreements on every class of nuclear weapon. How successful SDI would be in bringing about either objective was a matter of great controversy in the American and international political arenas.

Critics of SDI, who called the program "Star Wars," believe the concept to be too heavily dependent on uncertain technology to be fully or even reasonably effective. Since a full-scale missile attack might involve hundreds of missiles, thousands of

reentry vehicles, and tens of thousands of decoy reentry vehicles launched simultaneously at hundreds of targets, there is considerable doubt whether a tremendously complex system such as SDI would work in the first minutes of an attack.

Some experts were concerned that, if the technology were deployed and was even partially successful, it might have either provoked the Soviet Union into launching a preemptive nuclear strike or create the dangerous and false assumption among American leaders that a nuclear war could be survived and even won. Many European nations would have raised the question of how SDI affected the balance of power between the United States and the Soviet Union, since the Soviet Union had a large numerical superiority in conventional military forces.

Opponents of the program also pointed out that space-based weapons or weapons-support systems are highly vulnerable to attack by a variety of different methods. This weakness, the critics argued, could render SDI useless and leave the United States more vulnerable to nuclear attack. Satellites in space can be destroyed by the detonation of a nuclear bomb above the atmosphere. They can be attacked by specially designed hunter-killer satellites that can be placed in orbit near the original satellites and used at the first sign of hostilities. Moreover, the same laser or particle beam technology that can be used to destroy an ICBM would be equally effective against satellites in Earth orbit.

The vulnerability of space-based weaponry led SDI scientists and designers to include satellite defense and antisatellite weapons systems in the SDI blueprints. These systems, in turn, made the whole SDI system more complex and, its critics suggested, more unreliable, because of the greater potential for technical problems and hardware failure.

Proponents of SDI, however, believed that SDI technology would enable the United States and the Soviet Union to control massive defense spending and that it would reduce the threat of nuclear war. They believed SDI to be more rational and moral than mutual assured destruction as a nuclear strategy for the twenty-first century.

With the election of Bill Clinton to the presidency, SDI was essentially cancelled following the collapse of the Soviet Union. The Ballistic Missile Defense Organization, a scaled-back successor to SDI, is prohibited by treaty to practice its methods on real missiles and has rechanneled its resources into such research projects as Clementine.

Knowledge Gained

The knowledge gained from SDI falls into two categories: scientific and political. As a scientific endeavor, SDI significantly increased the funding and resources available to researchers working on directed energy (lasers and neutral and charged particle beams) and on satellite defense, communications, and support systems. Ways were found to create free electron lasers and nuclear powered X-ray lasers, to guide a charged particle beam with a laser, and to accelerate neutrons into neutral particle beams, all of which have potential commercial and civilian applications.

Advances in kinetic weaponry can help reduce the threat of nuclear war by making the effect and success of a first strike more uncertain. That, in itself, made SDI an important part of the defensive structure of both the United States and the Soviet Union.

"Star Wars" also changed the scope of political debate about the arms race. The Reagan Administration used it as a bargaining chip; some observers claim that it helped bring about the 1988 Intermediate Nuclear Forces (INF) treaty, an agreement between the two superpowers to destroy a whole class of intermediate range nuclear missiles. SDI is also credited with helping to advance talks on the reduction of other classes of nuclear missiles.

SDI, by virtue of its high visibility and President Reagan's open and vocal support of the program, also helped to increase public awareness of the arms race. Americans gained a greater understanding of how, and how well, the nation is prepared to defend itself in the event of a nuclear war. This understanding, and the sentiment it generated, may help to prevent a nuclear attack as effectively as the technology behind the SDI program.

Context

The Strategic Defense Initiative was spurred by the desire to reduce the threat of nuclear war. When President Reagan proposed SDI, ballistic missile defense technology, particularly surface-to-air anti-ballistic missiles, had been under development for many years. BMD was the impetus for the ABM Treaty of 1972 between the Soviet Union and the United States. This treaty, in turn, created the avenue for BMD research that led to President Reagan's announcement of the SDI program.

In a broader sense, SDI was an attempt to move the United States away from its reliance on mutual assured destruction — a primarily offensive strategy — and toward the defensive posture of SDI, which would ultimately eliminate the need for a first-strike capability.

From a political perspective, SDI was initiated in the context of a massive increase in defense spending on the part of both the United States and the Soviet Union. Some experts believed that SDI, as an effort to diminish the effect of nuclear weapons, would allow the two nations with the most nuclear weapons in their arsenals to reduce their stockpiles and divert the funds for those weapons systems to other, productive uses.

Bibliography

Adragna, Steven P. *On Guard for Victory: Military Doctrine and Ballistic Missile Defense in the U.S.S.R. Foreign Policy Report.* Cambridge, Mass.: Institute for Foreign Policy Analysis, 1987. An overview of Soviet military policies relating to the Soviet ballistic missile defense program. Written from a conservative viewpoint, but very informative.

Broad, William J., et al. *Claiming the Heavens: The New York Times Complete Guide to the Star Wars Debate.* New York: Times Books, 1988. A compilation of an exhaustive series of articles on SDI that appeared in *The New York Times*, this resource is essential reading for the beginner interested in learning the basics before going on to more complex aspects of the SDI controversy.

Codevilla, Angelo. *While Others Build: The Common Sense Approach to the Strategic Defense Initiative.* New York: Free Press, 1987. Codevilla, a noted expert on SDI, provides a strong, easy-to-read overview of the program and a critical look at its advantages and disadvantages from both a political and a technological viewpoint. The subject is presented mostly from a proponent's perspective, but the text is filled with interesting information.

Davis, Jacquelyn K., and Robert L. Pfaltzgraff. *Strategic Defense and Extended Deterrence: A New Transatlantic Debate.* Cambridge, Mass.: Institute for Foreign Policy Analysis, 1986. This booklet examines the debate among the United States' allies over the benefits and drawbacks of SDI from an international political perspective. It also examines the Soviet strategic defense program and the European defense structure. A scholarly work, yet understandable to the nonspecialist.

Godson, Dean. *SDI: Has America Told Her Story to the World?* Cambridge, Mass.: Institute for Foreign Policy Analysis, 1987. This brief treatise looks at how the United States presented the arguments for SDI to the world community, particularly its European allies. Somewhat scholarly and technical, the work nevertheless provides a good geopolitical overview of a complex question.

Mikheyev, Dmitry. *The Soviet Perspective on the Strategic Defense Initiative: Foreign Policy Report.* Elmsford, N.Y.: Pergamon Press, 1987. This book, written by a former Soviet physicist, looks at the Soviet government's views on the SDI program. Provides interesting insights into Soviet military and political policies regarding strategic defense. Written from a conservative perspective.

U.S. Congress. Office of Technology Assessment. *SDI: Technology, Survivability and Software.* Princeton, N.J.: Princeton University Press, 1988. This is an unclassified version of a classified report prepared by the Office of Technology Assessment on SDI progress and feasibility as of 1985. Although somewhat technical in nature, the report gives a clear picture of "Star Wars" technology and reviews some of the arguments surrounding SDI.

Eric Christensen

The Stratospheric Aerosol and Gas Experiment

Date: Beginning February 18, 1979
Type of satellite: Scientific

The Stratospheric Aerosol and Gas Experiment (SAGE) instrument was designed to measure the concentration of some of the constituents of Earth's atmosphere. The SAGE satellite was one of the first remote-sensing satellites to provide estimates of the global distribution of stratospheric aerosols, ozone, nitrogen dioxide, and water vapor.

Principal personages

M. Patrick McCormick, the principal investigator and coordinator for SAGE 1 and SAGE 2, Langley Research Center

William P. Chu, a researcher, Langley Research Center

Theodore J. Pepin, a researcher, University of Wyoming

Derek M. Cunnold, a researcher, Georgia Institute of Technology

Royce Lane, a researcher, University of Wyoming

Phillip B. Russell, a researcher, Ames Research Center

Gerald W. Grams, a researcher, Georgia Institute of Technology

Summary of the Satellites

The SAGE class of instruments was developed by the National Aeronautics and Space Administration (NASA) primarily through the efforts of M. Patrick McCormick of the Atmospheric Sciences Division at NASA's Langley Research Center in Virginia. The prototype to the SAGE experiments was conceived and built by Theodore J. Pepin of the University of Wyoming. This prototype was a small, manually operated instrument flown on the Apollo-Soyuz Test Project in 1975; it consisted of a handheld package containing a telescope, a Sun sensor, and an external electronics package. Its purpose was to measure the extinction of sunlight caused by aerosols in the atmosphere. The instrument was named the Stratospheric Aerosol Measurement (SAM) instrument and was operated by an astronaut as he pointed it directly at the Sun during a sunrise or sunset aboard a spacecraft.

It was shown that aerosols, which are small solid or liquid particles on the order of a micrometer in diameter, could be accurately measured in the atmosphere. After the successful results of SAM were revealed, SAM 2 was designed by NASA engineers and built at the University of Wyoming. SAM 2 was a much more sophisticated instrument than SAM and was built to fly aboard the unmanned Nimbus 7 spacecraft. SAM 2 was to measure atmospheric aerosols in the polar regions of Earth between the latitudes of 64 to 84 degrees south and 64 to 84 degrees north. SAM 2 began operation in November, 1978, and continues to operate long beyond its initial design lifetime.

The SAGE 1 and 2 instruments were even more sophisticated than the SAM 2 instrument. Their primary objectives were to determine the spatial distribution of not only stratospheric aerosols but also ozone, nitrogen dioxide, and water vapor on a global scale. The principle of operation of the SAGE instruments is identical to that of the SAM instruments. On a spacecraft that orbits approximately 600 kilometers above Earth, SAGE instruments receive sunlight that is focused onto a set of detectors. Depending on the point in the orbit, this sunlight enters the instrument unobstructed (full-Sun condition), enters the instrument partially obstructed by Earth's atmosphere (occultation), or does not enter the instruments because of Earth's being in the path between the instrument and the Sun (total occultation). Scientists call the sequence just described an "event."

The SAGE instrument makes measurements of solar irradiance during the full-Sun condition in order to obtain a reference value for the intensity of the Sun as it is viewed unobstructed by the atmosphere. The actual atmospheric measurement occurs during the time that the Sun's disk appears through the atmosphere. During this portion of the event, the instrument focuses on a very small area of the solar surface with a telescope. Simultaneously, a mirror sweeps the image of the Sun across the field of the telescope. The actual spatial extent of the solar disk as viewed from the instruments varies in size depending upon how high the Sun appears above Earth. As the Sun is viewed in the higher layers of the atmosphere, the vertical extent of the Sun is approximately 32 kilometers. As the solar disk appears to move lower into Earth's atmosphere, the disk becomes flattened because of the refraction by Earth's atmosphere. The entire event lasts approximately one minute.

Events are classified as being either sunrises or sunsets. Sunrise events are observed as the instrument orbits out from behind Earth and into the full sunlight, while sunset events are observed as the instrument orbits out of the full-Sun condition and into Earth's shadow. The small region on the surface of the Sun observed by the SAGE instrument during an event corresponds to a region in Earth's atmosphere that is approximately 1 kilometer high, 1 kilometer wide, and 200 kilometers long. This 200-kilometer-long volume of atmosphere lies along the line of sight joining the instrument and the Sun and, at any time during an event, is within 1 kilometer of a fixed altitude above the surface of Earth.

At the heart of the SAGE instruments is a grating, a flat plate on which are ruled closely spaced lines. This grating splits the incident sunlight into a rainbow of colors. Each color corresponds to a particular wavelength of light emitted from the Sun. The SAGE instruments measure the intensity of several of these wavelengths of light. As the sunlight is occulted by the atmosphere, each individual wavelength is diminished in intensity more or less independently of the other wavelengths being measured. The measure of intensity that a particular wavelength loses because of the intervening atmosphere is directly related to the type of aerosol or gas present in the atmosphere.

The presence of aerosols and gases in the atmosphere causes incident light to scatter or be absorbed. When a light ray is scattered, it is simply redirected from its incident direction into another direction. When light is absorbed, it is taken up by the molecules in a gas or aerosol and converted into energy. The conversion process may change the energy into molecular rotation or vibration, kinetic energy, or light at wavelengths different from the wavelength of the incident light. The processes of scattering and absorption are also called extinction. Very little absorption occurs for aerosols at the wavelengths where SAGE operates.

Basic principles of optics, atmospheric physics, and orbital mechanics are used to convert the relative instrumental quantities measured by the SAGE instruments into the absolute quantities of extinction and gas concentration. The information provided for experimenters by the SAGE 1 and SAGE 2 instruments comes in the form of vertical profiles of aerosol extinction and gas concentration. Each profile contains approximately sixty measurements, one measurement per kilometer, over an altitude range from cloud top to the top of the stratosphere. During any given day, fifteen sunrise and fifteen sunset measurements are made at equally spaced intervals around the world. Because the instruments take measurements at a local sunrise or sunset, the set of measurement profiles represent atmospheric conditions at the terminator, or the transition region on Earth that separates night and day.

On February 18, 1979, SAGE 1 was launched by NASA on the Applications Explorer Mission 2 (AEM 2). It operated successfully until November 18, 1981, when a spacecraft power problem caused the instrument to stop functioning. The mission provided more than thirteen thousand events from which atmospheric profiles of aerosol extinction, ozone, and nitrogen dioxide were retrieved. The latitudes over which

measurements were made ranged from about 72 degrees south to 72 degrees north.

The SAGE 2 instrument was launched on October 5, 1984, on the Earth Radiation Budget Satellite (ERBS). Along with SAGE 2, the ERBS platform contains the Earth Radiation Budget Experiment. The ERBS is a key part of NASA's climate program. The SAGE 2 instrument makes atmospheric measurements from approximately 80 degrees south to 80 degrees north, depending on the season. It has recorded more than thirty-four thousand events from which scientists can retrieve profiles of extinction and gas concentration.

Knowledge Gained

The long-term record of SAM 2 and SAGE 2 aerosol extinction data reveals periodic phenomena that occur in the Southern Hemisphere. The summertime aerosol extinction is relatively constant. During fall and winter, however, the temperatures decrease, and a large polar low pressure system with swirling winds at its boundaries, called the polar vortex, establishes itself over the Antarctic. When the stratospheric temperatures reach their lowest values in the year, polar stratospheric clouds form inside the vortex and persist throughout the winter and early spring. Such clouds are always detected by SAM 2 and SAGE 2 aerosol extinction sensors during the winter and early spring. A high degree of variability in the set of daily aerosol extinction values is recorded as SAM 2 measures inside and outside the vortex. This high variability is also observed in the SAGE 2 ozone, nitrogen dioxide, and water vapor measurements taken at this time. With the onset of springtime and the consequent evaporation of polar stratospheric clouds, the polar stratospheric aerosol extinction measurements reach very low values relative to the rest of the year. Furthermore, as the polar vortex is replaced by warmer springtime air, the values of aerosol extinction increase to levels more typical of the summertime.

During the periods immediately after the eruptions of volcanoes, the aerosol extinction data of SAGE 1, SAGE 2, and SAM 2 show high values of aerosol. Volcanic eruptions typically place large volumes of gas and dust high into the atmosphere. The newly injected material perturbs the background aerosol distribution, which is normally hemispherically symmetric with a maximum above the equatorial latitudes. For some months after the volcanic injections, the aerosol extinction values continue to rise and eventually reach a maximum. This phenomenon is attributed to the conversion of gas to particles. That is, the large gas quantity emitted by volcanic eruption is converted to aerosols.

The SAGE 1 and 2 data reveal that the distribution of global ozone concentration is hemispherically symmetric with a maximum above the equator and minima near the polar latitudes. The temporal variation of the ozone in the midlatitudes is characterized by a strong annual cycle. Below 25 kilometers at these latitudes the ozone attains its maximum value of the year in summer, while between 25 and 37 kilometers the ozone attains its maximum value in winter. The temporal variation of the ozone at the equator is dominated by a semiannual cycle. Below approximately 60 kilometers there is no significant difference between sunrise ozone and sunset ozone. Above this altitude, the sunrise ozone is greater than the sunset ozone because of photochemical reactions. These types of reactions are controlled by the presence of sunlight. The chemical bonds of certain gases in the atmosphere break apart when light is absorbed.

The SAGE data reveal that the distribution of global nitrogen dioxide is essentially hemispherically symmetric. The temporal variation of the nitrogen dioxide in the midlatitudes is characterized by a strong annual cycle. Over the altitudes from 20 to 45 kilometers at these latitudes, the maximum value of the year occurs in summer. Furthermore, the sunset nitrogen dioxide concentration is always greater than that for sunrise because of photochemical reactions.

Because the occultation method of measuring atmospheric constituents is dependent on sensing the solar disk as it rises or sets in the atmosphere, any cloud that might be along the line of sight from the satellite to the Sun affects the measurement. Thick clouds, such as cumulonimbus, tend to block the sunlight completely, whereas thin clouds, such as cirrus, may only partially obscure the sunlight. Thus, the global distribution of the frequency of occurrence of clouds and cloud altitudes may also be estimated from the SAM 2, SAGE 1, and SAGE 2 data.

Context

A phenomenon called the ozone hole has been observed during the Antarctic springtime for more than a decade. The ozone hole is a decrease in the abundance of Antarctic ozone as sunlight returns to the pole in early springtime. Minimum monthly values of ozone concentration are observed in October. The ozone hole occurs inside the polar vortex and is believed to be related to chemical reactions that take place on the surface of polar stratospheric clouds. Man's industrial activities have caused chlorine compounds to be released into the atmosphere. The unique atmospheric conditions that exist inside the polar vortex during the Antarctic springtime cause the chlorine from these compounds to be released in a reactive form that destroys ozone. The SAGE 2 measurements have been shown to agree with ozone measurements taken by balloon-borne instruments in the same region. The large gradients in ozone concentration as one moves from outside the vortex to inside the vortex are recorded in the SAGE 2 data.

The SAGE instruments measured ozone in the midlatitudes more frequently than near the polar regions. In 1985, a report, based on satellite data, was presented to the United States Congress stating that large global ozone decreases had occurred since 1978. The satellite data set discussed was that of the Solar Backscattered Ultraviolet (SBUV) instrument aboard Nimbus 7. It indicated that ozone had decreased by approximately 20 percent at 50 kilometers since 1978. An independent investigation of the SAGE ozone data sets over the same period revealed no trend at 50 kilometers and a small decrease (-3 percent) at 40 kilometers. The trend calculated from the SAGE data sets agreed more closely with theoretical predictions of ozone and temperature change than the trend computed from the SBUV data.

It is important to know the magnitude of any long-term decrease in the amount of ozone in the stratosphere because of the potential damage to life on Earth. The stratosphere contains most of the ozone in the atmosphere, with approximately 10 percent of the atmospheric ozone contained in the troposphere (even though man's activities generate ozone). Loss of ozone means that more solar ultraviolet light would penetrate the atmosphere to Earth's surface. Ultraviolet light is absorbed by stratospheric ozone and prevented from penetrating to the surface. The damage to humans, animals, and vegetation by unshielded solar ultraviolet light is potentially large.

The global distribution of aerosols mapped from the SAGE data shows that during volcanically active periods, gas and particles from eruptions are transported to different parts of the atmosphere by the general circulation. The data from the four aerosol extinction wavelengths measured by SAGE 2 have shown that it is possible to discriminate between the particle sizes injected into the atmosphere by a volcano. The aerosols emitted by volcanoes also act as tracers for the atmospheric circulation. In turn, researchers can further understand the dynamic motion of Earth's atmosphere.

The SAGE 1, SAGE 2, and SAM 2 satellite instruments have served as important tools in the understanding of Earth's atmosphere. Using their data, scientists have been able to map aerosol and ozone concentrations on a time scale shorter than major stratospheric changes; to locate stratospheric aerosol, ozone, nitrogen dioxide, and water vapor sources and sinks; to estimate long-term global ozone trends; to study the Antarctic ozone hole; to monitor circulation and transport phenomena; to observe hemispheric differences; and to investigate the optical properties of aerosols and assess their effects on global climate.

Bibliography

Anthes, Richard A., Hans A. Panofsky, John J. Cahir, and Albert Rango. *The Atmosphere.* 2d ed. Columbus, Ohio: Charles E. Merrill, 1978. A nonmathematical introductory book on meteorology. Many illustrations and photographs show examples of tropospheric phenomena. Suitable for a general audience.

Craig, Richard A. *The Upper Atmosphere: Meteorology and Physics*. New York: Academic Press, 1965. Most of chapters 2 and 3 of this work are nonmathematical in nature and deal exclusively with the stratosphere. The emphasis is on historical discoveries and observed phenomena. Many charts and graphs illustrate the subject matter.

Scorer, Richard S. *Cloud Investigation by Satellite*. New York: Halstead Press, 1986. Provides six hundred satellite images of cloud structures of the Northern Hemisphere. Written for a general audience.

Wallace, John M., and Peter V. Hobbs. *Atmospheric Sciences: An Introductory Survey*. New York: Academic Press, 1977. An excellent introduction to physical and dynamical meteorology, this college-level book assumes a first-year calculus and chemistry background. The authors alternate the subject matter in the chapters between dynamical meteorology and physical meteorology.

Watson, R. T., et al. *Present State of Knowledge of the Upper Atmosphere, 1988: An Assessment Report*. Reference publication 1208. Washington, D.C.: National Aeronautics and Space Administration, 1988. A publication geared toward those interested in the efforts of scientists to assess the ozone trend in 1987. All aspects of ozone measurement and trend estimation are discussed. Topics include satellite measurements, ground-based measurements, instrument calibration, the Antarctic ozone hole, stratospheric temperature, trace gases, chemical modeling, and aerosol distributions and abundances.

Young, Louise B. *Earth's Aura*. New York: Avon Books, 1979. Chapter 6 gives an interesting description of the ozone layer and the controversy concerning its destruction by man's activities. A historical account leading up to regulations on the use of certain chlorine compounds is given.

Robert Veiga

THE SURVEYOR PROGRAM

Date: May 30, 1966, to February 21, 1968
Type of program: Unmanned lunar probes

Surveyor 1 achieved the first lunar soft landing by a fully automated spacecraft. The Surveyor program developed the technology for soft-landing on the Moon, verified the compatibility of the design of the Apollo manned lunar-landing spacecraft with actual lunar surface conditions, and added to the scientific knowledge of the Moon.

PRINCIPAL PERSONAGES

> BENJAMIN MILWITZKY, Surveyor Program Manager at
>   NASA Headquarters, Office of Lunar and
>   Planetary Programs
> WALKER E. GIBERSON, Surveyor Project Manager at
>   JPL from its inception to 1965
> ROBERT PARKS, Surveyor Project Manager at JPL
>   from 1965 to 1966
> HOWARD H. HAGLUND, Surveyor Project Manager at
>   JPL from 1966 to 1968
> R. G. FORNEY, Spacecraft Systems Manager at JPL
> LEONARD JAFFE, Chairman, Surveyor Scientific
>   Evaluation Advisory Team
> EUGENE M. SHOEMAKER, Principal Investigator,
>   imaging
> A. L. TURKEVICH, Principal Investigator,
>   alpha scattering
> RONALD F. SCOTT, Principal Investigator,
>   soil mechanics surface sampler

Summary of the Program

In May, 1960, the National Aeronautics and Space Administration (NASA) approved a Surveyor program consisting of two parts: a lunar orbiter for photographic coverage of the Moon's surface and a lunar lander to obtain scientific information on the Moon's environment and structure. Before human beings could safely be sent to the Moon, the Surveyors were to provide spacecraft designers with information on the load-bearing limits of the lunar surface, its magnetic properties, and its radar and thermal reflectivity.

The Jet Propulsion Laboratory (JPL) was assigned project responsibility, and four Surveyor study contracts were awarded in July, 1960, to Hughes Aircraft, North American, Space Technology Laboratories, and McDonnell Aircraft. On January 19, 1961, NASA chose Hughes Aircraft's proposal for the Surveyor and began planning at JPL for seven lunar land-ing flights, the first of which was planned for launch on an Atlas-Centaur booster in 1963.

Because of development problems with the Centaur, early failures of the Ranger lunar impactor, and increasing demands for information on the lunar surface to support the Apollo program, the orbiter portion of Surveyor was dropped in 1962 and replaced by the Lunar Orbiter project, managed by NASA's Langley Research Center. Problems with the Centaur upper stage forced postponement of the first Surveyor launch and required a reduction in the spacecraft's weight — from an original 1,134 kilograms with a 156-kilogram payload to 953 kilograms carrying only 52 kilograms of instruments.

The Atlas-Centaur became operational in 1966, and Surveyor 1 was launched from Launch Complex 36A at Cape Kennedy, Florida, on May 30, 1966. Surveyor 1, the first test model of the series, carried more than one hundred engineering sensors to monitor spacecraft performance. No instrumentation was carried specifically for scientific experiments, but the spacecraft was outfitted with a survey television system and with instrumentation to measure the bearing strength, temperature, and radar reflectivity of the lunar surface.

After injection on a trajectory intersecting the Moon, the Surveyor spacecraft separated from the Centaur upper stage. Midcourse maneuvers, using vernier engines on the spacecraft, were performed to bring it within the desired target area. For the terminal descent, the main retro-engine was ignited, by command of an on-board radar altimeter, to provide most of the braking. After this retro-engine burned out, at about 10 kilometers above the lunar surface, it was jettisoned. A second radar, providing measurements of velocity and altitude, was used with the smaller vernier engines, in a closed loop under control of an on-board analog computer, for the final descent phase. To reduce the disturbance to the lunar surface, the vernier engines were extinguished by the computer when the spacecraft was about 4 meters above the surface and the descent rate was about 1.5 meters per second. At 0617 Greenwich mean time on June 2, 1966, Surveyor 1 soft-landed in the southwest portion of the Ocean of Storms, becoming the first U.S. spacecraft to soft-land on another celestial body.

During the next two lunar days (28 Earth days each) the

Surveyor returned some 11,000 photographs of the surrounding terrain to Earth. Surveyor 1 completed its primary mission on July 14, 1966, but engineering interrogations were conducted at irregular intervals through January, 1967.

Surveyor 2, launched on September 20, 1966, had essentially the same configuration as Surveyor 1 and was intended to land in the Central Bay, another area of the potential Apollo landing zone. When the midcourse maneuver was attempted on September 21, one of the three vernier engines failed to ignite, and the unbalanced thrust caused the spacecraft to tumble. Although repeated commands were sent in an attempt to salvage the mission, Surveyor 2 crashed on the Moon on September 22, 1966.

Surveyor 3, though similar to the two earlier spacecraft, was equipped with two fixed mirrors to extend the view of its television camera underneath the spacecraft. A remotely controlled surface sampler arm, capable of digging trenches and manipulating the surface material in view of the television camera, was also added. The Surveyor 3 spacecraft was launched on April 17, 1967, from Launch Complex 36B at Cape Kennedy — the first time the two-burn capability of the Centaur was used on an operational mission. After separation from the Atlas, the Centaur engine ignited and burned for approximately 5 minutes to place the vehicle into a 167-kilometer circular parking orbit. The Centaur coasted for 22 minutes, then reignited to place the spacecraft on a lunar-intercept trajectory. The use of a parking orbit greatly increased the fraction of the lunar surface to which the Surveyor could be targeted.

The midcourse correction maneuver and the firing and jettisoning of the retro-engine proceeded as planned. A few seconds before touchdown, however, the onboard radar lost its lock on the surface, apparently because of unexpected reflections from large rocks near the landing site. As a result, the spacecraft guidance system switched to an inertial mode, which prevented the vernier engines from extinguishing about 4 meters above the surface as planned. The spacecraft touched down with its vernier engines still firing, lifted off, touched down a second time, lifted off again, then touched down for a third time after receiving a command from Earth 34 seconds after the initial touchdown. The spacecraft had a lateral velocity of about 1 meter per second at the first touchdown, and the distance between the first and second touchdowns was about 20 meters, while there was a distance of about 11 meters between the second and third touchdowns.

The landing occurred on April 19, 1967, in the southeast part of the Ocean of Storms, a potential Apollo landing region. Surveyor 3 landed in a medium-sized crater and came to rest tilted at an angle of about 14 degrees. This location allowed the crater to be viewed from the inside, and the unplanned tilt permitted the camera to aim high enough to photograph an eclipse of the Sun by Earth, which would not have been possible if the landing had been on a level surface. The spacecraft returned 6,315 television pictures and operat-

ed its surface sampler for more than eighteen hours before transmissions ceased shortly after local sunset on May 3, 1967.

Surveyor 4, carrying the same payload as Surveyor 3 had, was launched on July 14, 1967. After a flawless flight to the Moon, radio signals from the spacecraft ceased abruptly during the final descent, approximately 2.5 minutes before touchdown and only 2 seconds before retro-engine burnout. Radio contact with the spacecraft was never reestablished, and Surveyor 4 crashed into the lunar surface, possibly after an explosion.

Surveyor 5 was launched from Cape Kennedy on September 8, 1967. Because of a helium regulator leak which developed during flight, a radically new descent technique was engineered, and the Surveyor 5 performed a flawless descent and soft landing in the Sea of Tranquillity on September 11, 1967. The spacecraft was similar to its two immediate predecessors, except that the surface sampler was replaced by an "alpha backscatter instrument," a device to determine the relative abundances of the chemical elements in the lunar surface material. In addition, a bar magnet was attached to one of the footpads to determine if magnetic material was present in the lunar soil.

During its first lunar day, which ended at sunset on September 24, 1967, Surveyor 5 took 18,006 television pictures, performed chemical analyses of the lunar soil, and fired its vernier engines for 0.55 second to determine the effects of high-velocity exhaust gases impinging on the lunar surface. On October 15, 1967, after exposure to the two-week deep freeze of the lunar night, Surveyor 5 responded to a command from Earth, reactivating, and transmitted an additional 1,043 pictures and data from the lunar surface.

The Surveyor 6 spacecraft, essentially identical to Surveyor 5, was launched on November 7, 1967, and landed on the Moon on November 10, 1967. The landing site, near the center of the Moon's visible hemisphere in the Central Bay, was the last of four potential Apollo landing sites designated for investigation by the Surveyor program. From landing until a few hours after lunar sunset on November 24, 1967, the spacecraft transmitted more than 29,000 television pictures, and the alpha backscattering experiment acquired 30 hours of data on the chemical composition of the lunar soil. On November 17, 1967, Surveyor 6's vernier engines were fired for 2.5 seconds, causing the spacecraft to lift off from the lunar surface and move laterally about three meters to a new location. Television pictures showed the effect of rocket firings close to the lunar surface. When combined with images taken from the earlier landing site, the new photographs provided stereoscopic data of the surrounding terrain and surface features.

Since the previous Surveyors had completed the Apollo landing site survey, Surveyor 7 was targeted at the scientifically interesting rock-strewn ejecta blanket of the ray crater Tycho. Launched on January 7, 1968, the Surveyor 7 spacecraft landed less than two kilometers from its target on

Surveyor spacecraft (top); lunar panorama near the crater Tycho (bottom). (NASA)

January 10, 1968. During the first lunar day, 21,046 pictures of the lunar surface were acquired, but the alpha backscatter package failed to deploy. The surface sampler arm was programmed from Earth to force the package into position and subsequently to move it to two additional sites. Laser beams from Earth were also detected by the television camera during a test. Forty-five more photographs and additional surface chemical data were obtained during the second lunar day of operation, before the spacecraft was deactivated on February 21, 1968.

Having met all the Apollo survey objectives, follow-on Block 2 missions were canceled because of budget constraints, and the Surveyor Project Office at JPL was closed on June 28, 1968.

Knowledge Gained

The surface sampler arms on Surveyors 3 and 7 made measurements of how much weight the lunar surface soil could bear before being penetrated as well as how depth affected the bearing capacity and shear strength in trenches up to twenty centimeters deep. The strength and density of individual lunar rocks were also determined. Strain gauges on the shock absorbers determined the loads on the legs of the spacecraft during the touchdown on the lunar surface. The radar reflectivity and dielectric constant of the lunar surface material were determined by the landing radar system. Thermal sensors determined the surface temperatures, thermal inertia, and directional infrared emission at all five landing sites.

At the mare landing sites, firings of the vernier rocket engines on Surveyors 3, 5, and 6 and the attitude control rockets on Surveyors 1 and 6 against the lunar surface provided information on the permeability of the surface to gases, the cohesion of the soil, its response to gas erosion, and its adhesion to spacecraft surfaces.

The alpha backscattering experiments on Surveyors 5, 6, and 7 determined the abundances of the major chemical elements from carbon to iron on six surface samples and one subsurface sample at two mare sites and one highland site. These analyses indicated that the most abundant element of the lunar surface is oxygen (57 atomic percent, or 57 atoms per 100 representative atoms), followed by silicon (20 atomic percent), and aluminum (7 atomic percent). These are the same elements, and in the same order, that are most common in Earth's crust. The major chemical elements at the mare sites are generally similar to those found in terrestrial basaltic rocks. The similarity to basalts, as well as the morphology of the surface features determined from the lunar orbiter images, provides strong circumstantial evidence that some melting and chemical separation of the lunar material had occurred in the past. This surface composition is significantly different from primordial solar system material. It is also different from most

known meteorites, indicating that the Moon is not a major source of the meteorites which hit Earth.

The magnets carried on Surveyors 5, 6, and 7 provided information on the magnetic particles at two mare sites and one highland site. The single highland site showed a lower abundance of iron and other chemically similar elements, demonstrating that the lunar surface is not homogeneous. These chemical differences were suggested as an explanation for the difference in "albedo," or surface reflectivity, between the highland and mare regions.

The five successful Surveyor spacecraft returned 87,700 television pictures of the lunar surface, Earth, the solar corona, Mercury, Venus, Jupiter, and stars to the sixth magnitude. The lunar surface images provided information on the size-frequency distribution of lunar craters ranging from a few centimeters to tens of meters in diameter. The observed distribution was consistent with the distribution predicted to be produced by prolonged bombardment by meteorites and allowed the size distribution of the incoming meteoroids to be inferred.

Two and a half years after its landing, Surveyor 3 was visited by the Apollo 12 astronauts, who returned the video camera, some camera cable, aluminum tubing, and a glass filter to Earth. These Surveyor samples were analyzed to determine the extent of lunar weathering, micrometeorite impacts, and ion bombardment from the Sun.

Context

Although the Soviet Union's Luna 9 spacecraft preceded Surveyor in soft-landing on the Moon, the Surveyors provided the first quantitative data on the bearing strength of the lunar soil, its radar reflectivity, and the variations in the lunar surface temperature. The Surveyors also demonstrated the existence of loose dust on the Moon.

Prior to Surveyor 1, the photographs of the lunar surface having the best resolution were taken by the Luna 9 spacecraft. Surveyor supplemented these with photographs from five additional sites, including one in the lunar highlands. A comparison of the highland surface at the Surveyor 7 landing site with that at the mare sites showed fewer craters larger than eight meters in diameter at the highland site, indicating that the Tycho rim material on which Surveyor 7 landed was much younger than the mare material. This provided a relative chronology for the Moon's different regions.

The computer-controlled soft landings of the Surveyor spacecraft on the Moon verified the soft-landing techniques planned for the Apollo lunar landing missions. Four of the Surveyors landed in potential Apollo landing regions and determined that the lunar surface was sufficiently strong to support the footpad loads planned for the Apollo spacecraft. They also provided information on the surface roughness and topography necessary for the Apollo flights.

Bibliography

French, Bevan M. *The Moon Book*. New York: Penguin Books, 1977. Describes the explanations for the origin of the Earth-Moon system and places the contribution of the Surveyor program into the context of the entire lunar exploration program. Includes a thorough description and illustration on the alpha backscattering experiment and its results. Suitable for general audiences.

Hibbs, Albert R. "The Surface of the Moon." *Scientific American* 216 (March, 1967): 60-74. A well-illustrated article describing the results from the first nine spacecraft, including Surveyor 1, which provided close-up photographs of the lunar surface. Describes the various theories for the origin of the Moon in the context of scientific knowledge as it existed immediately after the landing of Surveyor 1.

Newell, Homer E. "Surveyor: Candid Camera on the Moon." *National Geographic* 130 (October, 1966): 578-592. A comprehensive collection of early Surveyor photographs and a description of the preliminary scientific interpretations of the video data.

Scott, Ronald F. "The Feel of the Moon." *Scientific American* 217 (November, 1967): 34-43. The principal investigator for the soil mechanics surface sampler describes the Surveyor spacecraft, emphasizing the sampler arm and the experiments to determine the strength of the lunar surface. Includes Surveyor images of the lunar surface and diagrams of the spacecraft and the sampler arm.

————. "Report on the Surveyor Project." *Journal of Geophysical Research* 72 (January 15, 1967): 771- 856. In this series of articles, the scientists who participated in the Surveyor program report their original data and interpretations. Contains descriptions of the Surveyor instruments and a detailed account of the scientific results.

————. *Surveyor Program Results*. NASA SP-184. Washington, D.C.: Government Printing Office, 1969. This official program overview includes introductory articles summarizing each Surveyor mission and describing the principal scientific results. These are followed by individual articles on the imaging, surface mechanical and electrical properties, and temperatures experiments, and the chemical analysis. Document also contains a list of the program participants.

George J. Flynn

MARITIME
TELECOMMUNICATIONS SATELLITES

Date: Beginning February 19, 1976
Type of satellite: Communications

A network of satellites designed to upgrade the communications capabilities of commercial and military maritime vessels was first proposed in 1972. In 1979, the International Maritime Satellite Organization was formed to create such a network.

Summary of the Satellites

In 1972 a subgroup of the United Nations, the Intergovernmental Maritime Consultative Organization (IMCO), expressed an interest in using satellites in space for maritime purposes. IMCO represented the concerns of seafaring nations in areas such as distress systems, navigations and position determination, and operation of maritime mobile services. The tremendous growth of the international maritime industry after World War II indicated a serious need for improved communication methods. The American Institute of Shipping echoed this concern when it estimated that by 1980, the number of vessels on the high seas weighing more than 1.5 million kilograms could exceed fourteen thousand at any one time. Vessels were still using the inefficient "brass key" radiotelegraph method for their ship-to-shore communications.

In a report to the U.N. secretariat, IMCO proposed a satellite system that would allow the exchange of telegraph, telephone, and facsimile messages and improve navigation for maritime interests. A panel was formed to study the legal, financial, technical, and operational problems involved in creating an entity that would be responsible for such a system. An international convention was held in April and May of 1975, and delegates and observers from forty-five nations and fifteen international agencies attended. The conference concluded that an international organization was needed to administer a worldwide maritime satellite system, and by the end of the first session, the United States and thirteen other countries had agreed on the major elements of a system that would eventually be known as the International Maritime Satellite Organization (INMARSAT). It would be patterned after the International Telecommunications Satellite Organization (Intelsat), with some modifications for its unique maritime interests. INMARSAT would provide global telecommunications services for maritime commercial activities and safety.

After four years of study and debate, INMARSAT was formally chartered on July 16, 1979, in London. The organization comprised forty-three nation members, including the United States, Great Britain, Norway, Japan, the Soviet Union, and Canada. The system was scheduled to become operational by the early 1980's.

Between 1976 and 1979, maritime communications were enhanced through the U.S. maritime satellite (Marisat) system. The Marisat system consisted of three satellites built by Hughes Aircraft Company for the Communications Satellite Corporation (COMSAT). Launched in 1976, the satellites formed the first maritime communications system in the world. The Marisats provided rapid, high-quality communications between ships at sea and home offices; they greatly improved communications of distress, safety, search and rescue, and weather reports.

All three Marisats were structurally identical. At lift-off they weighed 655 kilograms and were approximately 231 centimeters long with their antennae. A panel containing seven thousand solar cells supplied the craft with primary power. A "Straight-Eight" Delta rocket launched the satellites from Cape Canaveral, Florida, into geostationary orbits, wherein they would travel around Earth once every twenty-four hours. The payloads consisted of three ultrahigh frequency (UHF) bands, reserved for U.S. government use, and L-band and C-band channels. The latter were kept entirely separate from the UHF channels and were used to translate ship-to-shore signals. Each Marisat had a life expectancy of five years.

Marisat-A was launched on February 19, 1976, and was renamed Marisat 1 when it became operational over the equator at longitude 5° west. It served maritime traffic in the Atlantic. Marisat-B, later Marisat 2, was launched on June 9, 1976, and became operational in July. It served the major shipping lanes of the Pacific Ocean and covered a 120-million-square-mile area. It was stationed slightly west of Hawaii at longitude 176.5° west. On October 14, 1976, Marisat-C, the final satellite in the Marisat system, was launched. It was stationed over the Indian Ocean at longitude 73° east and was renamed Marisat 3 when it became operational. Initially, it served only the U.S. Navy, but it also acted as an in-orbit backup for Marisats 1 and 2.

Although the U.S. Maritime Administration and the U.S. Navy were early users of the Marisat system, the Marisats' main function was to provide communications services to ships of all nations. The satellites were owned by COMSAT, and the company reimbursed NASA for all administrative and launch costs.

Commercial satellite service to shipping continued to develop and expand during this period. Eight thousand ships were expected eventually to be equipped with satellite communications terminals. On February 15, 1979, the United States signed the convention on INMARSAT, in accordance with the International Maritime Communications Act of 1978. That summer, the existence of INMARSAT became official.

INMARSAT was to become operational by the early 1980's. It was expected to rely heavily on Intelsat 5 spacecraft for its generation of satellites. Eventually, the system included the three U.S. Marisats, two Marecs leased from the European Space Agency (ESA), and four Intelsat 5's leased from COMSAT. All the craft were placed in geostationary orbits over the Atlantic, Pacific, and Indian Oceans.

Development of the Marecs portion of the INMARSAT system began in 1973 under the auspices of ESA. Initial funding came from Belgium, France, Italy, the United Kingdom, Spain, and West Germany. Later, the Netherlands, Norway, and Sweden joined in the effort.

The Marecs satellite consisted of a service module, a derivative of the European Communications Satellite design, and a payload module. The payload contained a C-band to L-band forward transponder and an L-band to C-band return transponder incorporating a Search And Rescue (SAR) channel. The system was capable of operating without continuous ground control.

Marecs 1 was launched December 20, 1981, from the French Guiana polygon by an Ariane rocket; it was sent into a geostationary equatorial orbit at longitude 26° west. Marecs 2 was lost at launch because of the failure of the Ariane rocket. Its replacement, Marecs B2, was launched from the French Guiana complex on November 10, 1985; it orbited at longitude 177 ° east. The Marecs satellites had a life expectancy of seven years. The estimated cost of the Marecs program was $359.8 million.

Four Intelsat 5 satellites made up the remainder of the INMARSAT configuration. Each had a box-shaped housing and payload support system with a pair of winglike solar arrays. The three-axis stabilized satellites weighed about 1,950 kilograms at launch and about 860 kilograms in orbit. They were capable of being launched by the Atlas-Centaur rocket, the NASA space shuttle, and ESA's Ariane rocket. Ford Aerospace and Communications was the prime contractor for the Intelsat 5 series. The satellites were launched from 1980 through 1984 and had an expected life span of seven years.

The INMARSAT system has greatly improved communications between ships at sea and their land-based offices. There are more than 3,700 users of the system, and INMARSAT reports that new stations are being activated at the rate of about forty per week. More than 32,000 vessels of all types from about sixty nations receive navigation and communications services from the INMARSAT system. The largest users are the United States, Great Britain, Norway, Japan, the Soviet Union, and Canada.

Each INMARSAT user organization purchases its own terminal for installation on a ship or at a remote site. The INMARSAT charter is limited to providing services to users in the "marine environment," which includes ships, offshore oil rigs, Arctic research stations, and some remote inland temporary construction sites. There have been discussions, however, of expanding the charter to include aeronautics users.

Knowledge Gained

INMARSAT has proved that the use of space-based satellites to enhance marine communications is both economically and technologically sound. Shipboard antennae link ships' telephone systems via INMARSAT satellites to coastal stations. Users can route their calls through any of the coastal stations within range of their transmissions. These Earth terminals are provided by the individual member countries; INMARSAT purchases or leases only the space-based portion of the system.

Although the agency does not provide the land-based stations for its members, it does control and monitor the specifications for the equipment at these stations. It therefore pays for technical consultancy services for its members. The need for these services is determined by the organization's monitoring commitments, which fluctuate. Purchasing the services allows INMARSAT to maintain its standards without the need for a large staff of employees.

The international agency is run along strictly commercial lines. There are no requirements as to the work share of each member nation, and countries other than members can sell their goods or services to INMARSAT if they meet the specifications and have the best price.

INMARSAT also spends about $1 million a year on research and development programs. One program is designed to install a distress beacon on every ship. The beacon is activated automatically or by a crew member. It reports the ship's location and details to rescue agencies on shore rather than at sea. It has been shown that shore-based rescue operations are better equipped to respond quickly and effectively to a maritime emergency than are other vessels at sea.

INMARSAT sees the program as a business venture; it will not purchase the Emergency Pointing/Indicating Radio Beacon Systems (EPIRB) but will assist in providing the technical specifications for the systems.

INMARSAT is also exploring the possibility of entering the television transmission and medical assistance fields. It is investigating a narrow-band video transmission system that would transmit a high-quality picture. INMARSAT has had some successes in a procedure whereby medical diagnosis is conducted from a remote location via television cameras and the INMARSAT system.

INMARSAT also conducts market research on future technical needs. One area of focus is the expansion of its technological services to include telex, computer links, remote control, automated reporting, and high-speed data transmissions.

The agency has also considered entering the aeronautical field, initially by providing air-to-ground communications. A change in the original charter that would permit INMARSAT to provide aeronautical and other nonmaritime services is being considered, since much of the technology developed by the agency can be applied to other fields.

Context

When Arthur C. Clarke, British scientist and science fiction writer, suggested in a technical paper published in 1945 that communications satellites were feasible, the Soviet launching of Sputnik 1 was still twelve years in the future. It would be fifteen years before the National Aeronautics and Space Administration (NASA) launched Echo, the silvery balloon that orbited Earth every 118 minutes and reflected radio signals back to the surface. Echo was a passive satellite, but two years later NASA launched Relay, an active satellite that received signals, amplified them, and returned them to Earth.

With the launches of the early satellites in the late 1950's, many scientists and industry leaders recognized the potential for practical applications of the space program, especially in the area of communications. Even before any formal effort was made to utilize space technology, maritime interests were beginning to reap the benefits of satellite technology. The orbiting spacecraft were providing weather forecasts, storm tracking data, and information on iceberg locations with a degree of accuracy never before possible.

By the early 1960's, private companies were producing their own communications satellites, and in 1962 Congress authorized the Communications Satellite Corporation. COMSAT later became the U.S. representative in and manager of the International Telecommunications Satellite Corporation (Intelsat). As the satellites' ability to transmit voice, picture, and computer data grew, maritime interests began to investigate the possibility of using these technologies to improve and expand the outdated radiotelegraph technology that still handled most ship-to-shore communications on the high seas.

INMARSAT, the maritime equivalent of Intelsat, was formally chartered in 1979 after four years of study by a large group of nations and international maritime agencies. The space-based maritime communications system was dedicated to developing and implementing the latest in satellite and communications technology to enhance the worldwide marine communications network and to ensure the safety and efficiency of the international maritime industry. In the future, INMARSAT plans to expand its original charter to include aeronautical activities, services to remote land stations, improved rescue systems and devices, and long-distance medical diagnosis services.

Bibliography

Caprara, Giovanni. *The Complete Encyclopedia of Space Satellites*. New York: Portland House, 1986. This volume presents short entries on both civil and military satellites of all nations. The book includes line drawings and an index. Suitable for general audiences.

Executive Office of the President. National Aeronautics and Space Council. *Aeronautics and Space Report of the President: 1979 Activities*. Washington, D.C.: Government Printing Office, 1979. A compilation in textbook format of all space-related activities carried on during 1979. Some technical data presented in general terms.

————. *Aeronautics and Space Report of the President: 1983 Activities*. Washington, D.C.: Government Printing Office, 1983. This book has the same format as the one above, but the information is for 1983. Suitable for general audiences.

Gregory, William H., ed. "Inmarsat." *Commercial Space* 1 (Summer, 1985): 55- 57. This is an article in a quarterly publication that focuses on the commercial applications of space technology. The magazine contains about a dozen articles per issue and includes high-quality color photographs. Suitable for general audiences.

National Aeronautics and Space Administration. *NASA: The First Twenty-five Years*. NASA EP-182. Washington, D.C.: Government Printing Office, 1983. A brief chronological and topical history of NASA and the U.S. space program during its first twenty-five years, this book is designed for classroom teachers and features charts, graphs, drawings, color photographs, and suggested classroom activities. Topics include tracking and data-relay systems, space applications, aeronautics, and manned and unmanned missions. Suitable for general audiences.

Ritchie, Eleanor H. *Astronautics and Aeronautics 1976: A Chronology*. NASA SP-4021. Washington, D.C.: NASA, Scientific and Technical Information Branch, 1984. A compilation of U.S. space activities during the year 1976, this book presents technical information clearly for a general audience.

Rosenthal, Alfred, ed. *Satellite Handbook: A Record of NASA Space Missions 1958 – 1980*. Washington, D.C.: Government Printing Office, 1983. Brief summaries and documentation of all NASA manned and unmanned space missions undertaken between 1958 and 1980. Technical data on each mission includes descriptions of the spacecraft and launch vehicle, the payload, the mission's purpose, project results, and major participants. Suitable for a general audience.

Sherman, Madeline, ed. *TRW Spacelog: Twenty-fifth Anniversary of Space Exploration, 1957–1982*. Redondo Beach, Calif.: TRW, 1983. This is a booklet in magazine format that focuses on international space-related activities and programs. Contains brief summaries of various topics. Suitable for a general audience.

U.S. Congress. Office of Technology Assessment. *Civilian Space Policy and Applications*. OTA-STI-177. Washington, D.C.: Government Printing Office, 1982. An official publication that discusses the different policies and agencies that make use of space technology and applications. Contains technical material.

Lulynne Streeter

PASSIVE RELAY
TELECOMMUNICATION SATELLITES

Date: August 12, 1960, to June 7, 1969
Type of satellite: Communications

The United States' Echo satellite project included the first passive relay communications satellite launched into space and the first cooperative space venture between the United States and the Soviet Union. The satellites were inflated Mylar balloons that bounced radio signals between ground stations on Earth.

PRINCIPAL PERSONAGES

WILLIAM J. O'SULLIVAN, JR., an aeronautical engineer
at the National Advisory Committee for
Aeronautics
JOHN R. PIERCE, a researcher at Bell Telephone
Laboratories

Summary of the Satellites

The Echo passive relay communications satellite project was the first practical application of space technology in the field of telecommunications. The satellites were aluminum-coated Mylar spheres that were inflated in space. The concept was simple. The spheres passively reflected electromagnetic radio waves directed to them from a ground-based station at one location on Earth to a ground-based station at another location. They also provided information about the density of the upper atmosphere.

Echo was the first communications satellite project of the National Aeronautics and Space Administration (NASA). It was rooted in an earlier project conceived by William J. O'Sullivan, Jr., an aeronautical engineer at the Langley Aeronautics Laboratory, which was part of NASA's predecessor, the National Advisory Committee for Aeronautics (NACA). In 1956, O'Sullivan proposed that studies to measure Earth's atmosphere during the International Geophysical Year (IGY) — a period of intense scientific research set for July, 1957, to December, 1958 — could be conducted more efficiently by using a low-density inflatable sphere that could be tracked optically. John R. Pierce of Bell Telephone Laboratories had proposed a similarly designed balloon in a 1955 article entitled "Orbital Radio Relays," but Pierce had wanted to use the orbiting inflatable spheres as reflectors for radio signals.

Several attempts to launch O'Sullivan's balloon with IGY payloads failed when the launch vehicles malfunctioned. Pierce proposed a cooperative communications experiment using O'Sullivan's inflatable spheres. According to O'Sullivan, it was a logical next step to consider using the balloons for communications purposes. In 1958, NACA's director, Hugh L. Dryden, told the U.S. Congress that the technology to orbit such a passive communications satellite existed — but certain design changes had to be made. It would be necessary to increase the size of the air-density balloon to provide a larger surface from which to bounce signals, and the surface would also have to be treated to increase its reflective capacities.

By 1959, the IGY project had become the newly formed NASA's passive communications satellite project. It was designated Project Echo. Technicians at NASA's Langley Research Center had three major requirements as they began designing a passive communications satellite that would inflate in orbit into a perfectly smooth-surfaced sphere: They needed a suitable material for the sphere, an inflation system, and a canister in which to launch the collapsed balloon.

It was decided that the balloon would be made from an aluminized polyester film manufactured by E. I. Dupont. Known as Mylar, the material was 0.5 millimeter thick. Another company cemented the Mylar into eighty-two flat gores that formed the Echo sphere. Benzoic acid was chosen as the inflating agent because it could change from a solid state to a gaseous state without going through the liquid stage. A spherical metal canister impregnated with plastic would carry the deflated balloon into space.

In October, 1959, the first test model was assembled and readied for testing. On the fourth attempt, in April, 1960, the balloon satellite was successfully inflated at an altitude of 375 kilometers. On May 13, 1960, a three-stage Thor-Delta rocket launched Echo A-10. During the vehicle's coast period, the attitude control jets on the second stage failed. The vehicle reentered the atmosphere and decomposed.

Three months later, on August 12, 1960, the world's first passive communications satellite was successfully launched from Cape Canaveral, Florida. Echo 1 was placed into space by a Thor-Delta three-stage rocket at an inclination of 47.28 degrees to the equator. The satellite measured 30.5 meters in diameter and weighed 76 kilograms. The inflatable portion was packed in a magnesium sphere and was released about two minutes after injection into orbit. In addition to the balloon, which was designed to transmit images, music, and voice signals from one side of the United States to the other,

Echo carried two small radio beacons that assisted in locating and tracking the satellite. The beacons were mounted on small disks and attached to the balloon.

A pretaped message by President Dwight D. Eisenhower was the first voice signal to be bounced off Echo. It traveled from NASA's facility at Goldstone, California, to Bell Telephone's station at Holmdel, New Jersey. The first known two-way voice communication was bounced off Echo 1 on August 13, 1960, between Cedar Rapids, Iowa, and Richardson, Texas. The first reported image transmission via Echo 1 occurred on August 19, 1960, again between Cedar Rapids and Richardson.

For the next four months, Echo 1 was utilized by Bell Telephone Laboratories in New Jersey and the Jet Propulsion Laboratory in California. Eventually, micrometeoroids damaged the balloon's sensitive "skin," and its orbit was affected by solar winds. Yet it continued to reflect a variety of communications signals to and from Earth at ground stations all over the world. Echo 1 reentered Earth's atmosphere on May 24, 1968.

Echo 1 performed successfully, but it was apparent that some modifications of the design were necessary. In January of 1962, several suborbital tests of a modified Echo inflation system were conducted. Later that same year, plans were announced for the launch of two Echos to determine how smooth the surface area of an advanced Echo had to be. In December of 1962, the United States and the Soviet Union agreed to cooperate in the upcoming experiments planned for Echo.

Early in 1963, however, NASA officials announced that because of the formation of the Communications Satellite Corporation (COMSAT) and the Department of Defense's decision to cancel its Advent project, NASA would cancel its own plans for advanced passive and intermediate altitude communications satellite projects. In August, a private contractor was selected to build three second-generation Echos. This project was also eventually canceled.

On January 25, 1964, Echo 2 was launched from Vandenberg Air Force Base in California. Launched by a Thor-Agena B, the balloon was successfully placed in orbit at an inclination of 81.5 degrees. With a weight of 256 kilograms and a diameter of 41 meters, Echo 2 was somewhat larger than its predecessor. It was also more durable. The sphere was made up of 106 gores of Mylar, three layers thick, and bonded between two layers of a soft aluminum foil alloy. Its outer surface was coated with alodine, and its inner surface was coated with India ink. The improved satellite could maintain its rigidity for a longer period of time. Pyrazole crystals were used as the inflating agent. The crystals were positioned such that upon ejection from the canister into the sunlight they would gradually expand. It was estimated that with this method it would take about ninety minutes for the balloon to become fully inflated. That would permit higher pressures and produce a stronger structure and better reflecting power. Echo 2 also carried two beacon transmitters powered by solar

cells and nickel–cadmium batteries.

Echo 2 had several objectives. It was to perform passive communications experiments with radio, telex, and facsimile signals, collect data concerning the spacecraft's orbital environment, and test the new inflation method. In addition, the Echo 2 spacecraft was used to conduct the first joint U.S.-Soviet space experiment, under the auspices of a 1962 cooperative space exploration experiment consisting of a communications link between stations at Jodrell Bank in Great Britain and the Zimenski Observatory at Gorki University, near Moscow. Echo 2 reentered Earth's atmosphere on June 7, 1969.

Knowledge Gained

The Echo project proved that an inflatable sphere coated with aluminum to increase reflecting abilities could be successfully launched into space and placed into orbit. The balloon could be inflated in space and remain orbiting as it transmitted radio communications and images between distant points on Earth. It also provided a means of measuring such things as the density of the atmosphere. Echo 2 also carried out experiments on the pressure created by solar radiation. During both Echo missions the orbital parameters of the satellites underwent continual variation because of the pressure effects of solar radiation.

Passive relay communications satellites were soon replaced by more efficient and technologically advanced "active" communications satellites. The Echo satellites, however, did offer two advantages over the more sophisticated communications satellites which followed. They were extremely reliable, because of the simplicity of their design and the lack of electronic equipment, and they had multiple access capabilities.

Context

Echo satellites constituted the first civilian telecommunications system set up in space. They were the beginning of a complex space-based communications network that could handle telephone, television, telex, facsimile, and radio signals. They represented the advent of communications systems that would eventually reach every portion of Earth, no matter how remote.

Passive communications satellites were not the most promising method of establishing space-based communications systems, and they were soon abandoned in favor of active systems. These active satellites were capable of receiving signals from Earth and then retransmitting them to another part of the planet. These satellites did not require the expensive ground-based stations necessary for the passive satellites.

Since it seemed clear that these so-called active-repeater satellites in synchronous orbits were more viable than the passive satellites for building a commercial communications system, NASA decided to direct its research to that area, and the agency abandoned further plans to upgrade the Echo project. In fact, Relay 1, an active-repeater satellite, was launched in

1962, two years before the flight of Echo 2.

At first, the use of space communications systems was an international effort through such organizations as the International Telecommunications Satellite Corporation (Intelsat), but soon more and more nations began launching their own satellites to reach remote areas. Such systems offer education, news, entertainment, and business and financial information to every citizen in possession of a receiver.

It is certain that as communications satellites become more advanced and efficient, their numbers will increase. It has even been suggested that widespread use will eventually result in overcrowding of frequencies and orbital positions. Yet whatever the future holds for communications satellites, the Echos, once visible to the naked eye as they traveled around Earth, are remembered as popular symbols of the peaceful and practical application of space research.

Bibliography

Branigan, Thomas, ed. "Echo." *TRW Spacelog* 4, no. 2 (1964): 8-9. The magazine in which this article appears reports on and describes current international space research, development, and missions on a quarterly basis. This article features technical information, photographs, and charts. Written for a general audience.

Caprara, Giovanni. *The Complete Encyclopedia of Space Satellites*. New York: Portland House, 1986. This volume contains a complete listing of every civilian and military satellite launched from 1957 to 1986. It contains color and black-and-white photographs, a bibliography, an index organized by country, a general index, and a table of contents.

National Aeronautics and Space Administration. *NASA: The First Twenty-five Years, 1958–1983*. NASA EP-182. Washington, D.C.: Government Printing Office, 1983. Designed for use by classroom teachers, this book features color photographs, charts, graphs, tables, and suggested activities. Topics include NASA history, programs, and missions.

Rosenthal, Alfred. *Satellite Handbook: A Record of NASA Space Missions, 1958–1980*. Greenbelt, Md.: Goddard Space Flight Center, 1981. This book contains detailed descriptions of major NASA satellite missions from 1958 to 1980. It includes a NASA launch record for those dates, a list of abbreviations and acronyms, an index, and photographs.

Van Nimmen, Jane, Leonard C. Bruno, and Richard L. Rosholt. *NASA Historical Date Book: 1958–1968*. NASA SP-4012. Springfield, Va.: National Technical Information Service, 1976. A complete history of NASA's programs and projects from 1958 to 1968. Contains much detailed information. Includes illustrations, charts, and tables.

Wells, Helen, et al., eds. *Origins of NASA Names*. Washington, D.C.: Government Printing Office, 1976. A brief report on the origins of NASA project names. Includes acronyms, abbreviations, and space terms. It also features the International Designation of Spacecraft, a list of major NASA launches from 1958 to 1974, reference notes, and an index.

Lulynne Streeter

PRIVATE AND COMMERCIAL TELECOMMUNICATIONS SATELLITES

Date: Beginning August 12, 1960
Type of satellite: Communications

Since the early 1960's, communications satellites have been designed and built by private corporations to serve the needs of their customers. These commercial ventures in space have contributed to U.S. technological developments.

PRINCIPAL PERSONAGES

RONALD REAGAN, the U.S. president who legalized a
partial privatization of space for U.S. industry
HOWARD HUGHES, the corporate initiator of the
Syncom satellite series
JOSEPH CHARTYK, a president of COMSAT
DEAN BURCH, Director General, Intelsat
GERARD K. O'NEILL, head of Geostar Corporation

Summary of the Satellites

The satellite communications industry was born shortly after the first satellite was placed in orbit in the late 1960's. The usefulness of communications via satellite was obvious: An orbiting satellite's altitude allows it to transmit signals over very long distances; ground-based transmissions are limited by Earth's curvature. Since the microwaves that make up radio, television, and telephone signals travel in a straight line, ground relay stations must be placed about every 55 kilometers to compensate. Before satellites came into use, therefore, worldwide communications costs were prohibitive.

Scientists experimented with two types of communications satellites: the passive reflector and the active repeater. The passive reflector is typified by Echo 1, a "balloon" satellite, whose large surface was used to reflect radio signals from the ground. This sort of satellite had two advantages: Any transmitter could bounce a signal of any frequency off the reflective surface, and there were no parts to malfunction. The disadvantage was that the received signal was extremely weak. The first Echo was launched on August 12, 1960, and was useful for a little more than four months.

In 1962 the Telstar and Relay satellites were launched. These were active repeaters that received transmissions from ground stations, amplified them, and relayed them back to Earth. Telstar was designed and built by Bell Telephone Laboratories, and Relay was built by RCA. The early satellites were placed in relatively low orbits and moved rapidly across the sky, requiring elaborate tracking.

Telstar could handle one television channel, the equivalent of six hundred one-way voice channels. Because the satellite's signal was weak even after amplification, the ground station was located in the shelter of a ring of low mountains to reduce radio interference. A horn-shaped antenna with an opening 1,100 meters wide was used to focus the signal, which even then reached only a billionth of a watt.

Besides their usefulness as experimental communications satellites, Telstar and Relay also were important as probes of the space environment. They carried instruments that collected data on the Van Allen radiation belts, regions that encircle Earth and contain radioactive particles. The belts' existence was inferred from data returned by the first orbiting satellite, Sputnik.

Syncom, built by Hughes Aircraft for NASA, represented the next generation of communications satellites. Launched in 1963, it was placed in an orbit with a twenty-four-hour period and therefore matched Earth's rotation. Syncom was the first "geosynchronous" satellite. The main advantage of a geosynchronous satellite is that ground control is greatly simplified. Because the satellite remains over one area of Earth's surface, expensive and complex tracking equipment is unnecessary. The satellite is accessible to any ground station within its line of sight. A geosynchronous satellite must be placed directly over the equator to maintain its position relative to the surface; complex firings of gas jets are required to maneuver it into position. From its high altitude of about 35,900 kilometers, a geosynchronous satellite can transmit signals to nearly a third of Earth's surface. With a network of three such satellites, communications service can be provided to all populated areas of Earth. Although Syncom 1 failed and the network was not completed at that time, Syncom was the first step toward a global communications system.

Three Syncoms were launched. The first was lost when a high-pressure nitrogen bottle aboard the satellite burst. Syncom 2 was successful, and it set new records in long-distance communication. Syncom 3 transmitted the first television program (a relay of the Olympic Games in Tokyo) ever to span the Pacific via a geosynchronous satellite. Syncoms 2 and 3 remained in service until 1969.

The successes of Relay, Telstar, and particularly Syncom clearly demonstrated that communications satellites were moving away from experimental projects and toward commercial ventures. In 1963, the Communications Satellite Corporation (COMSAT) was created as a private company that would develop satellite systems. Half the corporation's stock was to go to the public; the other half, to large communications companies. COMSAT is part of Intelsat, the International Telecommunications Satellite Corporation, which was set up to provide a global communications network. COMSAT's first commercial satellite was Early Bird, launched in 1965. Early Bird relied heavily on technology from the Syncom satellites. This and other early commercial satellites demonstrated the feasibility of using geosynchronous satellites for commercial communications.

Private users of the satellites were communications common carriers, broadcasters, news wire services, newspapers, airlines, computer services companies, and television companies. At first, the satellites were used simply to receive television signals or telephone communications and relay them to one or several ground stations. It was foreseen that multipurpose satellites would eventually be built to serve the needs of various customers.

Early users of communications satellites would lease the services from one of a very few satellite owners. As the industry expanded, some large businesses began to own and operate their own satellites and lease excess capabilities to smaller businesses. Satellite capacity grew rapidly with new satellite technology. Satellites with multiple transponders (devices which, when triggered by a signal, transmit another signal at the same frequency) set at different frequencies began to support many customers simultaneously. In the mid-1960's, companies began building basic "production line" satellites; customers could choose one and then refine it to meet specific needs.

Western Union developed the first U.S. domestic satellite communications system. Westars 1 and 2 were launched in April and October of 1974, and Westar 3 was launched in August, 1979. Companies wishing to operate domestic satellite systems must obtain permission from the U.S. Federal Communications Commission (FCC), and Western Union was one of the first such companies to be licensed. The Westars were designed to relay voice, video, and data communications to the continental United States, Alaska, Hawaii, and Puerto Rico. The satellite can handle twelve color television channels or seven thousand two-way voice circuits simultaneously. The new generation of Westars — Westars 4, 5, and 6 — were deployed by 1982 and had twice the capacity of the earlier satellites. They could also relay signals to the U.S. Virgin Islands.

Another domestic communications system, Satcom, was placed in orbit by RCA beginning in 1975. Satcom was the first satellite system devoted to relaying signals to cable television installations in the United States.

A third domestic communications satellite system was launched in 1976. Comstar is a telephone satellite system for long-distance calling throughout the United States; it was designed to handle increasing domestic telephone usage. Each Comstar satellite can relay more than eighteen thousand calls simultaneously. They are leased jointly by the American Telephone and Telegraph Company and GTE Satellite Corporation, a subsidiary of General Telephone and Electronics. In the early 1980's, AT&T launched a new series of Telstar satellites: Telstars 301, 302, and 303.

In March, 1981, Satellite Business Systems (SBS) began to offer satellite services for private business communications to large U.S. companies. The third SBS satellite to be deployed was the first commercial satellite launched by the U.S. space shuttle.

Although Hughes Aircraft had been designing and building satellites for other customers since the early 1960's, it was not until 1983 that it began launching the Galaxy series, owned by Hughes Communications. The Galaxy series was dedicated to the distribution of cable television programming. Hughes offered cable programmers the opportunity to buy, rather than lease, transponders on the satellite.

Knowledge Gained

The major gains in knowledge made possible by private communications satellites have stemmed from advances in satellite technology and from the increased ease and speed of global and domestic communications.

The first satellites provided valuable information about the environment of Earth orbit. Technology improved almost immediately as the first satellites reported on the new conditions. Solar cells on Telstar 1 were damaged by radiation in the Van Allen belts, and satellite manufacturers learned new ways of compensating for radiation. Telstar 1 also returned valuable scientific data on the Van Allen belts as it passed through them. The densities and energies of free protons and electrons were measured, and the temperature and pressure within the satellite was recorded.

In 1959, an engineer working for Hughes Aircraft developed a satellite design that revolutionized the communications industry. Tiny rocket boosters were incorporated into the satellite to position it in a geosynchronous orbit. The rockets' periodic thrusts stabilize the satellite and orient it so that its communications antenna continually aims at Earth's surface. Hughes also developed spin-stabilized satellites. Spin-stabilization previously had been used to improve rockets' accuracy.

New technology was devised for the Comstars' antenna arrangement. A vertically polarized and a horizontally polarized antenna were used. Polarization changes the form in which signals are received and sent without changing the frequency, which allows the capacity of the system to be doubled.

Another Hughes development is the shaped beam. The microwave beam transmitted by the satellite is shaped to the contours of the receiving area, allowing a more concentrated and powerful transmission. Along with the new technology,

Telstar 1 (top); Galaxy spacecraft (bottom). (NASA)

the improvement of existing technology has allowed for increasingly longer satellite lifetimes and increased power output.

In 1981, the space shuttle began to be used for launching commercial satellites. Cargo space in the payload bay is limited, so satellite designs were modified with telescoping solar panels and folding antennae. Starting from a stowed size of about three meters high by two meters wide, a satellite can expand to the height of a two-story building when deployed in orbit.

Communications satellites function as channels for information. Knowledge must be distributed to be effective. It would be difficult to gauge the spread of knowledge that has resulted from improved communications. Satellites have helped bring a variety of educational programs to persons who were previously beyond the reach of educational systems. Engineering and technological programs are broadcast instantaneously to universities across the country, and businesses use satellites to provide training seminars for their employees.

Context

The development of a country is related to its ability to communicate quickly and efficiently. Satellite communications have brought about a huge increase in the ability to transmit information economically and quickly. The transition from ground-based telecommunications to satellite telecommunications took place in less than twenty years. This growth is still taking place, not only in technologically advanced countries such as the United States, but all over the world. Many underdeveloped countries are using satellite systems to unite remote regions of the country. They are finding that the cost of employing satellite systems is far less than the cost of installing land-based relay stations. Educational systems can be improved and can compensate for teacher shortages.

Satellite usage primarily involves telephone and television transmission, but as the industry expands and costs become less prohibitive, new satellite uses evolve. Satellites regularly provide such services as computer-to-computer digital communications, video conferencing, monetary fund transfers, and air traffic control. Technology is being developed for mobile communications, and businesses are using satellite systems to handle electronic mail and data transfer.

The U.S. commercial satellite industry has been affected by unreliable launch systems. Because the shuttle program was suspended after the 1986 *Challenger* accident and the revitalized program was slow to accelerate, satellite users were forced to use other, conventional launch systems. Some of those have proved unreliable, and several multimillion-dollar satellites have been lost. Insurance costs increase dramatically as the risk associated with a launch increases, and orders for new satellites slow as launch capability lessens. Many companies have looked to European launch facilities for reliable and timely deployment of satellites.

Some companies have argued that the federal government has not offered enough support to commercial space ventures. In the late 1980's, government support began to grow along with the perception that the United States was falling behind the rest of the world in the utilization of space. One way to increase a nation's space use is through the mobilization of the private sector.

Debris in space is also a growing problem. Hundreds of satellites are in geosynchronous orbit and satellite overcrowding is possible. For many applications, there are only a few suitable locations for signal transmission. Furthermore, satellites that are less than two degrees apart in orbit risk radio frequency interference.

Despite problems, the communications satellite industry has grown. Further technological developments may lead to communications satellites that can transmit power, rather than information, to Earth. It is conceivable that by 2025, one hundred power satellites could meet 30 percent of the United States' electrical needs.

There are four planets in the solar system with ring systems surrounding them. The deployment of geosynchronous satellites may produce a fifth. Earth's "ring system," however, would not be composed of dust, rocks, or ice, but of thousands of artificial satellites.

Bibliography

Aviation Week and Space Technology 120 (June 25, 1984). This entire volume is devoted to prospects for the commercial use of space. It contains a series of authoritative articles that are readily understandable by laymen.

Braun, Wernher von, et al. *History of Rocketry and Space Travel*. New York: Thomas Y. Crowell, 1966. Reviews the entire history of spaceflight from the beginning of the concept to the sending of humans into space and to the Moon. Fully illustrated and easily understood.

Goldman, Nathan C. *Space Commerce: Free Enterprise on the High Frontier*. Cambridge, Mass.: Ballinger Publishing Co., 1985. Written for nonspecialists, this book provides a useful review of American commercial space enterprises. It is both synthetic and specific. Charts, tables, and a few photographs augment the text. Contains six appendices and an index.

Ordway, Frederick I., III, Carsbie C. Adams, and Mitchell R. Sharpe. *Dividends from Space*. New York: Thomas Y. Crowell, 1971. Explores the benefits to mankind that have directly resulted from the space program. Discusses products developed for mass consumption and advances in medicine, industry, research, and the study of Earth. Sparsely illustrated. Written for general audiences with an interest in space technology.

Paul, Günter. *The Satellite Spin-Off: The Achievements of Space Flight*. Translated by Alan Lacy and Barbara Lacy. Washington, D.C.: Robert B. Luce, 1975. A survey of the commercial scientific and communications applications that developed from the space research of the 1960's and early 1970's. Contains a comprehensive account of the early, politically charged days of Intelsat. This book is written from the perspective of the European community, and it is necessary reading for those desiring a broader understanding of Intelsat than might be available from the U.S. point of view.

Porter, Richard W. *The Versatile Satellite*. New York: Oxford University Press, 1977. This short book provides a fine introduction to the various uses of satellites. Chapter 4 deals with communications satellites.

Schwarz, Michiel, and Paul Stares, eds. *The Exploitation of Space: Policy Trends in the Military and Commercial Uses of Outer Space*. London: Butterworth and Co., 1986. Consists of highly readable and informed articles written by scholars but intended for laymen. Well illustrated. Notes and references conclude each article.

Divonna Ogier

AIR AND SPACE TELESCOPES

Date: Beginning 1608
Type of technology: Telescopes

Telescopes, first invented in 1608, have undergone enormous diversi-fication in the twentieth century, particularly as it has become possible to elevate them into and beyond Earth's atmosphere. They have played a vital role in the discovery and investigation of a wide variety of celestial bodies.

PRINCIPAL PERSONAGES

HANS LIPPERSHEY, the inventor of the telescope

GALILEO, the first person to use a telescope for study of the sky

SIR ISAAC NEWTON, the English physicist who made the first reflecting telescope

JOHANNES KEPLER, a German astronomer, the founder of modern optics

N. CASSEGRAIN, the inventor of the Cassegrain reflecting telescope

SIR WILLIAM HERSCHEL, the father of stellar astronomy

GEORGE ELLERY HALE, an astronomer who helped establish the Yerkes, Mount Wilson, and Mount Palomar observatories

ALVAN GRAHAM CLARK, the maker of many of the largest telescopes in the United States

GEORGE WILLIS RITCHEY, the director of photo-graphic and telescopic research at the U.S. Naval Observatory

KARL AUGUST STEINHEIL, a German physicist

CHRISTIAAN HUYGENS, a Dutch astronomer, mathe-matician, and physicist

HEBER DOUST CURTIS, Director of the Observatory of the University of Michigan

Summary of the Technology

Astronomy, one of the most ancient of sciences, was altered for all time by the invention of the telescope. The first such instrument was developed by Hans Lippershey of Middelburg, the Netherlands, in 1608; he discovered that a certain combination of concave and convex lenses worked to make distant objects seem nearer. Experimentation and refinement of technologies over the subsequent centuries made possible the proliferation of ground and space-based telescopes of the twentieth century.

Telescopes can be divided into two broad categories: opti-cal telescopes and radio telescopes. Optical instruments col-lect light energy from a distant source and focus it into an image that can be studied by a number of different tech-niques. Those satellite-borne telescopes that are aimed at detecting celestial X rays or ultraviolet radiation are included in this category, since they operate according to the principles of geometrical optics. Radio telescopes are used in astronom-ical research to detect and measure radio waves coming from various parts of the Galaxy. These instruments consist of three complementary parts: a large reflecting surface that col-lects and focuses incident radiation, an electronic receiver that detects and amplifies cosmic radio signals, and a device that displays the information.

The diversity of modern telescope technology has its ori-gins in the pioneering work of scientists such as Galileo, who in 1609 heard of Lippershey's "magic glass" and became fasci-nated with the idea of using it to study celestial bodies. The instrument that he developed consisted simply of a paper tube and a pair of appropriate lenses; it had the advantage of an erect image, but afforded only a small field of view. Johannes Kepler, Francesco Generini, and others experiment-ed subsequently with variations on this design. Naturally, a principal goal was to increase the instrument's power, or mag-nification, which varies with the ratio of the focal length of the front (objective) lens to the focal length of the rear lens, or eyepiece. These early optical instruments, with two lenses centered on the same axis, are known as refracting telescopes.

In 1668, Sir Isaac Newton experimented with the use of a small, concave mirror in what became known as the reflect-ing telescope. Though his instrument was not particularly successful, its basic concept proved durable. The Frenchman N. Cassegrain produced a different type of reflecting instru-ment in 1672, with a convex mirror. These two types of reflectors, termed Newtonian and Cassegrain, remain in use in many large reflecting telescopes.

The simple, versatile Newtonian reflecting telescope is the most universally used type of telescope. It bends light rays by reflecting them from the surface of a concave mirror. Light from the object being viewed travels down the tube assembly and strikes the primary mirror; the light is then reflected toward a focal point that lies just outside the tube assembly. The focused rays are intercepted by a secondary diagonal mirror, which bends the rays at a ninety-degree angle to the incoming light. The light rays are then brought into contact with the eyepiece, which lies outside the tube plane. As a result of the bending of the light rays, the image is seen upside down, with some loss of light.

In the Cassegrain configuration, light is reflected by a hyperbolic mirror inserted in front of the prime focus, passes through a hole in a large parabolic mirror, and comes to a new focus on the back of the mirror. The Cassegrain telescope improved upon the Newtonian design through its use of a shorter focal length in its system. Its more compact formation requires a less massive mount for the same degree of stability. In addition, the eyepiece is placed more conveniently. Unfortunately, parabolizing the primary mirror to the required degree of accuracy is an expensive process.

Variations on these reflecting telescopes have been produced, each with its strengths and weaknesses. The Dall-Kirkham telescope uses an ellipsoidal primary mirror and a spherical secondary mirror. Its field of vision, however, is quite limited, and like the Cassegrain and Newtonian telescopes, it produces coma effects (a distortion in which points of light appear comet- or fan-shaped). The Ritchey-Chrétien telescope uses a hyperboloidal primary mirror and an elliptical, convex secondary mirror. It reduces coma almost completely but suffers from astigmatism and a severe curvature of its field of vision. In general, because reflecting telescopes are not sealed, dust can accumulate on the mirrors; in addition, air turbulence within the tube can disrupt viewing.

Chromatic aberrations (false colors arising from the refraction of different wavelengths of light) are not a problem with reflecting telescopes, although low-expansion glass, which can resist the tendency of large mirror-surfaces to expand and contract differently in different areas, is a necessity. This type of telescope possesses considerable advantages in spectroscopic work and is particularly useful in the observation of faint objects, such as nebulae and galaxies.

Though for a time, with the success of reflecting-type telescopes their refracting counterparts became obsolete, eventual improvements in glass technology made possible a return to refracting instruments. By 1799, a Swiss craftsman had mastered the art of making flint glass that had the optical qualities required for refraction. Innumerable further refinements have led to the development of refracting telescopes that are mechanically simple, durable, and readily available.

The refracting telescope's closed-tube design eliminates the air currents that can degrade images. The objective lens is mounted at the far end of the apparatus; its diameter ranges from about 0.05 to 1 meter. This lens is focused on the image of a distant object, capturing light, which it then focuses onto a microscopic point. The eyepiece lens then concentrates that point of light onto the human eye. Refracting telescopes are especially useful for visual observations and astrometric measurements of stellar parallax and binary stars and for the visual and photographic determination of stellar positions.

The glass of the reflecting telescope's lens must be optically homogeneous and free of bubbles. The lenses require expensive material, and unless the correct combination of lenses and the proper focal length are used, chromatic aberrations arise. The flint lens of a refractor becomes increasingly opaque to wavelengths shorter than 4×10^{-8} meter; it becomes useless in this spectral region.

Catadioptric telescopes, developed in the twentieth century, combine the best features of reflecting and refracting telescopes. These instruments have closed tubes, thus eliminating image-degrading currents and making the optics almost maintenance-free. These telescopes are portable and free of chromatic aberrations, and their additional optical elements allow users greater facility in correcting foci and surmounting the faults of the mirrors.

Various other telescopes have been developed for specialized functions. A heliostat, for example, is a telescope used to study the Sun; its flat moving mirror captures sunlight and feeds it to a fixed telescope with a stationary focal plane. A similar instrument used for observing stars is known as a siderostat. The zenith telescope has a fixed lens that points toward the zenith — that is, directly overhead. Incoming light is reflected from a mercury pool onto a photographic plate placed at the focus of the lens. Such telescopes are used for the precise determination of time. Transit telescopes are refractors capable of moving in altitude but not in azimuth; they are particularly useful in making accurate observations of the stars as they cross the meridian. A chronograph is used to obtain photographs and film of the solar corona, even in daylight.

Despite all the refinements that increased the quality of the images produced by ground-based telescopes, astronomers realized that they suffered from an inherent limitation: the distortions produced by Earth's atmosphere. Earth's air absorbs essentially all ultraviolet radiation below a wavelength of 3×10^{-7} meter, as well as much of the infrared spectrum between wavelengths of 1×10^{-6} and 1×10^{-3} meter. Moreover, the thermal currents in Earth's atmosphere deflect light and thus limit the sharpness of most Earth-based astronomical photographs.

It was a desire to observe the Sun that motivated early efforts to send telescopes into air and space. A balloon-borne refractor sent aloft in 1956 and 1957 reached an altitude of between 6,096 and 7,620 meters and enabled astronomers from the University of Cambridge in England and the Meudon Observatory in France to obtain photographs of the granulation of the Sun.

Stratoscope 1, an American project, followed on the heels of the European refractor's launch. An unmanned balloon capable of lifting a payload of 635 kilograms to an altitude of 25,603.2 meters was designed. It carried a reflecting telescope that yielded an optical enlargement corresponding to an effective focal length of 60.96 meters. Several pairs of photodiodes acted as remote eyes to point the telescope toward the

Sun's disk. A total of five flights of Stratoscope 1 were undertaken. The last flight furnished excellent photographs of solar granulation and sunspots, taken from an altitude of about 24,400 meters.

The success of these balloon experiments encouraged investigators to devise ways of observing other solar phenomena. Astronomers from Boulder, Colorado, organized the launch of the Coronascopes, special balloonborne telescopes whose purpose was to photograph the solar corona in full daylight. Coronascope 1 carried a small coronagraph with an aperture of 0.033 meter. It recorded images of the Sun automatically on a red-sensitive 0.035-meter spectroscopic film. Coronal streams between two and five times the solar radius from the Sun's center were detected.

Stratoscope 2 carried elaborate instruments to an altitude ranging from 21,641 to 25,603 meters. The first flight was on March 1, 1963; it was successful in obtaining measurements of a part of the infrared spectrum of Mars. Stratoscope 2's second flight took place on November 26, 1963; it returned images of the infrared spectra of about nine red stars and new data on Jupiter and the Moon.

Balloons could carry telescopes, however, only to the threshold of space. Further penetration into space required vehicles with greater lifting power. In response to this need, the first Orbiting Astronomical Observatory was launched into space on April 8, 1966. It carried eleven telescopes, four of which were coupled with ultraviolet vidicons to obtain maps of the sky. The other seven telescopes, in conjunction with two spectrometers and five photometers, were used to obtain spectral and energy distributions of numerous celestial objects.

A series of Orbiting Solar Observatories were launched beginning in 1962. These satellites were designed to return measurements of ultraviolet, X, and gamma radiation from the Sun. The Orbiting Geophysical Observatories, whose launches began in 1964, obtained measurements of radio noise and bursts of cosmic ray protons from the Sun.

Beginning in 1969, various probes — Mariners, Lunas, Voyagers, Rangers, Orbiters, Surveyors — were being launched in space with telescopes as standard equipment. Each craft carried a different type of telescope related to its function. By the late 1970's, these telescopes were being controlled by microcomputer.

With the May 14, 1973, launch of Skylab, a new era of space research began. Skylab carried eight separate solar telescopes on its Apollo Telescope Mount: two X-ray telescopes, an extreme ultraviolet spectroheliograph, an ultraviolet spectroheliometer, an ultraviolet spectrograph, a visible light coronagraph, and two hydrogen-alpha telescopes. These highly complex instruments gathered data on the Sun across the electromagnetic spectrum.

The first High-Energy Astronomical Observatory was launched on August 12, 1977; a series of such orbiting observatories were aimed at returning data on high-energy astro-

physical processes. High-Energy Astronomical Observatory 2 carried the first X-ray telescope capable of providing focused images of X-ray objects in the sky.

In the meantime, scientists continued to find that telescopes mounted on aircraft could also return data of considerable value. The Kuiper Airborne Observatory C-141, for example, was mounted with a 0.90-meter telescope that was used to study the nature of extragalactic X-ray background radiation, distant quasi-stellar radio sources (quasars), X-ray sources in Galaxy M31, and X-ray emissions from other clusters of galaxies.

Certain Explorer satellites were designed to provide data on sources emitting ultraviolet radiation. The International Ultraviolet Explorer, an international venture undertaken by the National Aeronautics and Space Administration (NASA), Great Britain's Science Research Council, and the European Space Agency, was launched on January 26, 1978, into an eccentric geosynchronous orbit. It was mainly used to gather data on the transmission and absorption of radiation in the atmosphere of subluminous stars, in the interplanetary medium, and around other objects within the solar system.

The Infrared Astronomical Satellite, a cooperative project of NASA, the Netherlands, and Great Britain, carried a cryogenically cooled telescope system. It surveyed various celestial sources of infrared radiation.

The Solar Maximum Mission satellite, another telescope-bearing craft, was put into space to gather information on solar flares and the globular solar corona.

The first generation of orbiting telescopes — the Orbiting Solar Observatories, the Orbiting Geophysical Observatories, and the High-Energy Astronomical Observatory, for example — were automated telescopes revolving around Earth. Such telescopes function simply as search cameras or survey instruments. Eventually, plans were made for the design of manned orbiting telescopes, so that astronauts could go into space to do maintenance work on them; instruments could also be picked up by the space shuttle and returned to Earth for refurbishment and relaunch. After several years of study and refinement of the design principles, the Hubble Space Telescope became an approved NASA project. This high-resolution, 2.4-meter telescope was placed in orbit in 1990 as a joint venture of NASA and the European Space Agency. The Cassegrain telescope is accompanied by a high-resolution camera, a faint-object spectrograph, an infrared photometer, and guidance and protective equipment. The space telescope is designed to operate from wavelengths of 91.2×10^{-9} to about 1×10^{-3} meter — across the ultraviolet, optical, and infrared regions.

Knowledge Gained
Almost all that is known regarding the Sun, the Moon, planets, stars, galaxies, and other celestial bodies has been gained with the aid of telescopes, ground-, air-, or space-based.

Galileo and other early telescope builders produced drawings of the sunspots that they observed, showing the presence of umbras (dark central regions) rimmed by penumbras (lighter areas). Scientists were able to observe and measure the sunspots' drifting motion across the Sun's face and thereby to show that the Sun rotates. Eventually, it was understood that the Sun goes through peaks (maxima) and valleys (minima) of sunspot activity in cycles of approximately eleven years. Telescopes carried by the Orbiting Solar Observatories, the Solar Maximum Mission satellite, and Skylab returned vital information regarding the sunspot cycle, coronal holes and loops, the solar wind, and much more.

With the help of his simple refractor, Galileo was able to observe the mountains and craters of the Moon's surface. He also discovered Jupiter's four large moons: Io, Europa, Ganymede, and Callisto. A dozen smaller Jovian satellites have been discovered. Jupiter's cloud belts have been found to be discontinuous and composed of a multitude of streamers and festoons. The 1955 discovery of radio waves being emitted from Jupiter led to the discovery of that planet's radiation belts. Mars's surface features, such as Syrtis Major and Mare Erythraeum, have been found to be discontinuous. That planet is known to be covered with volcanoes, canyons, and craters; Mercury, too, is highly cratered. Mercury and Venus were discovered to have unusual patterns of rotation. Telescopic examination has shown that Saturn has twelve concentric rings, as well as a red spot like that of Jupiter. Saturn also has some seventeen satellites of widely varying natures. Uranus, which was discovered by Sir William Herschel, has been found to possess only five moons. Rings have also been discovered around Uranus. Unexpected aspects of Uranus' orbital path led to the discovery of the planet Neptune, and distortions in Neptune's path in turn helped astronomers to locate tiny Pluto.

Halley's comet, whose periodicity was determined by Edmond Halley to be seventy-six years, has been studied telescopically during its three recurrences since 1758. Much has been learned about stars as well through the use of telescopes — for example, that there are many types of stars, including double, variable, and exploding stars. Quasi-stellar objects discovered in the 1950's with the aid of radio telescopes were dubbed quasars. Pulsars, celestial objects emitting rapid radio pulses, have also been discovered. With the aid of telescopic instruments, almost all the known galaxies, including the Milky Way, have been grouped into clusters. Large numbers of asteroids have also been discovered; most of them were found to move in solar orbits between the orbits of Mars and Jupiter.

Infrared emission from the planets has provided important information about the structure and evolution of planetary interiors. Various nebulae, such as the planetary nebulae, extragalactic objects, and some quasars, have been discovered to emit unexpectedly large amounts of infrared energy. Numerous quasars have been found to have a large redshift; thus, astronomers have discovered the rapid movement of those objects away from the Galaxy.

Radio observations of the interstellar medium have led to the discovery of complex molecules such as ammonia and formaldehyde, along with hydroxyl ions and water molecules.

Context

The invention of the telescope provided mankind with a window on the universe. Subsequent refinements opened that window ever wider. Yet at the beginning of the twentieth century, much remained to be discovered. For example, knowledge of solar radiation extended only slightly beyond the visible-light spectrum, a range of 4×10^{-7} to 7×10^{-7} meter. Earth's atmosphere blocks out most of the shorter and longer wavelengths. Exotic, nonthermal radiations as well as the radiation emitted by very hot and very cold objects were beyond understanding.

Thus, the ability to raise telescopes into air and space led to a remarkable amplification of astronomers' knowledge of celestial bodies and their radiations. Observations of the planetary spectra by means of telescopes carried to Earth's stratosphere in jets, for example, are far superior to those obtained from ground-based facilities. In general, measurements obtained with Earth-orbiting telescopes are free from the interference from Earth's atmosphere. Deep space probes are able to carry telescopes even farther— and further open the window to knowledge of the universe.

Bibliography

Brown, Sam. *All About Telescopes*. 5th ed. Barrington, N.J.: Edmund Scientific Co., 1981. This volume describes in detail the qualities of a good telescope and explains how to use a telescope and mount it for best viewing of the finer details of the sky. Suitable for high school and college-level students.

Fimmel, Richard O., James Van Allen, and Eric Burgess. *Pioneer: First to Jupiter, Saturn, and Beyond*. NASA SP-446. Washington, D.C.: Government Printing Office, 1980. This overview of the Pioneers' missions and data includes a brief introduction to telescopic study of the planets.

Kopal, Zdenek. *Telescopes in Space*. London: Faber and Faber, 1968. This readable book covers the literature of the telescope. Illustrated.

Kuiper, Gerard P., and Barbara M. Middlehurst, eds. *Telescopes*. Vol. 1, *Stars and Stellar Systems*. Chicago: University of Chicago Press, 1960. This well-illustrated volume describes optical and radio telescopes and their accessories. There is a particular focus on the Hale telescope and the Lick Observatory.

Moore, Patrick, ed. *Astronomical Telescopes and Observatories for Amateurs*. New York: W. W. Norton and Co., 1973. This practical book for amateur astronomers contains detailed descriptions of many telescopes, along with illustrations.

Page, Thornton, and Lou Williams Page. *Sky and Telescope*. Vol. 1, *Wanderers in the Sky*, and vol. 4, *Telescopes*. New York: Macmillan, 1965 – 1966. Volume 1 describes various celestial bodies and defines key astronomical concepts. Volume 4, illustrated with more than 120 photographs, drawings, and diagrams, describes in detail the principles of telescope design and methods of fabrication. It also discusses some of the world's famous observatories and telescopes.

Pendray, Edward G. *Men, Mirrors, and Stars*. New York: Funk and Wagnalls, 1935. This history of the development of telescopes for space observation gives due recognition to the greatest names and foremost discoveries of astronomy.

Ronan, Colin A. *The Practical Astronomer*. New York: Macmillan, 1981. This book describes in detail the characteristics of the planets, their satellites, and other celestial objects. It is suitable for the general reader.

Sidgwick, J. B. *Observational Astronomy for Amateurs*. 4th ed. Hillside, N.J.: Enslow Publishers, 1982. This book presents observational techniques for various fields of amateur astronomy. It contains a vast amount of data collected by various deep space probes, suborbital vehicles, and ground-based facilities.

Traister, Robert J., and Susan E. Harris. *Astronomy and Telescopes: A Beginner's Handbook*. Blue Ridge Summit, Pa.: TAB Books, 1983. Traister and Harris survey optical telescopes, from the simple seventeenth century instruments to the most sophisticated telescopes of the 1980's.

Raj Rani

TETHERED SATELLITE SYSTEM

Date: Beginning 1966
Type of program: Space experimentation

During the early 1990's the National Aeronautics and Space Administration (NASA) conducted a series of experiments involving tethered satellite systems. Thin cables ranging in length from 20 kilometers to 100 kilometers were deployed between a main satellite and a sub-satellite to determine if it would be feasible to use tethered satellite systems as generators to provide electrical power to the space shuttle or to recharge failing satellite batteries. Scientists also suggested using tethered satellites to conduct upper atmosphere research.

PRINCIPAL PERSONAGES

KONSTANTIN TSIOLKOVSKY, pioneer in Russian rocketry who first proposed a space elevator in the 1890's

YURI N. ARTSUTANOV, Soviet engineer who first proposed a space elevator in the 1950's

PAUL BURCH, a British scientist who proposed a cosmic railway in the 1980's

GIUSEPPE COLUMBO, professor at the University of Padua and co-author of the 1974 "Skyhook" report

MARIO D. GROSSI, co-author of the 1974 "Skyhook" report

BILLY NUNLEY, NASA Tethered Satellite System mission manager

JEROME PEARSON, American scientist active in research into space elevators in the 1980's and 1990's

RAYMOND SCHWINDT, Martin Marietta's Tethered Satellite System program manager

THOMAS D. STUART, NASA program director

JEFFREY A. HOFFMAN, STS-46 mission specialist

Summary of the Program

Tethered satellite systems involve using a retractable cable, or tether, to connect satellites orbiting the Earth. Tethered satellite systems have been proposed for use in generating electrical power, for deploying instruments for upper atmosphere research, and even to serve as elevators for transferring materials from a planet's surface to a space station. For most intended applications, the simplest way to visualize a tethered satellite system is to picture someone walking a pet. In the case of tethered satellite system intended for upper atmos-

phere research, for example, the space shuttle or space station would be analogous to a pet owner circling the neighborhood with a dog on a leash. Walking steadily down the middle of the sidewalk the human maintains a constant distance from neighbors' homes while the dog on the leash explores their yards, with the owner shortening or lengthening the leash, or tether, depending on circumstances. Similarly, during a space mission devoted to upper atmosphere research, the space shuttle would maintain an orbit that was a constant distance from Earth while using a tether to deploy a small subsatellite containing instruments down to regions of the upper atmosphere that the shuttle could not safely descend to without losing orbital velocity and being forced into an early reentry and landing.

Proposals for various types of space elevators and tether systems date back to the late nineteenth century. Scientists from many nations suggested ideas that could exploit the gravity gradient, making use of the fact that the farther an object is from the Earth the less it apparently weighs, but they were frustrated by limitations in available materials and technology. For example, one of the pioneers of Russian rocket science, Konstantin Tsiolkovsky, proposed a space railroad in the 1890's. Tsiolkovsky theorized that immensely tall towers could be built that would serve as elevators to lift freight into space. A height of 36,000 kilometers (approximately 22,000 miles) would be required for the towers to reach the point where objects attain weightlessness. Tsiolkovsky envisioned eventually circling the Earth with a cosmic railway, but his ideas remained unrealized as even much shorter towers were beyond the capabilities of nineteenth century engineering. Almost one hundred years later engineers revived the concept of a space railroad, although in a different form than Tsiolkovsky had proposed.

As the space program progressed during the 1960's and 1970's, scientists and engineers began to discuss approaches to space elevators and space railways that would take advantage of natural forces such as gravity and the Earth's magnetic field. Launching spacecraft into orbit or deep space using rockets requires tremendous amounts of energy to lift the spacecraft beyond the attraction of the Earth's gravity. Using a tethered satellite system to serve as a skyhook to raise freight or passengers from the Earth's surface could be both more energy efficient and dependable than conventional rocket boosters. Ideas for possible space elevators ranged from relatively simple schemes to elaborate, globe-circling systems. Rather than building a tower from the ground, engineers suggested lowering a cable from an orbiting satellite or space station. While the simplest tethered satellite systems proposed, such as that advocated by Soviet scientist Yuri N. Artsutanov in the 1950's, generally involved placing a single satellite in a geostationary orbit above a fixed point on the Earth's surface, other plans were much more elaborate. In the 1980's a British scientist, Paul Burch, and an American, Jerome Pearson, both proposed using electromagnetic forces rather than rigid towers to hold a cosmic railway in place. Burch suggested launching a series of electromagnetic coils into orbit and then connecting them to form a tube circling the globe. By using a connecting wire that conducted electricity, a magnetic field would be created that would accelerate the wire to speeds above orbital velocity. This in turn would create a force pushing the tube away from the Earth's surface. Such a railway could be placed in a relatively low orbit for use as an intermediate platform to launch research satellites into higher orbits or space craft into outer space. Materials would be lifted from the Earth's surface to the railway using tethers.

The advantage to the ring suggestion was that it could be placed in a fairly low orbit, perhaps only one hundred kilometers or so above the Earth's surface, while the system employing a geostationary satellite would require that satellite be placed in a much higher orbit, 36,000 kilometers, in order for the satellite to remain in position above one spot on the Earth's surface. Many of the proposals for a geostationary tethered satellite system also involved a series of electromagnetic rings. Rather than circling the globe, the rings would be placed to form a tube leading straight out from the Earth, creating an elevator to the satellite. By the mid-twentieth century advances in knowledge and materials since Tsiolkovsky's time made such elevators to the stars seem more technologically feasible, but the incredibly high costs, the length of time needed for completion, and numerous safety and environmental considerations made it unlikely that such a project would ever be completed. Pearson estimated it would take thirty to forty years to build the globe-circling ring Burch had proposed.

In addition, the possibility for catastrophic damage on a global scale caused even the most fervent advocates of such tethered systems to admit that such systems would be best developed for use on the Moon or other planets, such as Mars. Several hundred kilometers of even microthin polymer cable falling from the sky could devastate a wide area around the ground station for a skyhook system. The 1974 Skyhook report co-authored by two Italian scientists, Giuseppe Columbo and Mario D. Grossi, suggested instead that an excellent application for a tethered satellite system would be to deliver and pick up cargo to and from the Moon's surface. Such a Skyhook system would eliminate the need for spacecraft to actually land on the Moon, theoretically reducing both the costs and the risks involved in lunar exploration and development.

Intriguing though these ideas for space elevators were, research in space with tethered satellites proved more feasible than exploring the elaborate schemes for Earth-anchored systems. Although early experiments involving spacecraft and short tethers revealed numerous technical obstacles to be overcome, engineers and scientists still believed tethered satellite systems could serve many different purposes in space. For example, engineers believed that applications might include using tethers to connect two spacecraft that would then revolve around each other to create artificial gravity.

Beginning in the 1960's both Soviet and American spaceflights involved experimentation with tethers. In 1966, during two manned missions, Gemini XI and Gemini XII, NASA used a 100-foot long Dacron strap to connect the Gemini spacecraft with the upper stage of an Agena rocket as part of research into docking methods in space. Astronauts reportedly could not get the tether to remain taut as both the Gemini capsule and the rocket upper stage tumbled erratically.

Other possible applications of tethered satellite systems include using tethers to connect storage tanks for fuel or other volatile substances a safe distance from a space station in orbit. Dangerous materials could be stored in canisters deployed well away from the space station with the canisters being reeled in to the station when needed. Fuel could be stored to be used by spacecraft that had launched from Earth and then needed to take on additional fuel to leave Earth orbit as part of research missions to other planets or to the Moon. The amount of fuel needed to boost a satellite or space craft out of orbit at 36,000 kilometers is extremely small compared to the amount of fuel required to lift it from the Earth's surface to that orbit to begin with. By using tethered satellites as storage, a space station could serve as a staging area for final assembly for both manned and unmanned explorations of the solar system.

Tethered satellite systems could also be utilized for conducting research in the upper atmosphere. Small, expendable sub-satellites could be connected to larger satellites or the space shuttle and lowered hundreds of kilometers into the upper atmosphere to record scientific measurements unattainable by other methods. These small subsatellites could either be retrieved by reeling them in or cut loose to burn up harmlessly as their orbit decayed on reentry. Such a tethered satellite system could provide data for a variety of scientists. Meteorologists might devise instrumentation for use in weather forecasting, while aerospace engineers have suggested using tethered satellites to obtain data for use in developing a space plane, that is, a space craft that could leave from the Earth's surface and fly into space rather than being lifted by external rockets as in the case of the space shuttle.

Another proposed use for tethered satellite systems was that of generating electricity. Electricity is created when a conductor, such as a copper wire, is passed through a magnetic field. Dragging a satellite connected to the Shuttle with a conducting wire many kilometers long through the Earth's magnetic field theoretically would have the same results that passing a wire through a magnetic field on the Earth's surface would. Engineers suggested that if a subsatellite was connected to a main satellite via a tether composed of a conducting material, the system would generate electricity that could be used to power a satellite or to recharge existing batteries. NASA conducted a number of experiments to test this theory in 1992 and 1993.

The first experiments involving NASA's Tethered Satellite System (TSS) took place aboard the Space Shuttle *Atlantis* during STS-46 space shuttle mission as part of a joint research effort with the Italian Space Agency. Using a deployer system and satellite developed by Martin Marietta for the Italians, astronauts attempted to deploy the satellite upward away from the *Atlantis*. The protocol for the experiment called for deploying the satellite on a tether extending twenty kilometers from the shuttle and conducting 5,000 volts to the orbiter. When the astronauts attempted to deploy the satellite, however, they were unable to do so. The tether snagged shortly after the initial deployment and never extended farther than 256 meters from the shuttle and conducted only 50 volts. The *Atlantis* crew succeeded in unsnagging the tether, but ground controllers canceled any further attempts at deployment. The satellite was successfully retrieved for re-use.

Following the unsuccessful attempts of the *Atlantis* crew to deploy a tethered satellite in 1993, NASA launched two unmanned tether experiments. The first employed a Small Expendable Deployer System (SEDS) and was designed to test deployer technology. The second, the plasma motor generator experiment, was meant to investigate the behavior of electrodynamic tethers and hollow-cathode plasma contactors in space. These missions, as well as the second SEDS mission the following year, were flown as secondary payloads on Air Force Global Positioning System missions.

The Air Force launched the Delta II ELV carrying SEDS-1 on March 30, 1993, as part of the Delta II GPS-31 mission. SEDS-1 was intended primarily to test the components of the deployment system such as the reel and the brake. SEDS-1 consisted of a deployer housing a light-weight spinning reel, a brake, ether turns counter, tether cutter, and electronics and an end mass payload connected to the deployer with a 20-kilometer-long tether. The tether, which was 0.75 millimeters in diameter, was made from an eight-strand braid of a high strength polyethylene polymer, Spectra 1000. The end mass payload carried three primary science sensors: a three axis accelerometer, a three axis tensiometer, and a three axis magnetometer. The end mass payload was designed to function independently of the deployer and so carried its own battery and electronics.

After achieving orbital velocity, the SEDS-1 deployer successfully completed the downward deployment of the end mass payload. A brake under the control of an on-board timer was applied to the free reeling deployment of the tether as it

neared its end and brought the subsatellite to a smooth stop. Following the successful deployment, the deployer's tether cutter severed the tether. The end mass payload transmitted data for approximately twenty-two hours before burning upon re-entering the atmosphere. NASA considers SEDS-1 to have been the first successful twenty-kilometer space tether experiment.

In June, 1993, NASA launched the plasma motor generator as a payload on the Air Force Navstar II-21 Delta II second stage. A subsatellite containing a hollow-cathode plasma contactor was connected to a hollow-cathode plasma contactor on the Delta with a 0.5 kilometer conductive tether. The two contactors provided a low-resistance path with the ionosphere for the tether current. The experiment was designed to test a possible system for use in grounding high power systems in space. Research has shown that interactions between high-voltage power systems, such as those on the space shuttle, and the ambient Earth orbit plasma, can result in arcing and sputtering, leading to severe structural damage. A tethered system using hollow-cathode plasma contactors could be used to prevent such damage as well as serving as a source of power or to alter an orbit. When current on the tether flowed in one direction, the tether could actually serve as a propulsive mechanism to increase the orbital velocity. If the current were reversed, it would have the opposite effect. The tether would then serve as a brake, slowing the spacecraft and eventually causing it to move closer to the Earth's surface.

SEDS-2, launched almost a year later on March 10, 1994, as part of the Delta II GPS-6 mission, also tested a downward deployment of a tethered satellite system. This mission was designed to test long-term tether dynamics. After a successful deployment, the end mass payload remained connected to the deployer with a twenty-kilometer tether consisting, like SEDS-1,of a 0.75 millimeter eight-strand braid of Spectra 1000 polymer. The SEDS-2 end mass payload transmitted data for approximately ten days before exhausting its batteries, although space debris or a micrometeorite severed the tether after only five days had elapsed.

Knowledge Gained

By the mid-1990's attempts at implementing tethered satellite systems had met with mixed results. The upward deployment of a tethered satellite system from the space shuttle *Atlantis* had proved unsuccessful. A NASA investigation concluded that the deployment mechanism had been insufficiently tested prior to launch. Last-minute changes to a crucial bolt caused the tether to snag, a problem that by itself need not have been fatal to the mission. However, when thrusters that investigators concluded were too small failed to provide adequate motive power to the satellite, slack in the tether contributed to binding in the reel and deployment halted at 256 meters. The *Atlantis* mission had been designed to test the theory that it was possible to generate electricity to recharge satellite and space shuttle batteries by dragging a conducting cable through the Earth's magnetic field. Because the tether stopped far short of its intended length of 20 kilometers, the theory remained untested. Crew members were able to conduct tests of various parts of the system, including seeing the effect normal shuttle operations, such as releasing cooling water overboard, might have on the conducting nature of the shuttle's surroundings, but many of the big questions regarding conductivity in space remained unanswered.

The successful deployment of SEDS-1 and SEDS-2, however, indicated that many of the problems scientists and engineers had feared would occur with tethered satellite systems were solvable. SEDS demonstrated the feasibility of post-deployment stabilization of tether dynamics by passive means, deployment control, and the long-term dynamics of satellites in space when deployed on a tether of twenty kilometers in length. While some observers of the tethered satellite system experiments claimed the failure of the TSS to successfully deploy from the *Atlantis* raised questions about the validity of the theories underlying the experiment, most scientists and engineers agreed that the problems lay in the hardware employed. NASA researchers remained confident that a tethered satellite system could be developed for future use in the space program.

Context

The small expendable deployer system and tethered satellite system experiments were important to the overall development of the United States space program for several reasons. As NASA moved forward with plans for the construction of a permanent manned space station, numerous technical hurdles remained. Tethered satellites held the potential to serve as research tools, as storage units for hazardous materials and for fuel, for generating electricity, and for use as microgravity laboratories. Connecting a subsatellite to the space station with a retractable tether could allow scientists to take advantage of the gravity gradient in conducting research. A subsatellite lowered even a few kilometers closer to the Earth than the space station itself would possess a different gravity. By moving along the tether, scientists could actually measure the differences in gravity at various points and perhaps answer some fundamental questions about the nature of gravity and its effects on both living and inanimate objects.

The possibility of using tethered satellite systems as an energy source also held great potential for the space program. In the 1990's all satellites and spacecraft that were in long-term orbits around the Earth or engaged in deep space exploration relied either on batteries, photovoltaic cells, or a combination of the two for power. Despite advances in battery technology, the power source for a satellite still contributed considerable mass to any payload. The successful development of tethered satellite systems for recharging satellite or spacecraft batteries could reduce the weight of a satellite's power plant, thus allowing engineers to incorporate more scientific instrumentation into a satellite or to use a smaller rocket for launching the satellite into space. Further, photovoltaic panels were vulnerable to damage in space from space debris and micro-meteors, which reduced their efficiency. A tethered satellite system with a thin conducting cable might withstand damage that photovoltaic panels could not.

Before tethered satellite systems could be incorporated into planning for a space station, however, engineers and scientists needed to know if those systems would behave as predicted by scientific theories. If it proved impossible to maintain tautness between a main satellite and a subsatellite, for example, scientists would know that all components of a research facility in space would have to be connected using only rigid materials. Thus, experiments like SEDS-1 and SEDS-2 that tested basic components of a tethered satellite system, such as the deployer mechanism, were a vital step in advancing the goals of the United States space program as a whole. The knowledge gained regarding post-deployment stabilization meant that further research into tethered satellite systems as part of the space station development was worthwhile.

Bibliography

Anderson, Loren A., and Michael H. Haddock. "Tethered Elevator Design for Space Station." *Journal of Spacecraft and Rockets* 29 (March-April 1992): 233-238. Good discussion of the theories behind tethered systems as well as the problems engineers must consider in designing tether systems. Contains some technical jargon, but is accessible to the general reader.

Asker, James R. "Atlantis to Evaluate Characteristics of Tethered Satellite." *Aviation Week & Space Technology* (July 20,1992): 40-44. A specific description of a proposed tethered satellite experiment that includes a good, easily understandable explanation of both the history and the potential of tethered satellite systems.

Baracat, William A. *Tethers in Space Handbook.* Washington, D.C.: The Administration, 1986. Description of tethered satellite proposals and experiments.

Beletskii, Vladimir V. *Dynamics of Space Tether Systems.* San Diego: Univelt, Inc., 1993. Discussion of theoretical considerations of tethered satellite systems. The advanced mathematical concepts may prove difficult for some readers.

Pearson, Jerome. "Ride an Elevator into Space." *New Scientist* 14 (January, 1989): 58-61. Excellent summary of the evolution of the concept of tethered satellite systems as well as the author's suggestions for future research. Includes illustrations.

Nancy Farm Mannikko

TIROS Meteorological Satellites

Date: April 6, 1960, to September 17, 1986
Type of satellite: Meteorological

A TIROS satellite provided the first television picture from space on April 1, 1960. Thereafter, the satellites in this series would provide high-altitude views that have increased meteorologists' capability to forecast weather.

Principal personages
WILLIAM G. STROUD, TIROS Project Manager
MORRIS TEPPER, TIROS Program Manager

Summary of the Satellites

The first Television Infrared Observations Satellite, known as TIROS 1, was the earliest of a series of weather satellites designed to provide environmental data for the 80 percent of the globe that was not covered by conventional means. The purpose of these satellites is to measure temperature and humidity in Earth's atmosphere, the planet's surface temperature, cloud cover, water-ice boundaries, and changes in the flow of protons and electrons near Earth. TIROS satellites have the capability of receiving, processing, and transmitting data from balloons, buoys, and remote automatic stations.

The TIROS series was the first of four generations of polar-orbiting weather satellites launched and operated by the United States. The series included ten successful launches from Cape Canaveral between April, 1960, and July, 1965. The first eight spacecraft—18-sided polygons 106 centimeters in diameter, 56 centimeters high, covered by solar cells, and approximately 12 kilograms in weight— were in orbits inclined 48 and 58 degrees to the equator. With TIROS 9, engineers made the first attempt to place a satellite in polar orbit from Cape Canaveral, using a series of dogleg maneuvers over twenty orbits to reach that orbit. Similar maneuvers were used to achieve polar orbit with TIROS 10.

Initially, the TIROS spacecraft were equipped with two miniature television cameras, a tape recorder for each camera, two timer systems for programming future camera operations as set by a command from ground stations, and sensing devices for determining spacecraft attitude, environment, and equipment operations. As the program moved ahead, the satellites had progressively longer operational times and carried infrared measuring instruments (radiometers) to study the amount of radiation received and reradiated from Earth.

TIROS 8 had the first Automatic Picture Transmission (APT) equipment, allowing pictures to be sent back to Earth immediately after they had been taken instead of being stored in tape recorders for later transmission. TIROS 9 and 10 were improved configurations; they led to the second generation of TIROS satellites, called ESSA after the government agency that financed and operated them, the Environmental Science Services Administration, later known as the National Oceanic and Atmospheric Administration (NOAA).

TIROS 1 through 10 enjoyed 6,630 useful days and provided 649,077 pictures. TIROS 1 proved television operation in space to be feasible. It had 89 days of useful life and transmitted 22,952 pictures. TIROS 2 studied ice floes and had 376 useful days, providing 36,156 pictures. TIROS 3 made the first hurricane observation and provided the first advance storm-warning. It had 230 days of useful life and sent back 35,033 pictures. TIROS 4 permitted the first international use of its weather data. It operated for 161 days and transmitted 32,593 pictures. TIROS 5 extended the coverage of weather satellites, operated for 321 days, and sent 58,226 pictures. TIROS 6 provided support for the Mercury 8 and Mercury 9 manned space missions, operated for 389 days, and provided 68,557 pictures.

TIROS 7 provided weather data to the International Indian Ocean Expedition, operated for 1,809 days, and sent back 125,331 pictures. TIROS 8 inaugurated the APT direct readout system, had a useful life of 1,287 days, and sent 102,463 pictures.

TIROS 9 and 10, with their improved configurations and near-polar orbits, improved global coverage. They had useful lives of 1,238 and 730 days, respectively, and sent back 88,892 and 78,874 pictures.

The ESSA satellites for the first time provided daily, worldwide observations without interruption in data. There were nine ESSA satellites. ESSA 1 was launched in February, 1966, operated for 861 days, and sent back 111,144 pictures. The average useful life of each of the nine ESSA satellites was nearly three years. Transmitting more than a quarter of a million television pictures, ESSA 8 lasted the longest, from December, 1968, until March, 1976. Altogether, ESSA 1 through 9 had 10,494 useful days and returned 1,006,140 television pictures.

In the 1970's, the third generation of meteorological satellites, known as the Improved TIROS Operational Satellite

(ITOS), was developed. ITOS 1, launched in January, 1970, was the first of the TIROS system satellites equipped with television cameras for daytime coverage of the sunlit portion of Earth and infrared sensors sensitive to temperatures of the land, sea, and cloud tops for both daytime and nighttime coverage. A single ITOS spacecraft furnished global observation of Earth's cloud cover every 12 hours, as compared to every 24 hours with two of the ESSA satellites.

A second ITOS satellite was launched successfully in December, 1970. Following a launch failure in October, 1971, four ITOS satellites were launched successfully, the last one being ITOS H in July, 1976. When the Environmental Science Services Administration became the National Oceanic and Atmospheric Administration, the satellites were renamed: ITOS A became NOAA 1; ITOS D, NOAA 2; ITOS F, NOAA 3; ITOS G, NOAA 4; and ITOS H, NOAA 5. The ITOS (NOAA) satellites were launched from the Western Test Range at Vandenberg Air Force Base in California.

The TIROS-N/NOAA series is the fourth generation of the meteorological satellites built by RCA. These carry a wide variety of new and considerably more advanced sensors. Besides providing weather imagery data, these satellites transmit atmospheric and sea surface temperatures, water vapor soundings, measurements of particle activity surrounding Earth, and data gathered from balloon-borne, ocean-based, and land-based weather-sensing platforms.

Sensor systems on the TIROS-N satellites include the TIROS Operational Vertical Sounder (TOVS), which combines data from three complementary sounding units to provide improved temperature and moisture data from Earth's surface up through the stratosphere; the Advanced Very High-Resolution Radiometer (AVHRR), which gathers and stores visible and infrared measurements and images, permitting more precise evaluation of land, surface water, and ice as well as information on cloud conditions and sea surface temperatures; the Space Environment Monitor (SEM), which allows the measurement of energetic particles emitted by the Sun over essentially the full range of energies and magnetic field variations in Earth's near-space environment (these measurements are tremendously helpful in determining solar radiation activities); and ARGOS, which permits the collection and transmission of environmental data from platforms on land, at sea, and in the air (ARGOS also determines the geographic location of platforms in motion on the sea surface or in the air).

The first satellite in the TIROS-N series was launched in October, 1978. It operated for twenty-eight months, twice its designed mission life. The second in the series, NOAA 6, was launched in June, 1979, followed by NOAA 7 in June, 1981. NOAA 8 was launched in March, 1983, NOAA 9 in December, 1984, and NOAA 10 in September, 1986. All these launches were from the Western Test Range in California, using an Air Force Atlas launch vehicle.

These NOAA satellites operate in a near-polar, circular orbit at an altitude of 870 kilometers. In their operational configuration, two satellites are positioned with a normal separation of 90 degrees. One of the satellites operates in an afternoon ascending orbit, crossing the equator at 3:00 P.M. local solar time. The second operates in a morning descending orbit with an equator crossing at 7:30 A.M. local solar time. The satellites take 102 minutes to circle Earth.

These TIROS series satellites include instruments from the United Kingdom and France. The United Kingdom, through its Ministry of Defense Meteorological Offices, provides the stratospheric sounding unit, one of the three atmospheric sounding instruments on each satellite. The soundings provide three-dimensional observations of the atmosphere from the surface to approximately 65.5 kilometers. The two TIROS-N satellites provide roughly 16,000 timely soundings a day. The Centre National d' études Spatiales of France supplies the data-collection and location system for the satellite (ARGOS) as well as the ground facilities to process data and make the information available to users. The French operate a receiving station for weather data at Lannion, on the English Channel, about 80 kilometers northeast of Brest.

The TIROS-N/NOAA satellites have been designed with sufficient space, power, and data-handling capability to allow additional instruments to be installed—including a search and rescue system made up of repeater and processor units, a solar backscatter ultraviolet radiometer for ozone mapping, and two Earth Radiation Budget Experiment instruments. NOAA 8, 9, and 10 have these instruments and are known as Advanced TIROS-N (ATN) satellites.

The search and rescue equipment installed on ATN enables detection of emergency transmissions from downed aircraft or ships in distress. The devices support an international search and rescue effort in which the principal partners are Canada, France, the Soviet Union, and the United States. From its beginning in September, 1982, until July, 1988, more than eleven hundred lives had been saved as a result of the program. NOAA, the National Aeronautics and Space Administration (NASA), the U.S. Air Force, and the U.S. Coast Guard are the principal U.S. agencies supporting this international program.

For more than a quarter century, these meteorological satellites have shown improvement in quality, quantity, and reliability. Since 1966, the entire Earth has been photographed at least once daily on a continuous basis. These data are considered almost indispensable for short-range weather forecasts by meteorologists and environmental scientists.

Knowledge Gained

TIROS 1 demonstrated the practicality of meteorological observations from space and the ability to make observations with television. Observations from TIROS 1 confirmed and, in some cases, shed new light on the large-scale organization

The TIROS 8 spacecraft is seen in this artist's conception. (NASA)

of clouds, showing that they were arranged in a highly organized pattern. The cloud vortex stood out dramatically. Meteorologists learned that these cloud vortices could be used to pinpoint accurately atmospheric storm regions. Of the nearly twenty-three thousand cloud-cover pictures transmitted by TIROS 1, more than nineteen thousand were usable for weather analysis. This opened a new era in weather observation by providing data that covered vast areas of Earth that previously had not been available to meteorologists.

Some storms were tracked for as long as four days, underscoring the value of television satellite pictures for weather forecasting. In one case, TIROS 1 observations of a storm near Bermuda in May, 1960, provided a series of pictures recording the degeneration of the storm. The weather maps drawn from conventional data were correlated with the pictures, affording a unique set of data on which to judge future pictures.

TIROS satellite data were used to establish, modify, and improve cloud analysis and to brief pilots on weather. Pictures of cloud cover, transmitted by teletype, were used to refine weather analysis in the region from Australia to Antarctica, and the Australian Meteorological Service used the data in conjunction with special research projects on storms over Australia.

Pictures of ice-pack conditions in the Gulf of Saint

Lawrence proved that weather satellites could locate ice boundaries in relation to open seas. TIROS 2 observations were used in selecting proper weather conditions for the suborbital flight of Alan B. Shepard in May, 1961—America's first manned spaceflight.

The TIROS 3 satellite provided observations of all six of the major hurricanes in 1961. It detected Hurricane Esther two days before it was observed by conventional methods.

In 1963, three TIROS satellites (TIROS 5, 6, and 7) were in operation simultaneously and provided for the first time pictures of the cloud cover of various parts of the world on an almost continuous basis. These pictures formed the basis for adapting satellite data to forecasting methods and the use of numerical prediction models. Special projects surveying the Atlantic and Indian oceans brought clues to the origin and development of tropical cyclones. TIROS satellites have detected sandstorms in Saudi Arabia and have provided storm advisories in many areas of the world.

As a result of their analyses of the satellite pictures, meteorologists have been able to categorize cloud vortex patterns and to determine from those patterns the structure of moisture fields, the cloud structure of frontal zones, and the relationship of cloud patterns associated with jet streams. A series of atlases has been developed and are being used with the satellite data by research meteorologists to investigate a wide variety of theoretical conclusions.

Context

Until the 1960's, when meteorological satellites made their debut, the science of weather prediction left much to be desired. Meteorologists could not forecast with any great degree of accuracy the weather four or five days ahead. Forecasts can now be made for a number of days ahead, most of the time with great accuracy. A number of technological factors have contributed to this improvement. One was the development of the computer, which permits the analysis of millions of bits of information at high speeds so that the physical laws governing atmospheric motion can be calculated more accurately. The development of new communications links, enabling the rapid dissemination of meteorological data over great distances, is another.

Another advancement, equally important in the eyes of many, was the development of meteorological satellites. The development and operation of TIROS and other meteorological satellites have opened the inaccessible areas of the world to observation.

Improved weather forecasting extends to many daily activities of society. For example, improved forecasts are of inestimable value to agriculture. Better forecasting means better crop management and protection. A reliable forecast during the fruit-growing season in California or Florida could save fruit growers millions of dollars.

Improved forecasting offers benefits to transportation, to construction, to manufacturing, and to recreation. Better forecasts allow the routing of aircraft so that safety and comfort are enhanced. Thus, delays are avoided and money is saved as a result of improved fuel consumption. In construction, improved forecasting permits contractors to plan and schedule construction work as environmental conditions dictate. Improved forecasting also permits manufacturers to plan production, particularly when sales are sensitive to weather conditions.

Satellite pictures provide unique assistance to many. Images of the polar ice have helped vessels that otherwise would have been trapped find a way through the ice and have been used to assist ships that were caught in ocean storms. Weather services for Antarctica and for offshore oil-drilling platforms make it possible for activities in these locations to be undertaken in comparative safety. The infrared sensors can scan the ocean surface and determine the temperature, providing extremely useful information to the fishing industry, for example — temperatures can help locate schools of certain kinds of fish.

Meteorological support for humanity's daily activities on land, on sea, and in the air underscores the importance of continued measurements from space. Such measurements are being carried out on a global scale by all seafaring nations. Canada, France, Japan, India, the Soviet Union, the European Space Agency, the United States, and others all have active weather satellite programs.

Bibliography

Grieve, Tim, Finn Lied, and Erik Tandberg, eds. *The Impact of Space Science on Mankind*. New York: Plenum Press, 1976. Contains edited summaries of papers presented at the Nobel Symposium in Spatind, Norway, in September, 1975. In a paper presented by Robert M. White, Administrator of the National Oceanic and Atmospheric Administration, the use of environmental satellites is described. The use of satellites and their benefits to mankind are outlined.

Jakes, John. *TIROS: Weather Eye of Space*. New York: Julian Messner, 1966. A nontechnical review of the significance of TIROS and the benefits of the meteorological satellites. Reviews history of meteorological research as well as the development of the TIROS payload. A list of accomplishments is given, underscoring the benefits of satellite observations.

National Aeronautics and Space Administration. *Significant Achievements in Satellite Meteorology, 1958–1964*. NASA SP-96. Washington, D.C.: Government Printing Office, 1966. This volume is one in a series that summarizes the progress made by NASA's Space Science and Applications Program from 1958 through 1964. It reviews NASA's meteorological program and its benefits. Describes TIROS, Nimbus, and the meteorological sounding-rocket program.

Rosenthal, Alfred. *The Early Years, Goddard Space Flight Center: Historical Origins and Activities Through December, 1962*. Washington, D.C.: Government Printing Office, 1964. Outlines the historical origins of the Goddard Space Flight Center and its development through 1962 and contains historical documents concerning the establishment of Goddard as a NASA center.

Widger, W. K., Jr. *Meteorological Satellites*. New York: Holt, Rinehart and Winston, 1965. As a member of the team that developed TIROS, the author provides an interesting overview of the TIROS program itself and a definitive account of the value of satellite information, the equipment that provides the satellite weather data, and early results of the scientific investigations obtained from TIROS.

James C. Elliott

TITAN LAUNCH VEHICLES

Date: Beginning December 20, 1957
Type of technology: Launch vehicles, expendable

The Titan series of rockets provided the United States with a means of launching satellites into orbit. During its early years, the Titan served as an intercontinental ballistic missile for the United States Air Force.

PRINCIPAL PERSONAGES

ALBERT C. HALL, director of
technical activities for the Titan ICBM
at Martin Marietta Corporation
OSMOND J. RITLAND, director of development for
many early ICBMs, including the Titan
ROGER CHAMBERLAIN, manager of the Commercial
Titan Program
J. DONALD RAUTH, Chairman and Chief Executive
Officer, Martin Marietta Corporation
LAURENCE J. ADAMS, President, Martin Marietta
Aerospace
CALEB B. HURTT, President, Denver Aerospace

Summary of the Technology

During the years immediately following World War II, American officials had predicted that the Soviet Union would not have enough information to create an atomic bomb for at least another three years, and probably for five. It was, therefore, a great blow when in August, 1949, the Soviet Union detonated its first bomb. The Soviets again proved that they were an emerging superpower by performing Intercontinental Ballistic Missile (ICBM) tests in 1957.

Although ICBMs were among the new weapons the United States decided to build in the postwar years, several factors prevented their production before 1954. The major cause can be traced to the lack of conviction on the part of many influential scientists that a rocket could possess enough accuracy, let alone enough strength, to make the system feasible. In addition, such an undertaking would have required enormous financial resources and skilled manpower, neither of which was available to the military until it had become clear that the United States was lagging behind the Soviets.

As a response to Soviet ICBMs, Americans designed several rockets. In this field the United States had a great advantage, for during the closing stages of World War II, many German rocket experts had surrendered to the Americans

rather than face capture by the Soviets; these researchers became American citizens, willing to help their new country advance its weaponry.

The first American ICBM was the Atlas. It stood 24.4 meters tall and had two stages. During the early part of the program, both stages were ignited at lift-off to avoid having the rocket fail once the first stage had exhausted its fuel. The special Air Force Strategic Missiles Evaluation Committee and the Research and Development (RAND) Corporation published reports in 1954 and 1955 in which they cited two reasons that Atlas should be replaced: the necessity of igniting both boosters at the same time and the unsuitable nature of a thin pressurized skin for high-acceleration lift-off.

President Dwight D. Eisenhower ordered the development of a new ICBM, to be called Titan, in mid-1955. The contract was awarded to the Martin Company (later the Martin Marietta Corporation). It was to have a range of approximately 11,000 kilometers. The total thrust produced at lift-off was to be 1,343 kilonewtons, while the weight of the rocket (including fuel) was to be 99,790 kilograms. Its height was to be 29.87 meters, its maximum diameter 3.048 meters.

Titan was to have two stages that would ignite separately, using a propellant of kerosene and liquid oxygen. Two huge tanks were located inside the first stage. The bottom tank housed the kerosene while the top housed the liquid oxygen. Two pumps pushed the liquids to the combustion chamber, where they were ignited. Once the fuel was consumed, the separation rocket fired and the first stage was freed. The second stage was much like the first, except that it contained, in addition to the main engine, four smaller rockets to guide it. Above the second stage were the nuclear warhead and the guidance systems.

After lift-off, Titan rose vertically for twenty seconds. It then arched into a curved trajectory for one hundred seconds, at a speed of 8,530 kilometers per hour. Its fuel exhausted, the first stage separated. The second stage ignited and propelled the rocket to a speed of up to 28,970 kilometers per hour. Four vernier engines (stabilizers) corrected velocity and trajectory. Once this had been accomplished, the second stage was released, and the nose cone fell toward Earth. As a missile, the Titan reached a height of 805 kilometers.

In the early 1960's, researchers realized that the Titan system had some flaws. For example, it was necessary to fuel the rockets immediately before the launch so that the liquid oxygen would not evaporate. Furthermore, the range of the Titan was considered insufficient. Martin Marietta and the National Aeronautics and Space Administration (NASA) were contracted to build a successor to Titan.

Titan 2 was completed in 1964. It stood 33.22 meters tall, with a diameter of 3.048 meters. It weighed 185,034 kilograms. For its fuel it used a combination of nitrogen tetroxide and Aerozine 50. The two stages together produced a total thrust of more than 2,223 kilonewtons.

When NASA built Gemini, a successor to the Mercury space capsule, the Titan was seen as the perfect launch vehicle, with one exception: The Titan suffered from what was known as the "pogo effect" (named for the "pogo stick"), violent, up-and-down motions as the rocket rose. These oscillations mattered little when the Titan carried warheads; for human passengers, however, the pogo effect would create unacceptable conditions. Designers were able to eliminate the worst of the pogo effect, and on April 8, 1964, using Titan 2, the first Gemini spacecraft was launched, followed on March 25, 1965, by the first manned launch.

The Gemini program as a whole was considered a success. The Gemini spacecraft carried two astronauts, an improvement over the crowded, one-man Mercury capsule. During this program several records were set. The first docking with another spacecraft occurred, and the first American walked in space. For a time it was suggested that a variation of the Titan 2 vehicle should be used for the Apollo program. This plan was dropped in favor of using the more powerful Saturn launcher. Titan 2 was also intended to launch the ill-fated Dyna-Soar, a predecessor of the space shuttle.

From 1965 onward, Titan 2 was replaced by Titan 3. Titan 3, the modern-day heavy launch vehicle, can propel into orbit loads of 15,875 kilograms and handle as much as 3,175 kilograms for interplanetary missions. The core section of the rocket, sometimes known as Titan 3A, has two stages and is commonly flanked by boosters. The third stage, or transtage, contains the payload and the final fuel. Titan 3 has four configurations commonly in use.

The Titan's first stage is 22.25 meters tall and provides approximately 2,356 kilonewtons of thrust. The second stage measures slightly more than 7 meters and is capable of more than 445 kilonewtons of thrust. Both stages use nitrogen tetroxide and a mixture of unsymmetrical dimethylhydrazine (UDMH) and hydrazine. The transtage is of varying heights and can be as tall as 15.24 meters when launching Air Force payloads. It uses the same type of fuel as the first two stages. The two boosters are 26 meters long and use solid propellant.

The Titan 3B core has two stages and permits a variety of upper stages to be fitted, the most common being the Agena. It is not launched with boosters. Its cargoes have included Air Force reconnaissance satellites. The Titan 3B uses the standard inertial guidance system, which enables the satellite to maintain a predetermined course.

Titan II. (Lockheed Martin)

The Titan 3C, first launched in 1965, has a takeoff thrust of 10,490 kilonewtons. Its height is 47.85 meters, and it carries two boosters that are 26 meters in length with diameters of approximately 3 meters. At takeoff, the two boosters are the only source of thrust. The core is ignited just before the boosters are released. The fuel consists of nitrogen tetroxide and Aerozine 50. Approximately 80 percent of the satellites launched by the Unites States into geosynchronous equatorial orbit use a Titan 3C rocket. A geosynchronous orbit is one that keeps precise pace with the rotation of Earth, thereby maintaining a stable position relative to any given site.

The Manned Orbiting Laboratory (MOL), which was to have undertaken reconnaissance projects, was slated for launch by the Titan 3C. Growing costs and the high performance of a previously launched spy satellite resulted in the cancellation of the project; yet, the Titan 3C has continued to launch signal intelligence (SigInt) satellites.

Although the Titan 3D is similar to the 3C in many respects, it does not have a transtage. Instead of depending on inertial guidance, it depends on radio guidance. Titan 3D is the rocket used to launch the so-called Big Bird satellite.

In February, 1974, the Titan 3E-Centaur, a modified version of the Titan 3D, was first tested. Although the test ended in failure, the Titan 3E-Centaur performed flawlessly when it launched the Viking spacecraft to Mars in 1975.

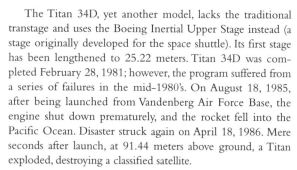

Titan IV. (NASA)

Launch of Voyager 1 aboard Titan/Centaur-6 on September 5, 1977. (NASA)

The Titan 34D, yet another model, lacks the traditional transtage and uses the Boeing Inertial Upper Stage instead (a stage originally developed for the space shuttle). Its first stage has been lengthened to 25.22 meters. Titan 34D was completed February 28, 1981; however, the program suffered from a series of failures in the mid-1980's. On August 18, 1985, after being launched from Vandenberg Air Force Base, the engine shut down prematurely, and the rocket fell into the Pacific Ocean. Disaster struck again on April 18, 1986. Mere seconds after launch, at 91.44 meters above ground, a Titan exploded, destroying a classified satellite.

The approximate cost of using a Titan 34D to launch a 12,700-kilogram satellite into a polar orbit of 185 kilometers is approximately $60 million if six are purchased at the same time. To place a satellite of 1,815 kilograms into geosynchronous orbit costs $125 million. In the event that only one launcher per year is ordered, the price would be raised by $121 million and $125 million, respectively, per flight. The average cost of the satellites that use the Titan 34D is more than $100 million.

In March, 1985, the U.S. Air Force awarded a $5 million contract to Martin Marietta to build a new and more powerful Titan rocket. The Titan 4's payload size is to be about the same as that of the space shuttle, and its height is to be 62 meters. The USAF's name for it is the Complementary

Expendable Launch Vehicle (CELV).

Knowledge Gained

The early Titan rocket served as a test vehicle. Various alternatives to the Atlas' thin casing were tested on the Titan, as were different fuels. Before a rocket is launched, thousands of modifications are made, ranging from the design of a better engine to the replacement of a first stage. The Atlas was designed so that both stages were ignited at once. When the Titan was built, a means of separating the first stage was devised. A system of igniting the second stage, after separation had occurred, was also designed.

In the process of choosing new propellants it was discovered that nitric acid and turpentine ignited upon contact. This fact proved to be very useful: Instead of having to reignite the fuel every time the rocket was started, it was possible, by mixing the fuel, to produce thrust once again. The combination of a fuel and an oxidizer that ignites spontaneously is called a hypergol.

For the Titan 2, a mixture of UDMH, hydrazine, and nitrogen tetroxide replaced the old mixture of kerosene and liquid oxygen. Before it could be placed in the rocket, many tests were performed. First, the pumps, which were designed to work well with liquid oxygen, had to be replaced with pumps that functioned at higher temperatures. The guidance system, containing some flaws, was replaced by an inertial system. Now that it is no longer necessary to add propellant a few minutes before launch, many of the ICBM silos could be simplified.

With the introduction of the even more powerful Titan 3, and the gradual phasing out of the Titan ICBMs in favor of more modern systems, Titan has become primarily a launch vehicle. As a launch vehicle, the Titan is required to do much more than propel a warhead into space and guide it back to Earth. For example, during the Gemini program the Titan had to boost the spacecraft into orbit, a job once likened to walking a mile with one's eyes shut and arriving no more than ten centimeters from one's destination. The job was made simpler during later flights by the use of an on-board computer that could be used to plot the flight path.

On interplanetary missions, such as the Viking mission to Mars, the Titan required a greater accuracy than on any of its previous missions. The distance from Earth to Mars is about 55,520,000 kilometers. The Titan vehicle benefited directly from the information gained from previous space missions.

When the Titan has reached the correct orbit with a satellite, a mechanism in the transtage allows it to separate from the satellite. The malfunction of this mechanism has resulted in the loss of many satellites. With the cost of spacecraft running into hundreds of millions of dollars, it has proved an expensive mistake.

Context

Titan was the second American ICBM to be produced. The Atlas and Jupiter rockets had been built not out of necessity but by scientists who were interested in doing something new. Engineers soon became content with traditional V-2

rocket design. It was only when the need for ICBMs arose that progress was made.

The Titan provided engineers with a chance to put into use some new practices. Previously it was considered too risky to ignite a second stage after the first had been jettisoned. The use of alcohol and oxygen, though somewhat safer, was thought to provide less thrust than other fuels. Experimentation resulted in the discarding of Titan's first stage before the second stage was ignited, the use of oxygen and kerosene, and finally the mixture of UDMH, hydrazine, and nitrogen tetroxide as fuel.

During the Gemini program, the Titan was found to suffer from the pogo effect. Previously, the condition, though recognized, was regarded as unfortunate but unpreventable. Using the Titan in the manned space program prompted a rocket redesign. When the solution was found, it was applied to other rockets as well.

Until the 1980's, the Titan 3 rockets were used often in the U.S. space program. Yet with the deployment of the space shuttle, an increasing number of companies showed an interest in the more advanced reusable orbiter as a launcher. The U.S. Air Force, however, had regarded the shuttle as too experimental. Expressing its confidence in the basic Titan design, the Air Force awarded Martin Marietta a contract for the successor to the Titan 3 series. With the explosion of the *Challenger* space shuttle in January, 1986, the Air Force's fears were realized, and the Titan was once again in demand for the launching of satellite payloads. After the *Challenger* explosion, all satellites that carry volatile fuel were banned from the shuttle. This ban created a new market for the Titan.

Bibliography

Binder, Otto O. *Victory in Space*. New York: Walker and Co., 1962. One of the best books written in its period, *Victory in Space* answers many key questions that were being asked during the early part of the space program. The author does not set out to write about the Titan rocket, but it is mentioned in the context of other subjects.

Emme, Eugene M., ed. *The History of Rocket Technology*. Detroit: Wayne State University Press, 1964. This collection of essays provides a useful survey of the U.S. space program as a whole. Essays on the Atlas, Thor, Titan, and Minuteman ICBMs are followed by a concluding essay on the Titan 3C. Includes a bibliography.

Ley, Willy. *Events in Space*. New York: David McKay Co., 1969. This book includes sections covering the beginning of satellites and rockets. Ley covers Titan, Titan 2, and Titan 3C. It should be noted, however, that the primary topic is space exploration, not the Titan series.

Newlon, Clarke. *One Thousand and One Questions Answered About Space*. New York: Dodd, Mead and Co., 1962. Newlon answers some of the questions most frequently asked about the Titan ICBM and Titan 2. Information is also given about Titan's contemporaries. Capsule biographies of important people in the field of rocketry are included.

Peterson, Robert A. *Space: From Gemini to the Moon and Beyond*. New York: Facts on File, 1972. This volume shows the involvement of the Titan launcher in the American space program. The Gemini project, MOL, and various other payloads are discussed. Suitable for high school and college reading.

Sobel, Lester A., ed. *Space: From Sputnik to Gemini*. New York: Facts on File, 1965. Gives additional information on the use of the Titan rocket as a launching vehicle during the beginning of the space program. The book ends with the introduction of the Titan 3C.

Taylor, John W. R., ed. *Jane's All the World's Aircraft, 1981–1982*. London: Jane's Publishing Co., 1981. Written for a general audience, this volume includes information on the later series of Titan rockets.

John Newman

TRACKING AND DATA-RELAY COMMUNICATIONS SATELLITES

Date: Beginning April 4, 1983
Type of satellite: Communications

Both the United States and Russia have developed and begun implementation of separate communications satellite systems, which are designed to improve tracking and data-relay capabilities from low-Earth-orbiting manned and unmanned spacecraft. Scheduled to be fully operational by the end of the 1990's, both systems have been hampered by payload losses, malfunctions, and delays.

PRINCIPAL PERSONAGES
DALE W. HARRIS, the Tracking and Data-Relay
Satellite System Project Manager at Goddard
Space Flight Center
CHARLES M. "CHUCK" HUNTER, TDRSS Deputy
Project Manager

Summary of the Satellites

As manned and unmanned missions in both the U.S. and the Russian space programs increased in number and complexity, it became apparent that former tracking and data-relay systems were becoming obsolete. Based on expensive, complicated networks of ground stations, communications ships, and outmoded equipment, the old systems only provided minimal contact with manned spacecraft. They also were unable to process and relay the enormous volumes of data generated by a new era of scientific applications satellites. Upgrading and expanding the old systems was economically unfeasible and inefficient for the task at hand.

In December of 1976, the National Aeronautics and Space Administration (NASA) contracted with the Space Communications Company (Spacecom) to develop and implement a tracking and data-relay system that would be able to transmit data efficiently and continuously from low-Earth-orbiting manned and unmanned spacecraft and maintain almost uninterrupted contact with manned crews. Spacecom, a joint venture between Continental Telecom and Fairchild Industries developed the Tracking and Data-Relay Satellite System (TDRSS). Its basic design consisted of two operational satellites, one in-orbit spare, and a ground station. The satellites would function as repeaters, neither processing nor altering any communications. The uplink channels would receive transmissions from the ground, amplify them, and transmit them to the spacecraft. The downlink channels would receive, amplify, and transmit signals from the craft to

the ground station. When completely operational, the system could provide almost continuous contact with manned space vehicles such as Skylab and the space shuttle. Data transmitting equipment could transfer the entire contents of a medium-sized library in seconds. It would be the largest and most advanced space tracking and data-relay system ever developed. TDRSS would form a vital link between Earth and near space.

The TDRSS ground terminal is located at White Sands Test Facility in southern New Mexico. TRW Defense and Space Systems Group designed, built, and tested the three satellites and integrated the ground terminal equipment. By 1983, the first satellite was ready for launch and the ground station, with its three 18-meter-diameter dish antennae, was completed. The New Mexico location was chosen for its arid climate and clear line of sight to the relay satellites. All transmissions to and from TDRSS pass through the White Sands station.

On April 4, 1983, at 1:30 P.M., eastern standard time (EST), space shuttle mission STS-6 was launched at Cape Canaveral, Florida. After the first of the three TDRSS satellites, TDRS-A, was launched from the *Challenger* payload bay and proper orbit was achieved, it became known as TDRS 1. The satellite, weighing 2,250 kilograms, measures 17 meters across at the two winglike solar panels. The panels provide the craft with electrical power. The design of the satellite consists of three modules. The equipment module, located in the lower part of the core hexagon, encompasses the subsystems that control and stabilize the satellite. It also contains the equipment that stores and manages the power supply and the machinery that operates the telemetry and tracking equipment. The communications payload module is in the upper portion of the hexagon, just below the multiple-access antenna elements. It contains electronic equipment that regulates the flow of transmissions between the various antennae and other communications functions. Finally, the antenna module consists of a platform holding various antennae, including thirty helices of the multiple-access phased array.

An Inertial Upper Stage (IUS) booster rocket was to place TDRS 1 in a geostationary orbit (an orbit in which a satellite revolves around Earth once every 24 hours). A malfunction in the IUS left the satellite tumbling in a 35,317-by-20,273-kilometer orbit, forcing ground controllers to separate the satellite from the IUS, and the craft stabilized itself in an elliptical orbit.

A NASA/TRW team devised a way to use the tiny onboard attitude control thrusters (designed for minor station-keeping duties) to nudge the satellite into its proper orbit through a series of controlled firings. The maneuver took extensive engineering and technical analysis and thirty-nine separate burns to get the spacecraft into its appointed orbit of 67 degrees west longitude on June 29, 1983. Following checkout, it was moved to its operational position of 41 degrees west. In its position 35,690 kilometers above the equator on the northeast coast of Brazil, TDRS 1 is considered the "east" station in the TDRSS network.

With the launch of TDRS 1, the new system became partially operational. The satellite, assisted by a few of the remaining ground stations from the Spaceflight Tracking and Data Network (STDN), began transmitting data to White Sands. Eventually, TDRS 1 would be replaced by another satellite and moved to 79 degrees west to function as the in-orbit spare for the system.

On January 28, 1986, at 11:40 A.M., EST, TDRS-B, which would have become TDRS 2 and the "west" link in the TDRSS network, was destroyed when the space shuttle *Challenger* exploded during launch at Cape Canaveral. The *Challenger* accident caused many delays in the U.S. space program, and these included the TDRSS launch schedule. It finally moved forward when TDRS-C (after launch, TDRS 3), designated to replace the destroyed satellite, was launched from the space shuttle *Discovery* on mission STS-26 on September 29, 1988. The satellite was placed in orbit at 151 degrees west for testing and support of shuttle mission STS-27. After STS-27 completed its mission, TDRS-C (TDRS 3) would be moved to a final position of 171 degrees west over the Pacific Ocean. It would become the "west" station in its geostationary orbit southwest of the Hawaiian and Gilbert Islands.

With the successful launch of TDRS-D (TDRS 4) from STS-29, and the replacement of TDRS 1 as an in-orbit spare, TDRSS would become fully operational. On January 13, 1993, the fifth TDRS (TDRS-F) was deployed from the cargo bay of *Endeavour*. It was the primary payload on the STS-54 mission and later successfully achieved geostationary orbit at 62 degrees west longitude.

It officially became known as TDRS-6 and assumed the role of backup to TDRS-East (TDRS-4). During STS-70, TDRS-G (TDRS-7) was deployed from *Discovery* on July 13, 1995. By the end of 1995, the TDRS constellation was in place. TDRS-1, partially operational and known as TDRS-Z, was moved to 139 degrees west longitude to support National Science Foundation investigations of the South Pole. TDRS-3, also partially operational, is at 85 degrees east, used for space shuttle/Mir rendevous missions and Compton Gamma Ray Observatory operations. TDRS-4 is located at 41 degrees west and is operating as TDRS-East. TDRS-5 is at

174 degrees west as TDRS-West. TDRS-6 is at 46 degrees west and serves as backup for TDRS-East. TDRS-7 is at 171 degrees west and is backup to TDRS-West.

In 1983, the same year the United States launched its first phase of TDRSS, the Soviets announced their plans to develop and implement an advanced satellite tracking and data-relay system known as the Satellite Data-Relay Network (SDRN). Confronted with many of the same problems that the United States was having with its outmoded system, the Soviets had the additional pressure of the sophisticated Mir space station and its proposed Kvant science module with which to deal. Both would collect vast amounts of data, and the old very high frequency (VHF) network of ground stations and communications ships was unable to handle the volume efficiently. The strain on manned space missions under the old system was becoming excessive. Opportunities for communications between Soviet manned missions and the ground were of short duration and irregular frequency under the old system. For example, communications both to and from Mir and the ground were only possible for about four hours each day. The Soviets considered this unsafe and a serious impediment to a smooth work pattern.

Apparently the system would be similar to TDRSS and feature two geostationary repeater satellites, one in-orbit spare, and a ground station. The first satellite was designated as "Luch." Observers believe that Luch, or Kosmos 1700, was launched sometime in early 1986. Listeners to the Mir VHF downlink detected a new signal that indicated that Mir was connected to a geostationary satellite. Later it was noted that Kosmos 1700 was in geostationary orbit over the equator at 95 degrees east. It had the capabilities of communicating with Mir for a forty-minute session during every orbit and had wideband microwave links capable of handling much more data than the very restricted VHF downlinks.

Observation showed that Kosmos 1700 provided television and data relay from Mir to the ground from March, 1986, to February, 1987. At that time, it was noted that the Mir crew had begun to use the old VHF system again, and certain signals indicated that the Mir/satellite hookup was no longer operational. It is believed that the satellite failed and drifted off-station. As of July, 1987, it appeared that the Soviets had abandoned trying to transfer the tremendous amounts of data collected by the Kvant science module until they could get the satellite functioning again.

It also appeared that the Soviets were having other problems implementing SDRN. It is suspected that the system's deployment was further delayed when the Soviets lost the payload launched on a Proton rocket on January 30, 1988, from the Baikonur cosmodrome. Observers believe that the satellite was stranded in a short-lived, low orbit because of a failure of the upper stage of the launch vehicle. Soviet authorities, however, have not reported the loss of any satellite with a Kosmos designation.

In spite of their losses, the Russians are expected to press

Artist's conception of one of the Tracking and Data Relay Satellites which comprise the space segment of NASA'S communications relay system. (NASA)

forward with the development and deployment of a sophisticated tracking and data-relay network for Mir, Kvant, and subsequent low-Earth-orbiting missions.

Knowledge Gained

Before the launch of TDRS-C, with only one satellite in orbit and the ground station operational, TDRSS had significantly influenced U.S. space research and operations. Alone, the first satellite and ground station had stretched communications between Earth and the space shuttle from 15 percent to 50 percent of mission time. Augmented with fifteen STDN ground stations, TDRS 1 and its ground station allowed the astronauts on STS-9, launched in November, 1983, to be the first U.S. shuttle crew to enjoy almost continuous communications with Earth. The European-built orbital research module Spacelab, carried in STS-9's cargo bay, was able to transmit volumes of data to scientists on Earth instantaneously. Researchers were able to respond to project results

while many experiments were still ongoing. On that mission, more data were retrieved through space-to-ground communications than on all other previous U.S. spaceflights combined because of the partially operational TDRSS. Fifty times more information was transmitted on Spacelab's ten-day mission than on Skylab's twenty-four-day mission a decade before. TDRS 1 also returned outstanding pictures from the Landsat 4 satellite, in what was the first satellite-to-satellite-to-Earth data relay.

The potential of TDRSS is vast. The completed system was designed to allow almost uninterrupted voice and data exchange between Earth and U.S. manned spacecraft. The only ground station involved is the New Mexico TDRSS station. All other STDN stations, except those used during shuttle launches and those used to track satellites incompatible with TDRSS, have been shut down. The fully operational TDRSS enables almost continuous command and telemetry communications between the ground control centers and unmanned, automatic research and applications spacecraft orbiting several thousand kilometers above Earth. Its ability to transmit vast amounts of data also eliminates the necessity to send additional bulky equipment aboard these craft for data storage.

A "typical" message from a low-orbiting research spacecraft would be transmitted to the proper TDRSS satellite. The TDRS would then relay that data instantly to the ground station at White Sands. From there, it might be sent back up to a commercial communications satellite and relayed to Goddard Space Flight Center (GSFC) in Greenbelt, Maryland. TDRSS is under the aegis of GSFC, and all data are transmitted there for processing and further routing or storage.

At GSFC, the message might again be relayed to another commercial satellite and sent to an appropriate data center such as the Payload Operations Control Center (POCC). This is where researchers can obtain instant results from their research equipment in space. Many POCCs are located at the home laboratories of researchers on university campuses or in research centers in the United States and abroad. The amount of information transmitted and the information gained under the partially operational TDRSS is more than impressive. The completed system will revolutionize the United States' abilities to increase and develop the commercial, scientific, and industrial potentials of space.

Certainly, the Russians have gained some information and data from their short-lived Kosmos 1700 satellite. The potential gains that their system will provide once it is operational mirror those of the United States.

Context

The Goddard Space Flight Center has played a pioneering role in both manned and unmanned spacecraft tracking and data communications since the earliest days of the U.S. space program. From the first minitrack system to the Spaceflight Tracking and Data Network, GSFC has played a prominent role in keeping track of and communicating with U.S. spacecraft. Until the mid-1970's, these systems consisted of a number of complex, sometimes overlapping networks of ground-based antennae, tracking stations, and communications ships.

In 1972, the STDN was founded when the Space Tracking and Data Acquisition Network (STADAN) merged with the Manned Space Flight Network (MSFN). The move to update and streamline the old system resulted in a network of fifteen international stations. Twelve of the stations tracked

manned and unmanned Earth orbital and suborbital missions, and three were used to support the infant space shuttle program. During the 1970's, STDN was continually upgraded to provide greater data processing capabilities and increase manned flight communications time. Most of the STDN equipment had been installed in the mid-1960's to support the Apollo program. Obsolescence and maintenance difficulties continued to increase.

With the advent of the Space Transportation System program and its schedule of frequent flights and the increasing sophistication of unmanned applications and research satellites, it became apparent that either STDN would have to go through a major upgrade or a new system would have to be developed. Studies revealed that it would be less expensive and more efficient to devise a new system using the latest technological advances than to attempt to refurbish the outmoded STDN. Plans were initiated for a new Space Network that would provide global coverage for U.S. spacecraft using a low-Earth orbit. The system would be able to support the ambitious projects scheduled for the future. The key component of the Space Network would be TDRSS. The system would be cost effective, provide almost uninterrupted coverage, and be able to transmit mountains of data. It would also eliminate the need to go through the political maneuvers necessary to maintain or set up additional ground stations in foreign countries.

The full deployment of the TDRSS and the completion of the $18.5 million ground terminal backup facility has made the Space Network a vital link between Earth and space. Each TDRSS satellite has a functioning life expectancy of about ten years and costs about $100 million.

Some of the projects that benefit from the Space Network include shuttle missions, space stations, commercial satellites, research and data-gathering satellites, and the Hubble Space Telescope. Although the Air Force makes some use of TDRSS, the system does have some disadvantages from a military point of view: The lack of security equipment and the fact that NASA only leases the satellites are two reasons for the armed forces' reluctance to make use of them.

Plans for a Russian geostationary tracking and data-relay system grew out of similar problems with an outmoded and inefficient system and the need to have a network that would handle the increased activity from its manned and unmanned programs.

Bibliography

Branegan, John. "Mir Communications in 1987." *Spaceflight Magazine* 30 (March, 1988): 108–112. The article reports observed communication signals both to and from the Soviet space station Mir. It also relates some background on the old Soviet tracking and data system and the new geostationary Satellite Data-Relay Network. College-level material.

Froelich, Walter. *The New Space Network: The New Tracking and Data Relay Satellite System*. NASA EP-251. Washington, D.C.: Government Printing Office. A booklet published by NASA that describes the Tracking and Data-Relay

Satellite System and the Space Network. Presents technical information in terms comprehensible to the layman. Includes color photographs, graphs, charts, and illustrations. Suitable for general audiences.

Karas, Thomas. *The New High Ground: Strategies and Weapons of Space-Age War.* New York: Simon & Schuster, 1983. An in-depth look at the latest in space-age military weapons and systems and their proposed deployment in space. The book briefly discusses the role of TDRSS in military applications. Some technical material but written for a general audience. Contains an index, notes, and acknowledgments.

National Aeronautics and Space Administration. *Goddard Space Flight Center.* Washington, D.C.: Government Printing Office, 1987. A pamphlet describing the history and programs of the Goddard Space Flight Center.

———. *NASA: The First Twenty-five Years, 1958–1983.* NASA EP-182. Washington, D.C.: Government Printing Office, 1983. A chronological history of NASA and the U.S. space program during its first twenty-five years. Designed for use by teachers in the classroom, it features color photographs, charts, graphs, tables, and suggested classroom activities. Topics include tracking and data-relay systems, applications, aeronautics, manned and unmanned spacecraft, and missions. Suitable for a general audience.

Rosenthal, Alfred, ed. *A Record of NASA Space Missions Since 1958.* Washington, D.C.: Government Printing Office, 1982. Brief summaries and documentation of all NASA manned and unmanned space missions since 1958. Technical data include descriptions of the spacecraft and launch vehicle, payload, purpose, project results, and major participants and key personnel. Suitable for a general audience.

Thomas, Shirley. *Satellite Tracking Facilities: Their History and Operation.* New York: Holt, Rinehart and Winston, 1963. A somewhat dated text, but useful for its history of early spacecraft tracking and data-relay systems and the relationship to the early U.S. space program. Contains an index and footnotes. Suitable for a general audience.

Lulynne Streeter

ULYSSES
SOLAR-POLAR MISSION

Date: Beginning October 6, 1990
Type of mission: Unmanned deep space probe

Ulysses is a joint mission between the National Aeronautics and Space Administration (NASA) of the United States and the European Space Agency (ESA) and is the first spacecraft to explore interplanetary space out of the plane of ecliptic, the imaginary plane above the Sun's equator where all of the planets orbit.

PRINCIPAL PERSONAGES

D. EATON, ESA Project Manager

W. G. MEEKS, NASA Project Manager

JOHN L. PHILLIPS, Principal Investigator, Solar Wind Observations Over the Poles of the Sun (SWOOPS) experiment

M. K. BIRD, Principal Investigator, Coronal Sounding (SCE) experiment

J. GEISS AND G. GLOECKLER, Principal Investigators, Solar Wind Ion Composition Spectrometer (SWICS) instrument

A. BALOGH, Principal Investigator, Magnetic Field (HED) experiment

LOUIS J. LANZEROTTI, Principal Investigator, Heliospheric Instrument for Spectra, Composition and Anisotropy and Low Energies (HI-SCALE)

E. KEPPLER, Principal Investigator, Energetic Particles Composition (EPAC) instrument

ROBERT G. STONE, Principal Investigator, Unified Radio and Plasma Wave (URAP) experiment

J. A. SIMPSON, Principal Investigator, Cosmic rays and solar particles (COSPIN) instrument

M. SOMMER, Principal Investigator, Solar X rays and Cosmic Gamma-Ray Bursts (GRB) instrument

EBERHARD GRÜN, Principal Investigator, Cosmic Dust (DUST) experiment

B. BERTOTTI, Principal Investigator, Gravitational Waves experiment

Summary of the Mission

The Ulysses mission is unique in that it is the first spacecraft to explore the region around the Sun out of the plane of the ecliptic. The plane of the ecliptic is the imaginary plane above the Sun's equator where all of the planets, including Earth, orbit the Sun. Before the Ulysses mission, all manmade spacecraft were confined to this plane, and our under-

standing of the region around the Sun, the heliosphere, was limited to what we could observe from within the plane of the ecliptic.

In 1977, the ESA added an "Out of Ecliptic" mission to its scientific program, later renamed the International Solar-Polar Mission (ISPM). Originally, it consisted of two spacecraft, one provided by ESA and one provided by NASA, but NASA cancelled its portion in 1981 due to budgetary restraints. This decision caused anti-NASA feelings in ESA and cancelled a primary mission objective of simultaneously observing both solar polar regions. In 1984, the project was named Ulysses. In the Inferno, Dante tells the story of the legendary Greek hero Ulysses, King of Ithaca, who, restless for adventure, decided to explore beyond the known world, which at that time ended with Gibraltar. He challenged his companions "To venture the uncharted distances...of the uninhabited world beyond the Sun...to follow after knowledge and excellence."

The primary objectives of the mission are to study the properties of the heliosphere at high latitudes away from the plane of the ecliptic. Specifically, Ulysses is to study the properties of the solar corona (the hot, thin outer regions of the solar atmosphere), the solar wind (the flow of high speed charged particles from the Sun), the heliospheric magnetic field, solar energetic particles, galactic cosmic rays (very high energy charged particles), and solar radio bursts. A Jupiter flyby also provided an opportunity to study the Jovian magnetosphere (the region surrounding Jupiter dominated by the planet's magnetic field). Additionally, radio-science investigations were conducted at specific times during the mission using the spacecraft and ground communications to conduct scientific measurements.

NASA provided the spacecraft power supply, launch vehicle, mission operation center at the Jet Propulsion Laboratory (JPL) in Pasadena, California, and tracking via the Deep Space Network (DSN). The ESA provided the spacecraft and is responsible for its operation. The scientific teams responsible for analyzing data collected by instruments on Ulysses come from both Europe and the United States. Ulysses was originally scheduled to be launched in 1983 but was delayed until 1986. The *Challenger* accident further delayed the mission until it was launched from the space shuttle *Discovery* on October 6, 1990, at 1:48 P.M., eastern daylight time, using a two-stage solid-fuel rocket. It was traveling at 11.3 kilometers

Artist's concept of the Ulysses spacecraft as it loops around Jupiter to gain a gravity assist from the giant planet in February, 1992. (NASA)

per second when it left Earth, making it the fastest interplanetary spacecraft ever launched. But even this high speed was not sufficient to achieve a solar-polar orbit. In fact, no man-made rocket could provide the velocity to reach the high solar latitudes. This is because the Earth orbits the Sun at thirty kilometers per second, so any rocket would have to cancel out this motion in addition to provide the necessary velocity to achieve the solar-polar orbit. So Ulysses was sent to the planet Jupiter, where that planet's large gravitational field could accelerate Ulysses out of the ecliptic plane.

The spacecraft weighed 367 kilograms at launch, including 33.15 kilograms of hydrozine rocket fuel and 55 kilograms of payload which included nine instruments. Because the spacecraft operates at a great distance from the Sun, solar panels would not supply sufficient amounts of power. Instead, power for the spacecraft is supplied by a Radioisotopic Thermoelectric Generator (RTG) that produces electricity via the nuclear decay of radioactive isotopes. Ulysses' most visible features are a 1.65 meter, Earth-pointing high-gain antenna for communications and a 5.6 meter boom mounted

opposite of the RTG, which carries several scientific sensors. The purpose of the boom is to minimize electromagnetic and radiation contamination of the instruments by the RTG. Most of the remaining scientific instruments are mounted on the main body as far removed from the RTG as possible. There is also a 7.5 meter boom and a 72 meter dipole wire boom that serve as electrical antennae. The spacecraft rotates five times per minute (5 rpm) with the HGA on the axis of rotation.

The HGA communicates with Earth with 20 watt X-band and 5 watt S-band transmitters. All commands from Earth are carried on the S-band, while the satellite sends telemetry data with the X-band and ranging codes with the S-band. The downlink bit rates can be selected at either 128 bits per second, or at 8,192 bits per second. Continuous coverage by ground stations was considered impossible for such a long duration mission, so data is stored on two on-board redundant tapes for sixteen hours, then replayed over an eight hour period, with real-time data interleaved in with the recorded data. There are also systems on-board for the detection of system failures and for safe reconfiguration. This is required for expected periods of nontracking and because of the long period of time it takes for a signal to travel between the spacecraft and Earth. There is also a search mode to reacquire the Earth if no signal is received within thirty days.

The nine instruments on-board are: the Vector Helium Magnetometer/Flux Gate Magnetometer (VHM/FGM) for studying the Sun's magnetic field around Ulysses; a solar wind plasma instrument for Solar Wind Observations Over the Poles of the Sun (SWOOPS); the Solar Wind I on Composition Spectrometer (SWICS) for measuring the composition of ions in the solar wind encountered during the mission; the Unified Radio And Plasma wave experiment (URAP) which measures the electric field in the vicinity of the spacecraft; the Heliosphere Instrument for Spectra, Composition and Anisotropy and Low Energies (HI-SCALE) for measuring the interplanetary ions and electrons throughout the entire Ulysses mission; the Energetic PArticles Composition instrument (EPAC), which was used to study energetic particles in interplanetary space; Cosmic Rays and Solar Particles (COSPIN) for studying high energy charged particles encountered by the spacecraft; the Cosmic Dust experiment (DUST) for direct measurement of particulate matter in interplanetary space; the Gamma Ray Burst (GRB) instrument for studying those unexplained phenomena; and the radio antennae which will be used in the study of gravity waves and for coronal soundings. The NASA spacecraft was to carry a camera, so no cameras were included in the instrument package on the ESA spacecraft. All instruments were turned on shortly after launch and the spacecraft was declared in commission in January, 1991. All instruments were functional throughout the mission.

Closest approach with Jupiter occurred at 12:02 Universal Time on February 8, 1992, when Ulysses was about 5.3 astronomical units (A.U.) from the Sun, where an A.U. is the average distance between the Earth and the Sun and is equal to about 150 million kilometers. Measurements by instruments on board Ulysses provided a wealth of information concerning Jupiter's magnetosphere. While similar information had been provided by Pioneer 10 in 1973, Pioneer 11 in 1974, and Voyagers 1 and 2 in 1979 when they passed Jupiter, Ulysses took a path through a region of Jupiter's magnetosphere that was not observed by the four previous spacecraft. Specifically, Ulysses passed through the dusk sector of the magnetosphere, as well as the doughnut shaped ring of highly charged ions encircling Jupiter in the orbit of the moon Io, known as the Io Plasma Torus.

The spacecraft trajectory and the powerful Jovian gravity field worked together to insert Ulysses into a six year orbit perpendicular to the ecliptic plane that first took the spacecraft over the Sun's southern polar regions and provided us with the first view of the regions over the Sun's poles. The maximum southern latitude of 80.2 degrees was achieved on September 13, 1994, at a range from the Sun of 2.3 A.U.

Ulysses then crossed the plane of the ecliptic at a distance from the Sun of 1.3 A.U. in February, 1995, and reached its maximum northern latitude of 80.2 degrees, 2.1 A.U. above the Sun, on July 31, 1995. These maximum latitudes were limited to about 80 degrees due to the requirement that it should not pass Jupiter closer than 6 Jupiter radii in order to avoid that planet's high radiation fields.

The nominal mission ended on September 30, 1995, when the spacecraft left the northern solar high-latitude region to return to Jupiter. The spacecraft and all instruments are functioning well, however, and Ulysses is expected to return to the southern solar high-latitude region in September 2000. It is expected that the spacecraft will detect a very different environment the second time due to the fact that the Sun was in the quiet period of its eleven-year solar cycle for the first orbit but will be in its active period for the second orbit.

Knowledge Gained

Once off the ecliptic plane, the pattern of the solar wind velocity followed a twenty-six-day cycle that was close to the rotation rate of the Sun. This pattern lasted approximately one year, during which time the solar wind speed varied between about 400 kilometers per second and 750 kilometers per second. After this pattern ended, the solar wind speed was a relatively constant 750 kilometers per second, with only small changes from that value. This pattern persisted as the probe went over the southern pole of the Sun and returned towards the ecliptic plane. The twenty-six-day cycle was not re-encountered. As the velocity of the solar wind increased, the density decreased. As a result, the product of solar wind velocity times density remained nearly constant. Solar wind measurements in the northern hemisphere mirrored measurements in the southern hemisphere.

While going southward, Ulysses encountered north and south magnetic fields equally, until it reached the point where the twenty-six-day pattern was detected in the solar wind. After this point the magnetic field was mainly a southward magnetic field, which continued until the point where the solar wind became constant. After this, the magnetic field was only southward in nature. It was expected that Ulysses would find the magnetic field strength to increase over the poles because all magnetic field lines would be funneling in to the surface of the Sun through this region, but measurements showed that the field was actually uniform throughout the polar region. Regular fluctuations in the number of charged particles were also detected. These fluctuations are caused by changes in the magnetic field strength, which are in turn caused by periodic oscillations that originate deep within the Sun. This was the first detection of these oscillations, which had been predicted by theory but never detected on Earth.

It was expected that Ulysses would find larger numbers of cosmic ray particles in the high latitudes, but this did not occur. An unexpected and unexplained observation was the detection of a twenty-six-day cycle in the number of cosmic rays detected by the spacecraft that matched the observed twenty-six-day solar wind cycle. But the observed cycle in the number of cosmic rays continued even after the solar wind cycle ceased and the solar wind became steady. Ulysses also made the first ever detection of neutral particles arriving from outside of our solar system.

Jupiter is a very large radio source and during Ulysses' flyby, instruments were able to measure individual radio signals and identify their sources. Instruments designed to measure the composition of the solar wind were able to measure the composition of the plasma within Jupiter's magnetosphere. Material was identified from three separate sources: the solar wind, Jupiter itself, and the volcanically active moon Io. Plasma from the solar wind and Io were detected in all regions investigated by Ulysses.

Context

Ulysses, a joint project between the ESA and NASA, placed an instrument laden spacecraft in a six-year orbit with a gravitational assist from Jupiter. It is the first spacecraft to go into a solar-polar orbit out of the ecliptic plane and investigate the heliosphere at all latitudes of the Sun.

Ulysses' measurements of the solar wind velocity off the ecliptic plane are considerably different than measurements near the ecliptic plane where the solar wind velocity is highly and irregularly structured: the solar wind velocity averages about 450 kilometers per second and gusts as high as 1,000 kilometers per second. Southern hemisphere measurements were mirrored in the northern hemisphere, indicating the solar wind is faster at latitudes higher than about 20 degrees, and slower and more structured at latitudes less than about 20 degrees. Ulysses measurements also indicated that there is more than one kind of solar wind, supporting previous theories that had suggested different kinds of solar wind with the differences being in the characteristic ion composition, velocity patterns, and density.

Measurements from Ulysses during the Jupiter encounter indicate how the magnetosphere can change in response to changing solar wind conditions, as well as the importance of large-scale current systems in determining the configuration and dynamics of the magnetosphere. Data on the composition of the plasma within Jupiter's magnetosphere provided new information concerning the origin and physical cycle of the Jovian plasma. The ability to identify the sources of particular regions of plasma shows that solar wind plasma is able to penetrate deep into the magnetosphere, while the Io plasma is able to move outwards. This is valuable data for any model of plasma circulation within the magnetosphere.

Changes in the solar wind and related magnetic disturbances give rise to changes in the Earth's magnetosphere which propagate down to the atmosphere and cause effects that impact our society. Some of the effects are radio communication problems, interference with radar signals, satellite communications, and loss of power in large power grids. These effects are most severe in high-latitude regions but can be experienced at much lower latitudes. The study of Jupiter's magnetosphere provides us with insights into the behavior of Earth's magnetosphere. Obviously the study of the solar wind is of great importance to our increasingly technology-dependent society.

Ulysses was very successful in making measurements of Jupiter's magnetosphere and exploring the high latitude regions of the heliosphere. It has provided valuable information concerning several theories, but it has also provided some interesting, and unexpected findings. All instruments are healthy and the second six-year orbit is highly anticipated during which the spacecraft will pass over the poles of the Sun during the active period of the eleven-year solar cycle when the solar atmosphere is expected to be radically different.

Bibliography

Covault, Craig. "European Ulysses Fired to Jupiter, Sun as *Discovery* Returns to Space." *Aviation Week and Space Technology* 133 (October 15, 1990): 22. Provides interesting details for general audiences.

Hathaway, David H. "Journey to the Heart of the Sun." *Astronomy* 23 (1995): 38. Provides an overview of what we know about how the Sun works, including solar oscillations and solar wind. Suitable for general audiences.

Mecham, Michael. "After Long Delay, Ulysses Mission Begins 5-year Voyage to Expand Solar Data Base." *Aviation Week and Space Technology* 133 (October 22, 1990): 111. Discusses the mission and the instrument package in more detail. Suitable for general audiences.

Talcott, Richard. "Seeing the Unseen Sun." *Astronomy* 18 (1990): 30. Provides a good overview of the mission objectives and expected results. Suitable for general audiences.

"Ulysses Spacecraft Prepared for Long-delayed Mission to the Sun." *Aviation Week and Space Technology* 131 (November 6,1989): 25. A good review of the program and its problems in the years leading up to its launch. Suitable for general audiences.

Gordon A. Parker

THE UNITED STATES SPACE COMMAND

Date: Beginning September 23, 1985
Type of organization: Military space agency

Space technology has developed to the point that systems first devised for exploration have become essential to national defense, providing critical functions. A unified command across all armed services now provides the needed operational focus, consolidating control of space assets and activities in support of nonspace missions.

PRINCIPAL PERSONAGES

RONALD REAGAN, President of the United States, 1981-1989

CASPAR W. WEINBERGER, Secretary of Defense

KEN KRAMER, Representative from Colorado

Summary of the Organization

Military space systems are critical elements in national defense. The military applications of space technology were recognized very early in the space age. As American policy expanded to include national defense as well as scientific concerns, the need became apparent for a space command, unified across all armed services, to provide an operational focus. The need intensified as plans for the Strategic Defense Initiative (SDI) became more concrete.

The U.S. Air Force had consolidated its space-related efforts in the Air Force Space Command (AFSC), formed in September, 1982. The parallel Naval Space Command was established in October, 1983. President Ronald Reagan's announcement in March, 1983, that he would endorse plans for the development of SDI once again focused public attention on the military uses of space. The Joint Chiefs of Staff began to consider the formation of a unified space command. After studying a June 7, 1983, proposal from the Air Force Chief of Staff, the Joint Chiefs recommended the establishment of a unified space command to Secretary of Defense Caspar Weinberger on November 8, 1983. Also, Representative Ken Kramer presented a letter, signed by fifty-three congressmen, to President Reagan on November 18, 1983, recommending that a unified space command be a vital part of the SDI organization.

On November 26, 1983, Secretary Weinberger presented the Joint Chiefs' proposal to the National Security Council, recommending that President Reagan approve it. The Joint Chiefs created the Joint Planning Staff for Space (JPSS) in February, 1984, to plan the transition. On November 20, 1984, President Reagan announced activation of a unified space command by October, 1985. The JPSS worked from late 1984 through 1985 establishing organizational roles and relationships and assigning missions for the U.S. Space Command.

One concern of the planners was the North American Air Defense Command (NORAD). The United States and Canada had formed NORAD to provide early warning of invasions by bombers and centralized operation of air defense forces. As the Soviet danger evolved to ballistic missile and antisatellite capabilities, NORAD's atmospheric early-warning line and jet interceptors became less valuable. (Canada participates in the U.S. missile warning and space surveillance system but declines to participate in missile defense research and development.)

The Joint Chiefs addressed Canada's concerns by specifying that the U.S. Space Command would not be a component of NORAD but would provide NORAD with space surveillance and missile warning capabilities. NORAD is responsible to the National Command Authorities of both the United States and Canada for warning of any aerospace attack on the North American continent but does not have any responsibility for ballistic missile defense.

The Joint Chiefs of Staff approved the NORAD commander in chief as the commander in chief of the U.S. Space Command. NORAD headquarters is located at Peterson Air Force Base, with key warning operations inside Cheyenne Mountain, sixteen kilometers to the west, near Colorado Springs, Colorado. The U.S. Space Command headquarters is also at Peterson. The commander in chief is directly responsible to the President through the Joint Chiefs and the Secretary of Defense. The Joint Chiefs of Staff completed assignment of responsibilities and missions in August, 1985. On August 30, President Reagan gave final approval, with the proviso for a one-year review.

The U.S. Space Command, activated September 23, 1985, has three components: the Air Force Space Command, the Naval Space Command, and the Army Space Command. The Air Force Space Command operates most military space systems for the U.S. Space Command, with headquarters at Peterson and other facilities at Cheyenne Mountain Air Force Base and the Consolidated Space Operations Center at Falcon Air Force Base, fourteen kilometers east of Peterson. The Air Force Space Division, another subgroup, is not part of the U.S. Space Command but part of the Air Force

Systems Command. It serves the Department of Defense, planning activities concerning the space shuttle's military use. Its staff also researches, develops, and performs in-orbit testing of Air Force satellites prior to their operation by the Air Force Space Command.

The Naval Space Command, based in Dahlgren, Virginia, has a senior liaison staff at the U.S. Space Command headquarters and two major space-oriented units: the Naval Space Surveillance System (NSSS), based in Dahlgren, and the Navy Astronautics Group (NAG), based in Point Mugu, California. The NSSS functions as a dedicated space-tracking sensor forming an electronic "fence" extending 1,609 kilometers out from both the Pacific and Atlantic coasts, and 4,827 kilometers across the Gulf Coast. It reaches 24,139 kilometers into space and disseminates information to forces at sea. The NSSS also acts as the Alternate Space Defense Operations Center and Alternate Space Surveillance Center. The NAG operates Transit, the oldest operational satellite system.

The Army Space Command, activated on August 1, 1986, as an agency (and as a command in April, 1988), is based at Peterson. It provides military satellite communications and space surveillance support to the U.S. Space Command and support to Army field forces.

The staff of the U.S. Space Command spent most of its first year organizing and hiring new personnel, initiating its planning and operations responsibilities, and establishing relationships with its component commands. These activities included transferring to it the space mission areas of Aerospace Defense Command before the latter's deactivation (NORAD was assigned the air defense responsibilities).

The Joint Chiefs conducted the required review in October, 1986. Besides approving the continuation of one commander in chief for both the U.S. Space Command and NORAD, the Joint Chiefs authorized the U.S. Space Command's deputy commander in chief to be NORAD's vice commander in chief. This ensured continuity of responsibility in supporting NORAD.

On August 14, 1987, the commander in chief reported that the U.S. Space Command Center had reached initial operational capability. The center monitors space events globally, around the clock. It transmits to the National Command Authorities both information and warnings. In the fall of 1987, the Army Space Agency took operational control of space-tracking radar functions and the Defense Satellite Communications System, beginning its transformation into the Army Space Command. Also in 1987, the Army Space Command trained personnel in the use of Global Positioning System (GPS) terminals. These employees received data on position and navigation that allowed speedy, precise response.

During that same year, the Air Force Satellite Control Facility was transferred to the Air Force Space Command from the Air Force Space Division. The satellite control network consists of the Consolidated Space Operations Center (CSOC); the Satellite Test Center in Sunnyvale, California;

seven worldwide tracking stations; and worldwide ground stations for the Navstar GPS, Defense Meteorological Satellite Program, and Defense Support Program. Automation of the Air Force Space Command tracking stations, completed in 1988, allows CSOC and satellite test centers to control the stations remotely.

The U.S. Space Command employs fewer than twelve thousand men and women. The U.S. Space Command has responsibility in three broad areas: space operations, space surveillance and warning, and ballistic missile defense planning. Space operations covers space control, directing space support operations, and operating space systems in support of other commands.

Space control, the primary mission of space operations, includes ensuring interference-free access to and operations in space; denying an enemy the use of space-based systems supporting hostile forces; protecting space-related assets; and surveillance of space objects.

Ensuring access to space includes the U.S. Space Command's effort to devise requirements for a space launch infrastructure. Denying an enemy the use of hostile space-based systems occurs when needed. Research and development in Antisatellite (ASAT) systems technology is ongoing.

Protection of the space-related assets of both the United States and its allies involves passive protection, including such measures as adding maneuvering fuel and hardening electronics, and the U.S. Space Command's Space Defense Operations Center, or Spadoc. Spadoc, located in the Cheyenne Mountain Air Force Base, monitors space activities, detects and verifies potentially hostile acts, and warns space systems owners and operators so they can take appropriate defensive measures if necessary.

The surveillance of space objects is managed by the Space Surveillance Center, also at Cheyenne. The center makes daily observations, predicts approximately where and when man-made objects will reenter the atmosphere, and warns of possible collisions between space objects. With its Space Surveillance Network, it makes thirty to fifty thousand observations daily, detecting, identifying, tracking, and cataloging more than seven thousand man-made orbiting objects. The network uses Earth-based sensors such as nonmechanical phased-array radars that track multiple satellites simultaneously, cameras with telescopes that can detect satellites more than 32,000 kilometers away, and the ground-based electro-optical deep space surveillance system, linked to video cameras to enable transmission to the Space Surveillance Network in minutes.

The second function under the category of space operations, directing space support operations and supporting launch and in-orbit requirements, consists of satellite support and terrestrial support. Satellite support includes both action in support of space-borne forces and operation of satellite systems. Actions consist of launch; telemetry, tracking, and commanding; in-orbit maintenance; crisis operations planning; and recovery. Operations incorporates the transit satellite operation, the Defense Meteorological Satellite Program

(DMSP), and the Satellite Early-Warning System (SEWS). Transit's successor is the Navstar GPS, a twenty-one-satellite radio navigation network designed to provide precise navigation and positioning information for both civilian and military use. The Air Force Space Command operates those GPS satellites already in place. The DMSP, a two-satellite network operated by the Air Force Space Command, provides oceanographic, meteorological, and solar-geophysical data to the Navy, the Air Force, and worldwide weather terminals. SEWS monitors the oceans and known ballistic missile sites worldwide for missile attacks.

Terrestrial support, parallel to satellite support, involves command control communications; surveillance from space; navigation; warnings and indications, including the ballistic missile early-warning system of large stationary radars and phased-array radars; and environmental monitoring for all sea, air, and land forces.

Another aspect of space operations, operating space systems in support of other commands, embraces "force-enhancement" support of communications, navigation, and surveillance for both U.S. and allied ground-based forces.

Space surveillance and warning, the second major responsibility of the U.S. Space Command, involves support to NORAD through providing space surveillance and missile warning data and to commanders in chief needing warnings of attacks for areas outside North America. The third responsibility of the command is ballistic missile defense planning. The U.S. Space Command develops requirements and plans for engaging attacking missiles.

The staff of the U.S. Space Command acts as both operational support and headquarters management. For example, the systems integration, logistics, and support director has important acquisition functions in supporting the Air Force Space Command and ensures operation of command electronics and communications. The intelligence director provides intelligence support to both command headquarters and the space operational intelligence watch crews, who identify space objects and supply strategic-launch warning information. The operations director supports several operational centers in Cheyenne and the U.S. Space Command Center, also providing for the worldwide component operational plans, coordination, and guidance.

The U.S. Space Command performs many other functions. Its deputy director for ballistic missile defense planning devises requirements and interacts with the SDI organization. The Center for Aerospace Analysis provides the U.S. Space Command and other defense agencies with analyses of space systems, air-breathing defense, and missile warning and defense. The command enhances the support provided to operational commanders by space systems and improves support to combatant commanders by codifying operational requirements for such areas as communications, wide-area surveillance, precision navigation, and environmental information.

The U.S. Space Command has established a space annex

for other commands' contingency and operational plans and developed procedures to allow theater commanders to request tactical data support. It has devised worldwide requirements for missile warning information and created procedures to distribute this information quickly. The command has also deployed a mobile command control system designed to survive an attack and provide jam-resistant tactical warning and assessment, as well as data on missile events.

Since its inception in 1985, the command has integrated its components. For example, GPS operational crews come from the Army, and Naval Transit experience shapes GPS operational techniques. Also, critical early-warning sites include Navy personnel. In short, the U.S. Space Command is well established to protect the United States in space.

Context

President and former Army General Dwight Eisenhower formulated the first space policy, maintaining that space should be used for peaceful, scientific activities. He favored using space satellites for "open skies" reconnaissance, instead of high-flying U-2 aircraft, which the Soviets successfully attacked. Eisenhower's emphasis on peace led to the United States' early separation of military from civilian space programs. The first proposal for a unified command of military space activities came in 1959; it had been developed by the Chief of Naval Operations but was basically ignored by the other services.

Each service developed different space interests. The Army focused on launch and booster vehicles; the Navy, high-altitude rockets and early satellite technology, growing into space-based fleet-support communications and navigation systems; and the Air Force, intercontinental ballistic missiles at first, and then extensive satellite applications in communications, surveillance, meteorology, and navigation.

President John F. Kennedy did not want the deployment of reconnaissance satellites publicized, and he had classified military launches by 1962. Yet all services were allowed to conduct space research; the Air Force became responsible for research and development and testing of all Department of Defense space projects.

The United States ratified three space treaties under President Lyndon B. Johnson: the Nuclear Test Ban Treaty, prohibiting nuclear explosions in space; the Outer Space Treaty, banning the orbiting of mass-destruction weapons; and the Astronaut Rescue and Return Agreement. To further his Great Society programs, programs aimed at expanding government's role in social welfare, Johnson emphasized the commercial and domestic benefits of space. Better capabilities in communications and meteorology resulted, aiding both civilian and military uses. The Air Force received some funding (in short supply because of the demands posed by the Vietnam War) for research on the Manned Orbiting Laboratory, a potential surveillance post which was later canceled.

Major space policy changes began with President Jimmy

Carter's Presidential Directive DD-37. It stated the objective of cooperating with other nations to ensure the freedom to pursue activities in space that enhance mankind's security. It also shifted policy by pursuing the survivability of space weapons systems, creating a program to identify which civilian space resources would be incorporated into military operations during national emergencies, and researching ASAT capabilities permitted by international agreements.

President Reagan built on Carter's foundation with National Security Decision Directive (NSDD) 42, issued in July, 1982. It stated that the United States would conduct national security space activities (including communications, navigation, surveillance, warning, command and control, environmental monitoring, and space defense), develop and operate ASATs, and deny enemies the use of space-based systems in support of hostile forces. Reagan also supported the space shuttle, giving priority to national security missions.

President Reagan's NSDD 85, issued on March 25, 1983, announced his plans for the development of SDI, with a goal of defending against missile attack. Reagan promulgated a national space strategy on August 15, 1984. The shuttle would be the primary launch vehicle for national security missions. The Department of Defense was required to implement the 1982 policy and SDI, ensure access to space by supplementing the shuttle with expendable launch vehicles, and emphasize advanced technologies to provide new capabilities and improve space-based assets. With the collapse of the Soviet Union, newly elected president Bill Clinton essentially cancelled SDI.

Bibliography

Covault, Craig. "Ground Troops to Benefit from Army Space Command." *Aviation Week and Space Technology* 128 (April 25, 1988): 80-82. Discusses the formation of the Army Space Command and its work with the U.S. Space Command. Also discusses the Army astronauts at Johnson Space Center.

———. "New Space Operations Center Will Improve Threat Assessment." *Aviation Week and Space Technology* 126 (May 25, 1987): 50, 52. This article offers a description of Spadoc's facilities and operations.

———. "NORAD: Space Command Request System for Surveillance of Soviet Weapons." *Aviation Week and Space Technology* 126 (April 6, 1987): 73-76. Discussion of the request for development of space-based intelligence and surveillance systems to counter a range of new Soviet strategic weapons.

———. "Space Command: NORAD Merging Missile, Air, and Space Warning Roles." *Aviation Week and Space Technology* 122 (February 11, 1985): 60- 62. Includes explanations of how NORAD and the U.S. Space Command will co-operate the new strategies to be implemented, and what the command will control.

———. "U.S. Space Command Focuses on Strategic Control in Wartime." *Aviation Week and Space Technology* 126 (March 30, 1987): 83-84. Descriptions of various operations centers in the U.S. Space Command.

———. "USAF Initiates Broad Program to Improve Surveillance of Soviets." *Aviation Week and Space Technology* 122 (January 22, 1985): 14 -17. This article describes the Air Force Space Command before the U.S. Space Command was formed and the surveillance systems the Air Force operated. Offers information on Soviet space activity as monitored by the Air Force.

McDougall, Walter A. *The Heavens and the Earth: A Political History of the Space Age*. New York: Basic Books, 1985. This scholarly text provides the history, especially the political history, of space exploration. The military uses of space are thoroughly covered.

Michaud, Michael A. G. *Reaching for the High Frontier: The American Pro-Space Movement, 1972–1984*. New York: Praeger Publishers, 1986. The prospace movement (more than fifty advocacy groups, involving more than 200,000 Americans) has developed since the end of the Apollo program. Michaud traces key groups, identifying their origins and goals and telling how they have influenced space policy—however subtly. Includes a bibliography, with many sources on military space activity.

"Several U.S. Military Spacecraft Operating on Final Backup Systems." *Aviation Week and Space Technology* 126 (March 30, 1987): 22-23. A description of what the U.S. Space Command is doing to maintain its surveillance in the light of several military satellites operating without backups.

Smith, Bruce A. "Air Force Supports Demonstration of Surveillance Technology in the 1990's." *Aviation Week and Space Technology* 128 (March 14, 1988): 93, 95. Two examples of space-based surveillance technologies are surveillance of space objects from a space platform and a space-based radar system. The Air Force Space Division is working with U.S. Space Command to define operational requirements. In this article, details are given on these and other ongoing similar projects.

———. "USAF Readies Vandenberg: Colorado Center for Military Shuttle Operations." *Aviation Week and Space Technology* 122 (March 18, 1985): 125-126. Discussion of what would be done at the U.S. Space Command once construction was completed and how it would interact with Vandenberg Air Force Base.

"Space Command Completes Acquisition of Pave Paws Warning Radar Installations." *Aviation Week and Space Technology* 126 (May 18, 1987): 128-129. A description of Pave Paws (Phased-Array Warning System), which tracks intercontinental and submarine-launched missiles as well as space objects for the U.S. Space Command.

Patricia Jackson

UPPER ATMOSPHERE RESEARCH SATELLITE

Date: September 12, 1991, to September, 1994
Type of mission: Meteorological satellite

The Upper Atmosphere Research Satellite (UARS) was deployed to measure chemical composition and changing conditions as they relate to and effect changes in the Earth's environment, especially the loss of ozone over Antarctica.

PRINCIPAL PERSONAGES

CHARLES E. TREVATHAN, UARS Project Manager
　　located at the Goddard Space Flight Center
CARL A. REBER, UARS Project Scientist located at
　　the Goddard Space Flight Center
JOHN O. CREIGHTON, STS-48 Commander
KENNETH S. REIGHTLER, JR., STS-48 Pilot
CHARLES D. "SAM" GEMAR,
JAMES F. BUCHLI, and
MARK N. BROWN, STS-48 Mission Specialists

Summary of the Mission

Atmospheric ozone depletion is recognized as a serious environmental problem. Chemical reactions occurring primarily in the stratosphere, approximately fifteen to fifty-five kilometers above the surface of the Earth, both generate ozone by the interaction of oxygen with solar radiation and, in the presence of other chemical reagents, deplete the ozone which is formed. Chlorine containing compounds, notably chlorofluorocarbons commonly used as refrigerants, are among the most serious of the ozone depleting chemicals. Production of these man-made chemicals is generally prohibited with the hope of lessening their contribution to ozone depletion. Other chemicals containing chlorine and nitrogen, some of which are naturally occurring, also contribute to ozone depletion.

In 1976 the United States Congress directed the National Aeronautics and Space Administration (NASA) to develop an Upper Atmospheric Research Satellite to study the chemical and physical properties of the upper atmosphere, namely the stratosphere, with special emphasis on ozone and ozone depleting chemicals. Prior studies directed towards this goal were of a limited nature. Instruments designed to measure chemical and physical properties in the atmosphere were contained aboard conventual aircraft. With their limited altitude capability only partial observations could be made relatively near the surface of the Earth. Instruments mounted aboard

rockets were also limited as they gave adequate height profiles of the various measurements but only over a relatively narrow land mass. The Upper Atmospheric Research Satellite was developed to obtain data over the entire circumference of the Earth and at a height suitable to map upper atmospheric chemical and physical changes at distances from ten to sixty kilometers above the Earth's surface. The satellite was designed to provide data for three years and was built at a cost of $740 million dollars.

The Mission to Planet Earth (MTPE) program, conceived in the late 1980's, was a worldwide cooperative venture to study global problems. The first United States contribution to this program was the launch of the UARS as part of STS-48 aboard the spaceship *Discovery* on September 12, 1991, with John Creighton acting as commander. The satellite itself was placed into circular orbit 600 kilometers above the Earth on September 15, 1991. It circled the Earth every ninety-seven minutes with observations covering an area from 80 degrees N to 80 degrees S latitude. The spacecraft had a total mass of 6,800 kilograms, which included 2,500 kilograms of scientific instruments.

The scientific instruments aboard UARS were designed to measure specific aspects of the upper atmosphere, namely chemical composition, temperature, wind speed and direction, intensity of solar radiations, and particulates. Data from these observations would then be tabulated, correlated with observations from other sources (satellites and Earth observations stations), and used to construct a model of the chemical and physical changes occurring above the Earth. From this model, predications of future changes could be made. Following is a brief description of each of the instruments aboard the Upper Atmospheric Research Satellite.

Four spectrometers aboard UARS measure the chemical composition of the upper atmosphere. The first of these is the Cryogenic Limb Array Etalon Spectrometer (CLAES), an infrared sensing instrument recording data over a wavelength range of 5.5 to 12.7 micrometer from infrared energies. It is capable of simultaneously recording twenty data points over a vertical distance of fifty kilometers. It detects the presence of nitrogen and chlorine compounds, ozone, water vapor, methane, and carbon dioxide. The Improved Stratospheric and Mesospheric Sounder (ISAMS) is another type of infrared instrument employing telescopes which allow it to

Upper Atmosphere Research Satellite (UARS). (NASA)

measure in both vertical and horizontal planes over the wavelength region of 4.6 to 16.6 micrometers. It monitors the presence of nitrogen compounds, ozone, water vapor, methane, and carbon monoxide. The Halogen Occultation Experiment (HALOE) instrument is a third type of infrared spectrometer capable of recording wavelength data from 2.4 to 10.5 micrometers. It is designed to measure hydrogen fluoride, hydrogen chloride, nitrogen compounds, ozone, water vapor, methane, carbon dioxide, and indirectly the gas density within the stratosphere. The Microwave Limb Sounder (MLS) instrument records signals in the microwave region of the electromagnetic spectrum rather than in the infrared region. Responding to energies at 1.46, 1.64, and 4.8 millimeter wavelengths, it monitors the concentrations of chlorine oxide, hydrogen peroxide, water vapor, and ozone.

Knowledge of the movement of chemical substances in the stratosphere is essential in understanding the interactions of the various chemical species present and in relation to various events on Earth that alter the natural abundance of these chemicals. Indirect knowledge of the movement of chemicals in the stratosphere is available from the observed displacement of various chemicals over time as recorded by the spectrometers described in the preceding paragraph. Two instruments aboard UARS, however, give, for the first time, direct measure of stratospheric winds. This is accomplished by monitoring the Doppler energy change in the electromagnetic radiations observed from selected molecules and unstable chemicals in the upper atmosphere. Recall that energy emitted by a stationary object is constant, all else being equal, but when the object and/or the device sensing the energy are moving either towards or away from each other the frequency at which the energy strikes the detector varies. This variation in frequency of the electromagnetic radiation emitted by particles propelled by stratospheric winds when measured from the moving satellite is used to calculate the wind velocity. The instrument capable of detecting slight changes in the frequency of electromagnetic radiations is the interferometer. Two interferometers are aboard the UARS. They are the High Resolution Doppler Imager (HRDI) and the Wind Imaging Interferometer (WINDII). These instruments are capable of measurement in both vertical and horizontal directions relative to the satellite. As the satellite circles the Earth, a three dimensional profile of stratospheric wind direction and speed is generated.

Energy distribution within the stratosphere arises from solar radiation and from charged particles, the former as ultraviolet radiation and the latter as high energy charged particles, namely electrons and protons, originating from solar flares. The Solar Ultraviolet Spectral Irradiance Monitor (SUSIM) responds to ultraviolet radiation from the Sun over the energy range from 120 to 400 nanometer. The Solar/Stellar Irradiance Comparison Experiment (SOLSTICE) also monitors changing ultraviolet energies over the wavelength range from 115 to 430 nanometer. For a fixed energy source to which changing ultraviolet energy levels can be compared, one of its detection channels focuses on a distant blue star as a constant source of ultraviolet energy. The Particle Environment Monitor (PEM) detects and records data from electrons and protons present in the stratosphere and, in addition, monitors the production of X rays produced by high energy electrons as they pass through the upper atmosphere. Electron energy detection is over the range from 1 to 5 million electron volts (MeV), protons are detected over the energy range from 1 to 150 MeV, and X rays are monitored in the 2 to 50 kilo electron volt (KeV) range. An electron volt is the energy gained by one electron as it accelerates in an electric field of one volt potential difference. One electron volt (eV) is equivalent to 1.2×10^{-19} joule of energy.

Data from the various instruments aboard URAS are collected from on-board computers. From these they are transferred to a Central Data Handling Facility (CDHF) at the Goddard Space Flight Center. From Goddard, the information is passed to the appropriate Remote Analysis Computer (RAC) located at the research site of each principal investigator responsible for refining and interpreting the data from a particular measurement instrument for which she or he is responsible. These sites are located at Georgia Institute of Technology, Jet Propulsion Laboratory, Lawrence Livermore National Laboratory, NASA Goddard Space Flight Center, NASA Langley Research Center, National Center for Atmospheric Research, National Oceanic and Atmospheric Administration, State University of New York-Stony Brook, United Kingdom Meteorological Office, University of Colorado, and University of Washington. This arrangement allows for very rapid data

transfer to the destination where it is intended and where it can be addressed in the minimum amount of time.

There was a brief period in June, 1992, when eight of the ten instruments aboard UARS were inoperative because of improper adjustments in the solar array panels. This was corrected by ground control. In December, 1994, three months past the expected shutdown date for receiving transmission from UARS, eight of the ten instruments aboard the satellite were still operational sending data back to Earth. It was anticipated at that time that some data reception could continue for several more years although at a reduced level.

Knowledge Gained

Data from the Upper Atmosphere Research Satellite provided information regarding the chemical composition of the stratosphere, measuring concentrations of both naturally occurring components and, more important, of artificially produced chemicals released into the air as a result of human activity. In addition, solar wind direction, intensity, and seasonal variations provided information about the migration of these chemicals and locations where they could be expected to concentrate at various times of the year. Temperature measurements gave insight into the reactivity of the various chemical species, certain chemical changes being favored over specific temperature ranges. Data on the daily fluctuations in solar winds, as measured by the High Resolution Doppler Imager (HRDI) from 90 to 105 kilometers above the Earth and varying both north and south from the equator were recorded. Recorded also were data pertaining to the Polar Stratospheric Clouds (PSCs) as monitored by the Improved Stratospheric and Mesospheric Sounder (ISAMS). Differences in the formation, composition, and duration of these clouds over the Antarctic as compared to those over the Arctic were used to explain the differences in extent of ozone depletion above the Antarctic as compared to that in the northern hemisphere. UARS gathering data from both regions at considerable height variations and for an expended period of time provided the necessary information to distinguish this difference. The type, intensity, and seasonal variation of radio activity were followed, measurements being recorded from both particle and x-ray emissions. Particulate monitoring capability of UARS proved also to play a significant role in devising theories to explain, for example, ozone depletion over Antarctica. It was fortunate that UARS was operational when the volcanic eruption of Mt. Pinatubo on the island of Luzon in the Republic of the Philippines occurred. It was estimated that 30 million metric tons of ash was forced into the atmosphere as a result of this eruption. UARS and other satellites were able to monitor the presence of this increased particulate concentration above the Earth.

The measurements obtained by UARS correlated and extended those of earlier studies from ground based observation and satellites launched previously. Perhaps most importantly, UARS data gave a rather complete three-dimensional overview of all variables recorded above the entire Earth surface through various seasonal changes.

Content

Ozone is formed in the upper atmosphere by the interaction of molecular oxygen and atomic oxygen, the latter produced by photo decomposition from the ultraviolet radiations emitting from the Sun. Ozone depletion results naturally by several processes including photo decomposition, interaction with nitrogen oxide present in the atmosphere, and interaction of hydrogen oxygen radicals also present. With the wide spread use of man-made chlorofluorocarbons (numerous compounds containing carbon, chlorine and fluorine) the destruction of ozone has markedly increased. The chlorofluorocarbons eventually find their way into the atmosphere where they are decomposed by ultraviolet radiation from the Sun. The liberated chlorine atoms rapidly attack ozone molecules forming chlorine oxide and oxygen molecules. Only through formation of a nonreacting chlorine compound can the conversion of ozone to oxygen be stopped. The presence of sulfuric acid, too, can promote chlorine oxide formation and thus ozone depletion.

By the late 1980's ozone loss over Antarctica was a recognized and growing concern. The Upper Atmospheric Research Satellite and numerous other probes launched to study the conditions causing this loss provided data from which theories were developed and steps proposed to reverse the depletion. The data also greatly increased understanding of the composition and changes taking place in the stratosphere and at other atmospheric elevations. Polar stratospheric clouds residing over the polar regions at an altitude of about twenty kilometers and extending to a height of ten to one hundred kilometers are sometimes called nacreous, or mother of pearl, clouds because of their appearance due to their iridescence from light reflecting off ice crystals.

Polar stratospheric cloud formation over Antarctica is favored during the austral (southern hemispheric) spring and winter because of the extremely low temperatures present during that time. It is during this time of extreme cold that chlorine oxide formation from chlorofluorocarbons reaches its maximum. Similar cloud formation over the Arctic region during its coldest period is less because of the relatively warmer temperatures. UARS data indicates a more dense cloud formation over Antarctica. The greater amount of particulate matter provides a greater area upon which ozone depleting reactions can occur. Cloud formation in Antarctica also persists for a longer period of time than in the relatively warmer Arctic region. Particulates contained within the clouds, ice crystals and others, serve as sites for chemical reaction between ozone and chlorine containing compounds. The unfortunate eruption of Mt. Pinatubo in 1991 with its massive thrust of particles into the atmosphere greatly accelerated this ozone depletion.

Bibliography

"NASA Launches Mission to Planet Earth." *Sky & Telescope* 83 (1992): 9. This brief note describes the Upper Atmosphere Research Satellite, its purpose, and its capabilities for studying ozone depletion above the surface of the Earth.

Cramer, Jerome, "A Mission Close to Home." *Time* 138 (September 16, 1992): 53. Background details leading up to and through the launch of UARS are presented including some controversy over the significance of space probes and their importance to national well being.

McCormick, M. Patrick, Larry W. Thomason, and Charles R. Trepte. "Atmospheric Effects of the Mt. Pinatubo Eruption." *Nature* 373 (1995): 399-404. The dramatic increase in airborne particulates and their consciences relative to polar stratospheric cloud formation and subsequent ozone depletion are described.

National Aeronautics and Space Administration. *Upper Atmosphere Research Satellite (UARS): A Program to Study Global Ozone Change.* NASA N92-11067. Linthicum Heights, MD: NASA Center for AeroSpace Information. This NASA publication details the purpose of the UARS project. It describes in detail each of the on-board instruments and it details the manner in which information from these instruments is relayed back to Earth for distribution.

Rodriquez, Jose M.. "Probing Stratospheric Ozone." *Science* 261 (1993): 1128-1129. This brief overview provides a summary of the several efforts made to monitor polar stratospheric clouds from 1987 through 1993 by various satellites.

Rowland, F. Sherwood. "Stratospheric Ozone in the 21st Century: The Chlorofluorocarbon Problem." *Environmental Science and Technology* 25 (1991): 622-628. This article written before the launch of UARS describes the manner in which various chlorofluorocarbons interact to destroy ozone in the upper atmosphere.

Singh, Hanwant B., Editor. *Composition, Chemistry, and Climate of the Atmosphere.* New York: Van Nostrand Reinhold, 1995. This recent compilation in book form summarizes the most current findings and theories regarding the physical and dynamic characteristics of the atmosphere.

Taylor, F.W., A. Lambert, R.G. Grainer, C.D. Rodgers, and J.J.Remedios. "Properties of Northern Hemisphere Polar Stratospheric Clouds and Volcanic Aerosol in 1991/92 from UARS/ISAMS Satellite Measurements." *Journal of the Atmospheric Sciences* 51 (1994): 3019-3026. This research paper details the data transmitted from UARS relating to polar stratospheric cloud formation and the role they play in ozone depleting reactions especially over Antarctica.

Webster, C.R., R.D. May, D.W. Toohey, L.M. Avallone, J.G.Anderson, P. Newman, L. Lait, M.R. Schoeberl, J.W. Elkins, and K.R. Chan. "Chlorine Chemistry in Polar Stratospheric Cloud Particles in the Arctic Winter." *Science* 261 (1993): 1130-1134. This article explains the theories proposed to account for the difference in ozone depletion over the Arctic pole as compared with the greater depletion over the Antarctic region.

Gordon A. Parker

VANDENBERG AIR FORCE BASE

Date: Beginning October 4, 1958

Type of facility: Military base and launch site

Vandenberg Air Force Base has served as the site for more than five hundred orbital and one thousand nonorbital launches of American rockets and ballistic missiles. Between 1965 and 1969, a launch complex for the Manned Orbiting Laboratory was built at Vandenberg. From 1979 to 1986, these facilities were expanded to form a West Coast launch complex for the space shuttle.

Summary of the Facility

Because of its ideal position for launching satellites into polar orbits, Vandenberg Air Force Base (VAFB) has become a prime launching site for orbital payloads. As a military facility, Vandenberg has seen the launching of most American reconnaissance satellites and the firing of ballistic missiles for test purposes. In the early 1980's, the base underwent massive construction for a planned West Coast launch and landing site for the space shuttle.

In January, 1956, the U.S. Department of Defense (DOD) had decided that the United States needed a facility to train the men who were handling Intercontinental and Intermediate Range Ballistic Missiles (ICBMs and IRBMs). On June 7, 1957, the DOD allocated to the U.S. Air Force the northern two-thirds of Camp Cooke, an inactive World War II Army training camp located along forty kilometers of the Pacific Coast just eighty-eight kilometers northwest of Santa Barbara, California. The portion of the land south of the Santa Ynez River fell to the Navy, which installed the Naval Missile Facility, Point Arguello. The Air Force made its new base the headquarters of the First Strategic Aerospace Division and integrated it into the Strategic Air Command.

On October 4, 1958, the Air Force's land was officially dedicated as Vandenberg Air Force Base, and on December 16, 1958, the first Thor missile was launched, thus making Vandenberg the first operational ICBM facility in the United States.

The Air Force, however, had even more ambitious plans than the testing of missiles, since the location of Vandenberg made it ideal for launching satellites into polar orbit. Flying from pole to pole, a satellite will pass over every part of Earth while the planet rotates beneath it. Because VAFB is situated on a promontory jutting west from the California coastline into the Pacific, a rocket which has been launched toward the South Pole from VAFB will not fly over land until it reaches Antarctica. The risks involved for such a launch are low, because fallout from failed missions will not hit inhabited areas.

In cooperation with the National Aeronautics and Space Administration (NASA) and the United States Navy, the Air Force immediately constructed control centers and pads for the launching of satellites from Vandenberg. The first of these facilities was erected on a small, round peninsula extending west into the Pacific Ocean, situated near the base's airport. Vandenberg, together with the Navy's Point Arguello, became the Western Test Range of the Space and Missile Test Organization (SAMTO), an umbrella organization for Pacific and Atlantic aerospace test ranges.

On February 28, 1959, the first successful lift-off of a launch vehicle took place at Vandenberg: A Thor-Agena A rose from launchpad 75-3-4 (now Space Launch Center 1 West, or SLC 1W) and ejected its payload, Discoverer 1, the first American satellite to reach polar orbit. Soon, the first of the powerful Atlas launch vehicles arrived at Vandenberg, where facilities grew and military and scientific personnel began to populate the village of Lompoc, east of the base.

The Discoverer series carried data capsules, which were designed to fall back to Earth to be recovered and analyzed by Air Force specialists. It took one and a half years before the capsule ejected from Discoverer 13 was recovered from the Pacific Ocean west of Vandenberg. Eight days later, the Air Force succeeded in retrieving the data capsule of the next Discoverer in midair. From that point onward, the military was able to launch payloads at Vandenberg or Point Arguello and safely recover the data packs; throughout the years, about three-quarters of all the midair recovery attempts have been successful.

In 1960 and 1961, a new class of reconnaissance satellites was launched from the Western Test Range. Enthusiastically promoted as "spy-in-the-sky-satellites" by General Bernard A. Schriever, then Head of Air Force Systems Command, this hardware forced a closer cooperation between the Air Force and the Central Intelligence Agency (CIA) at Vandenberg. As a military base, Vandenberg had a higher-level security classification than did other American space facilities. Yet the introduction of the new payloads of spy satellites was not kept a secret at first. Both the failure of the first Samos (Satellite and Missile Observation System) satellite, which was launched from Point Arguello on October 11, 1960, and the relative success of Samos 2 were publicized. Similarly, the MIDAS, or Missile Defense Alarm System, satellites were announced publicly at first.

Late in 1961, however, a shroud of secrecy descended on Vandenberg. Military reconnaissance satellites were no longer given names, but a CIA code— KH (for "keyhole") and a number— was applied to them. Discoverer 38 was the last named spy satellite to reach orbit from Vandenberg, in March, 1962. Thereafter, Air Force and Navy officials would not release any information other than the standard statement, "A classified payload went into orbit today."

In 1963 and 1964, the Air Force developed plans for a military Manned Orbital Laboratory (MOL). Riding atop a new Titan 3C (later revised to Titan 3M) rocket, the modified Gemini capsule would be launched from VAFB and circle Earth for thirty days. The main goal of the MOL was to test how well manned spacecraft could perform military space operations. The predecessors of the launch vehicle for the MOL, the Titan 1 and 2 series, had already been test-fired from Vandenberg in their military applications as ICBMs.

On February 13, 1969, President Richard M. Nixon established his Space Task Group, which recommended that MOL be canceled and the facilities that had already been erected be mothballed. This left SLC 6 with a launchpad, a mobile service tower, and a launch control center in the immediate vicinity.

Throughout the 1960's, Vandenberg had served as the prime launching site for both military and scientific satellites. ICBMs were tested and stored in silos at the northwestern edge of the base. Some of the older pads were decommissioned as advanced facilities were constructed. The Air Force cooperated closely with NASA for many successful launches of rockets delivering communications and scientific satellites such as Echo 2 and the Explorer series.

The 1970's brought new excitement after the frustration over the cancellation of MOL. On June 15, 1971, the first Big Bird (KH 9) Air Force reconnaissance satellite was successfully launched atop the first known Titan 3D-Agena rocket. Before the end of the program in 1984, Vandenberg saw nineteen launches of this type of classified satellite. The base also facilitated launching the more recent KH 11, the maiden launch of which occurred on December 19, 1976.

To equip Vandenberg with a landing strip for the shuttle orbiter, the runway at Vandenberg's airstrip was expanded to 4,500 meters. To transport the orbiter on the ground, a mobile orbiter lifting frame was developed. Finally, a special maintenance facility was built. In this facility, the shuttle is placed on seismic jacks, designed to protect the orbiter from the effects of an earthquake (since VAFB lies in an earthquake-prone area). Curtains hanging from the ceiling provide a clean environment (known as the "clean room") from the orbiter's nose to the rear end of its cargo bay. A sixty-ton bridge crane was designed to lift payloads from the shuttle

into a pit sixteen meters below the ground.

Because of cost overruns, Vandenberg was only equipped to perform minor repairs or service the shuttle after an aborted launch. Normally, the shuttle is prepared for launching at the Kennedy Space Center.

Plans to build a facility for the refurbishment and subassembly of the solid-fueled rocket boosters, and a related installation for the refurbishment of their parachutes, never reached the construction stage. Also, the building reserved for processing of the external tank has only been used as a storage area; one empty tank was stored there in 1984.

From its checkout stand, the refurbished orbiter is placed on a special seventy-six-wheel, self-leveling transporter and sent on a three- to four-hour journey of 25 kilometers to the launch complex. There, in a procedure different from that at Kennedy Space Center, the final assembly of shuttle, payload, external tank, and solid-fueled rocket boosters occurs directly on the launchpad.

Payloads for a shuttle mission are prepared in the payload preparation room, from which they are lifted into the mobile payload "changeout" room. This structure stands 52 meters tall and can roll the 250 meters to the shuttle assembly building. There, payloads are put into the cargo bay of the shuttle.

Once the shuttle is fueled on the launchpad, warm air is blown over the liquid hydrogen section of the external tank to prevent icing. The air comes from two jet engines which are housed in a concrete shack thirty-three meters away. The hot air is ducted around and away from the tank, warming its surface but not contaminating it. During lift-off, the exhaust gases from the boosters and the liquid fuel engines escape through two exhaust ducts.

Since 1979, the launch control center of the abandoned MOL program has been transformed into a modern facility of 13,800 square meters. Its outer walls consist of a solid structure sixty-five centimeters thick. Electronic equipment is protected from possible seismic shocks and potential overpressure during launch.

In January, 1985, Vandenberg's shuttle launch facilities underwent a verification check with the test flight orbiter *Enterprise*. The *Enterprise* was guided through the various stations, from the orbiter lifting frame to the launchpad. While still hoping for a more active role in an actual shuttle mission after the tragic explosion of the *Challenger* space shuttle in January, 1986, the base suffered from a new accumulation of financial and technical difficulties. A design flaw was discovered in the two exhaust and flame ducts at the launchpad. In those ducts enough hydrogen could be trapped to cause an explosion that could destroy the orbiter on the pad. The safest redesign required radical alteration of the ducts at a cost of millions of dollars. During the ensuing shuttle program delays, more safety concerns were expressed. Fog and occasional temperatures of below 10 degrees Celsius at Vandenberg could stress the reliability of some critical systems of the shuttle. Some quality control personnel believed that the distance of 350 meters between the pad and the launch control center was not enough to safeguard the delicate electronic equipment from vibrations during lift-off, or even protect the center itself in case of a shuttle explosion on the launchpad.

Finally, in July, 1986, after an intense debate about the need for a West Coast launch center for the shuttle program, the Air Force recommended mothballing Vandenberg's shuttle complex. President Ronald Reagan adopted the most extreme option outlined by Congress: All shuttle-related facilities at Vandenberg were to be shut down until at least the mid-1990's.

In the meantime, launches of orbital payloads as well as missile tests continued. In January, 1988, Congress cut the budget for the shuttle complex at Vandenberg to $40 million and recommended cancellation of the Air Force's shuttle plans. Instead, some of the existing facilities at Vandenberg would be used for the Strategic Defense Initiative (SDI), the Advanced Launch System (ALS, an unmanned alternative to the space shuttle), the Titan program, and minor NASA missions.

By 1988, forty-eight different types of launch vehicles, sounding rockets, and missiles had been launched from fifty-one different pads at Vandenberg. The base houses more than one thousand different buildings, which are connected by more than 830 kilometers of roads. After layoffs in 1986, it was estimated that more than ten thousand workers, both military and civilian, were employed there. While most launching activity has been redirected to South Vandenberg, North Vandenberg harbors eighteen vertical silo launchers for Minuteman 3 and Titan 2 ICBMs.

In 1989, the Space Shuttle Launch Complex at Vandenberg was placed in mothballs as plans for a West Coast launch facility were cancelled. In May, 1992, the Western Commercial Space Center (WCSC) was formed as a non-profit corporation to address issues of commercial operation at Vandenberg. WCSC plans to build its own launch pad and assembly facilities immediately south of SLC-6. The old SLC-6 will be used to support the Lockheed Launch Vehicle 1 (LLV-1), which will be launched from the former space shuttle launch pad.

Context

Vandenberg Air Force Base has served well in its dual function as testing site for ballistic missiles and as launching base for military and scientific satellites. Because of its geographical position, the base has a natural edge over Kennedy Space Center, where solid land to the north and south prohibits direct launches into a polar orbit. Thus, during the 1960's, more satellites were launched from Vandenberg than from Cape Kennedy (later, Cape Canaveral), and the American space program achieved splendid results from the California launches.

Despite their being military satellites, the Discoverers, the first satellites launched from Vandenberg, conveyed a series of important scientific findings about the dynamics and mechanics of atmospheric reentry and space radiation. These findings were crucial to the discovery of the Van Allen radiation belts and the commencement of the Mercury manned orbiting program, a program that put the first American into space.

Together with their Soviet counterparts, the Kosmos series satellites, the reconnaissance satellites that were launched mostly from Vandenberg have helped to make the world a safer place by providing each superpower with the means of gathering more exact knowledge of the other's military and nuclear capabilities. Samos 2, launched from Vandenberg on January 31, 1961, stayed in orbit for one month and proved with its photographs that the United States had vastly overestimated the so-called missile gap between itself and the Soviets and that there was far less to fear from Soviet ICBM superiority.

Also, during grave international crises such as the Arab-Israeli wars and the gulf war between Iraq and Iran, both superpowers rapidly launched reconnaissance and surveillance satellites to gather reliable information about the areas of concern. Then, the Big Birds launched from Vandenberg helped American politicians and military officials to make informed decisions and avoid haphazard guesswork. In terms of verification of arms accord treaties, spy satellites were equally helpful in the years before on-site inspections became politically possible.

In cooperation with NASA, Vandenberg has made possible the launch of the Landsat and Seasat satellites. Both programs have delivered invaluable data about the geography of Earth and have made remote sensing a reliable tool for the geological sciences. Agricultural projects can be assessed more easily, flood warnings can be served very quickly, and the effects of natural and man-made changes on the face and structure of Earth can be studied from a sharp, bird's-eye view.

In the future, Vandenberg's geographical advantages will guarantee the base its share of launch traffic. Furthermore, missile testing will remain an integral part of the base's function as a part of the Strategic Air Command of the United States. As home to the headquarters of the First Strategic Aerospace Division of the Air Force, Vandenberg's military and scientific contributions to the exploration of space will continue into the next millennium.

Bibliography

De Ste. Croix, Philip. *Space Technology*. London: Salamander, 1981. An exhaustive look at space exploration, with many cross-references to specific points of interest such as spy satellites, launch vehicles, and ballistic missiles. Places

Vandenberg in the context of the U.S. space effort. Informative, ideal for a general audience, with many color and black-and-white photographs and a detailed bibliography.

Diamond, Edwin. *The Rise and the Fall of the Space Age*. Garden City, N.Y.: Doubleday and Co., 1964. An early critique of the military's role in space and the relationship of NASA to the military-industrial complex. Describes and criticizes the programs situated at Vandenberg and places them in a national and international context. Argumentative but informative and readable. Contains no illustrations, but a few tables.

Klass, Philip J. *Secret Sentries in Space*. New York: Random House, 1971. Promilitary, this text emphasizes the technology behind spy satellites. It stresses the importance of these satellites for global safety. Klass is very good at placing Vandenberg in the broader context of an international espionage race. Includes photographs of the satellites, related hardware, and what they detect. For the technically inclined reader.

Sharpe, Mitchell R. *Satellites and Probes*. Garden City, N.Y.: Doubleday and Co., 1970. Close description and analysis of the international development of unmanned spaceflight. Stresses the contribution of the Vandenberg launch facilities to the success of U.S. military and scientific satellites. Compares this facility with its worldwide counterparts. Includes color and black-and-white photographs.

Shelton, William Roy. *American Space Exploration: The First Decade*. Rev. ed. Boston: Little, Brown and Co., 1967. Chronicles the history of U.S. spaceflight and provides good background information about the first decade at Vandenberg. Full of relevant anecdotes and biographies of persons important to the space effort. Written in a very readable, journalistic style, this book includes illustrations.

Sloan, Aubrey B. "Vandenberg Planning for the Space Transportation System." *Astronautics and Aeronautics* 19 (November, 1981): 44-50. A detailed description of the original, ambitious plan for the Vandenberg shuttle complex. Written by the man who was largely responsible for overseeing the development of this facility in its early stages. Good history of the decision-making process involved in bringing the shuttle complex to Vandenberg. Supplemented with illustrations, diagrams, and a useful bibliography for further, more specialized studies.

Stockton, William, and John Noble Wilford. *Space Liner: "The New York Times" Report on the Columbia Voyage*. New York: Times Books, 1981. A journalistic account of the space shuttle program from its conception to the first flight of the orbiter *Columbia* in April, 1981. Delineates the planned role of VAFB for further missions and talks about the decision to create a shuttle program with two major facilities in the eastern and western regions of the United States. Anecdotal and easy to read, with some fine black-and-white illustrations.

Reinhart Lutz

THE VANGUARD PROGRAM

Date: September 9, 1955, to September 18, 1959
Type of program: Launch vehicle and unmanned scientific satellite

Destined to be remembered for its failed attempts to launch the first U.S. man-made satellites, the Vanguard program generated important developments in rocket propulsion, satellite design, and satellite telemetry and tracking and eventually succeeded in launching three Vanguard satellites.

PRINCIPAL PERSONAGES

JOHN P. HAGEN, Vanguard Program Manager
JAMES A. VAN ALLEN, Professor of Physics, Iowa State
    University
WILLIAM H. PICKERING, Director, Jet Propulsion
    Laboratory
MILTON W. ROSEN, Vanguard program Technical
    Director
DONALD J. MARKARIAN, a Vanguard program engineer
    with the Glenn L. Martin Company
T. K. GLENNAN, NASA Administrator

Summary of the Program

Project Vanguard consisted of fourteen multistage launches, including test vehicles, with the stated purpose of placing the United States' first man-made satellite into Earth orbit.

In the early part of the twentieth century, Robert H. Goddard designed, tested, and successfully launched both liquid- and solid-fueled rockets. Rocket designers in the United States, the Soviet Union, Germany, and Austria were busy throughout the 1920's, 1930's, and 1940's developing the skills and technology that would later be used by rocket scientists of the post-World War II era. After World War II, a global awareness of the effective use of rockets forced the U.S. military to alter the scope and direction of its ballistic missile research. After the May, 1945, surrender to the Allies of roughly 120 German rocket scientists at Peenemünde—led by Wernher von Braun — the academic research community, industry, and military of the United States became engaged in dissecting, modifying, and eventually using German V-2 rockets for basic research. The U.S. Army, Air Force, and Navy were independently developing sounding rockets (rockets capable of suborbital flight) and spacecraft capable of orbital flight (research satellites). Because of the outbreak of the Korean War, however, military research was largely aimed at accurate delivery of nuclear or conventional weapon payloads.

On September 9, 1955, with the backing of President Dwight D. Eisenhower, the Department of Defense authorized the Naval Research Laboratory (NRL) to administrate a far-reaching program to design, build, and launch at least one artificial satellite during the International Geophysical Year, or IGY—a period of a year and a half running from July 1, 1957, to December 31, 1958. The United States' participation in the IGY, an international peacetime research effort, was problematic, because only military agencies had the hardware, financial backing, and manpower necessary to launch a satellite. Nevertheless, with the support of the Department of Defense, the Bureau of Aeronautics, the Office of Naval Research, the National Academy of Sciences, and the National Science Foundation, the NRL began implementing Project Vanguard. The NRL was backed by the Glenn L. Martin Company (GLM), Aerojet General Corporation, Grand Central Rocket Company (GCR), Allegany Ballistics Laboratory, General Electric (GE), International Business Machines (IBM), Minneapolis-Honeywell, and myriad U.S. universities and research facilities.

The planning and construction of the Vanguard launch vehicle's main stage was facilitated with testing in the late 1940's of the Viking rocket built by GLM for the NRL. Nurturing a working association with one of the nation's largest rocket builders, it was only natural that the NRL turn to GLM for help with the development and deployment of Vanguard test vehicles and satellite launch vehicles. Unfortunately, GLM also won a contract with the Air Force to construct the Titan missile, thus diluting the manpower it could expend on the Navy's Project Vanguard.

Despite these problems and others involving questions of responsibility and decision making, the NRL and GLM effort was successful. Rocket design specifications were completed by February, 1956, but again, disagreement between the NRL and GLM resulted in amendments and delay. The early Vanguards, Viking M-15's manufactured by GLM, were modified Viking M-10 missiles with a GE first-stage motor, an Aerojet Aerobee-Hi liquid-propellant second stage, and a GCR solid-fueled upper stage. The specter of the military dissipated, as the Viking missile was a renowned research vehicle that had long been used for atmospheric sounding.

Kerosene and liquid oxygen were used for the first stage

The Vanguard satellite displayed at the National Air and Space Museum. (NASA)

of the Vanguard rocket, which generated 120,096 newtons of thrust. The fuel and oxidant were supplied to the engines by a hydrogen peroxide decomposition technique, which produced superheated steam and oxygen to drive turbine-driven fuel and liquid oxygen pumps. Helium gas supplied pressure for the fuel tanks. The second stage produced 33,360 newtons of thrust with a mixture of white fuming nitric acid and unsymmetrical dimethylhydrazine, an explosive rocket fuel. The third stage was powered by a solid-propellant motor and generated roughly 13,344 newtons of thrust.

With the development of the transistor and miniaturized electronic circuitry, satellite instrumentation design programs at the NRL, Jet Propulsion Laboratory (JPL), and university campuses swung into full gear. Satellites were built under the directorship of the National Academy of Sciences. Meanwhile, NRL telemetering and tracking systems were in advanced stages of development. The deployment and success of the Minitrack system was a result of rigorous testing and research; this highly accurate tracking system formed the backbone of satellite tracking during and after the Vanguard era. GLM developed computer programs that balanced weight against anticipated flight trajectory. IBM offered the NRL free computer time at the Massachusetts Institute of Technology. Because of loans and contracts, both optical and electronic tracking and telemetry stations were built and manned largely by civilian personnel.

As originally planned, Vanguard test vehicles—numbered TV 0 through TV 5—would precede the production of Vanguard model satellite launch vehicles that would be used for missions SLV 1 through SLV 6; there would be a total of twelve launches. Because of engineering changes in payload shape, weight, and size, a moderate degree of launch failure, and the globally transmitted 20- to 40-megahertz beep of Sputnik 1, the original Project Vanguard firing schedule was accelerated and ultimately expanded to fourteen attempts with the launches of two backup test vehicles, TV 3BU and TV 4BU. In all, three highly successful satellites were placed into Earth orbit, and abundant new geophysical, atmospheric, and near-space data were gathered for processing and analysis.

The first Vanguard launch, TV 0, occurred on December 8, 1956, and successfully tested the Viking 13 first stage and the telemetry and tracking systems, which reached an altitude of 203.5 kilometers and a range of 157 kilometers. TV 1, or Viking 14, followed on May 1, 1957, and after a test of third-stage propulsion reached a range of 726 kilometers with a 195-kilometer peak altitude. Sputnik 1, launched October 4, 1957, interrupted the proposed Vanguard test launch schedule; TV 2 restored confidence in the project with a better-than-expected performance, involving the three-stage Vanguard prototype with inert second and third stages, on October 23, 1957. The Soviet Union launched the dog Laika into space on Sputnik 2 on November 3, 1957. The Vanguard TV 3, a three-stage missile complete with the United States' first artificial satellite, was fired on December 6, 1957, but to the amazement of all spectators, it toppled and exploded on the launchpad. This failure was a crushing blow to American pride, but it was also a stimulant for more careful engine system tooling and rocket construction techniques by GE and GLM, respectively, in preparation for future Vanguard firings.

In the interim, the Army Ballistic Missile Agency (ABMA) at the Redstone Arsenal in Huntsville, Alabama, had been researching modified German V-2 rockets with the assistance of the California Institute of Technology's Jet Propulsion Laboratory. On January 31, 1958, the ABMA-JPL team launched the Juno 1, a four-stage version of the Jupiter C rocket, which carried the first U.S. satellite, Explorer 1. Project Vanguard's second attempt, with TV 3BU, ended in failure on February 5, 1958, when a control system problem resulted in loss of attitude control and eventual breakup of the TV 3BU Vanguard after fifty-seven seconds of flight.

Finally, on March 17, 1958, TV 4 placed Vanguard 1 into orbit with an apogee (the point farthest from Earth) of 3,966 kilometers and a perigee (the point nearest to Earth) of 653 kilometers. Vanguard 1 was designed to measure Earth's shape and atmospheric density. It is expected to orbit, with its third-stage motor casing, for more than two centuries. The 25.8-kilogram, 16.26-centimeter spacecraft fulfilled the Project Vanguard goal of launching an artificial satellite within the IGY. The TV 5 launch attempt on April 28, 1958, failed to orbit a 50.8-centimeter, spherical, 9.8-kilogram X-ray and environmental satellite because of second-stage shutdown problems.

The first production version of a satellite launch vehicle, SLV 1, was, on May 27, 1958, to carry a satellite nearly identical to that destroyed during the TV 5 attempt. Again, during second-stage burnout, attitude control problems arose that resulted in firing of the third stage at an angle unsatisfactory for orbit. On June 26, 1958, SLV 2 encountered second-stage propulsion system shutdown after eight seconds. Failure to achieve orbit also plagued the SLV 3 launch on September 26, 1958, when the second stage underperformed.

With passage of the Space Act of 1958, the civilian National Aeronautics and Space Administration (NASA) was created to direct the U.S. space program. Vanguard 2 was successfully placed into orbit by NASA on February 17, 1959, with the launch of SLV 4. Weighing a total of 32.4 kilograms, the 10.7-kilogram payload and 21.7-kilogram third-stage motor casing attained an initial orbital apogee of 3,319 kilometers and a perigee of 556.7 kilometers. The Vanguard 2 payload, destined to orbit for roughly two hundred years, is a 50.8-centimeter spherical satellite that measures cloud distribution and the terrestrial energy cycle budget.

SLV 5 was fired on April 13, 1959; it attempted to launch a 33-centimeter satellite magnetometer and a 76.2-centimeter expandable sphere. Problems arose during the separation of the second stage and resulted in an aborted flight. SLV 6 was fired on June 22, 1959, but again, a second-stage failure sabotaged the launch with a rapid drop in fuel tank pressure, faulty ignition, and explosion of the helium tank because of overheating. The 50.8-centimeter satellite on board, designed to measure solar radiation and its reflection from Earth, failed to orbit.

NASA decided to use a spare backup launch vehicle, TV 4BU, in a final attempt to launch a third Vanguard satellite. Using a new solid-propellant third-stage motor, built by Allegany Ballistics Laboratory, the rocket successfully launched Vanguard 3 on September 18, 1959. The final Vanguard weighed 42.9 kilograms and included a 23.7-kilogram payload and 19.2-kilogram motor casing. The initial orbital apogee was 3,743 kilometers, and the perigee was 510 kilometers. The payload included a magnetometer, an X-ray device, and environmental measuring systems. In large part, Vanguard 3 fulfilled the agenda of previous unsuccessful Vanguard launch attempts.

Knowledge Gained

With the end of NRL control of Project Vanguard in late September, 1958, and the creation of NASA in October, many key Vanguard personnel joined the NASA staff. As such, the knowledge gained during the Vanguard era was applied toward all subsequent U.S. space ventures. Growth in the fields of vehicle engineering, construction, fueling, and launch were predictable outcomes of Project Vanguard. The three Vanguard satellites that were successfully launched investigated energy fields in the boundary between Earth's atmosphere and space and carried out Earth-directed research.

Vanguard 1 achieved a high-apogee orbit and provided a tracking signal until 1965, thanks to its pioneering use of solar cells. By analyzing changes in orbital acceleration, the satellite detected a bulge in the atmosphere caused by solar heating and recorded a bulge in the Southern Hemisphere of Earth itself, thus confirming that Earth is nonspherical and Earth's interior is inhomogeneous.

Vanguard 2 performed an experiment that measured variations in cloud-top reflectivity. The results were not conclusive, but they contributed some data and perfected meteorological techniques used in later missions. Vanguard 3 was a 50.8-centimeter sphere with sensors for solar X-ray and Lyman-alpha radiation measurements, environmental sensors, and a 66-centimeter projection supporting a magnetometer. Because of the Van Allen radiation belts, the radiation detectors were overloaded and failed to provide accurate data, but accurate temperature monitoring was accomplished over seventy days. Measurements of interplanetary cosmic dust showed a variable but significant influx of particulate matter estimated at 9,072,000 kilograms per day. The magnetometer provided accurate measurements of Earth's magnetic field, plus data on magnetic disturbance events and upper atmosphere lightning ionization.

Context

The technological impact of Project Vanguard has permeated every aspect of manned and unmanned space exploration and discovery. First and foremost, Project Vanguard developed budgeting, command, and scheduling techniques for effective launching of missiles. Advances in missile guidance, tracking, telemetry, and antennae systems, developments in electronic miniaturization, and solar cell and mercury cell use in satellites were made possible by Project Vanguard. The use of fiberglass casings and the eventual design of the Air Force Thor-Ablestar booster and NASA's highly successful Delta and Atlas launch vehicles are all direct descendants of Viking and Vanguard technology.

In addition, the Vanguards provided new views of Earth's geologic cycles, thus promoting environmental awareness. The subsequent research in electronics, computers, communication, and optics has changed the quality of human life. Finally, Project Vanguard demonstrated that the American military-industrial-academic complex was capable of far-reaching outer space missions.

Bibliography

Bergaust, E., and W. Beller. *Satellite!* New York: Hanover House, 1956. Details for the layperson the planning for satellite launches during the International Geophysical Year (IGY) and includes detailed drawings of the Vanguard missile. Dated by post-IGY satellite development, this volume discusses the visionary goals of project scientists for the Vanguard and early Explorer missions.

Braun, Wernher von, et al. *History of Rocketry and Space Travel.* Rev. ed. New York: Thomas Y. Crowell, 1969. Includes a comprehensive, superbly illustrated history of post-World War II rocket research and abundant tables of data on missiles, missions, satellites, and manned spacecraft. It contains a detailed bibliography and is recommended to rocketry enthusiasts. A new edition, entitled *Space Travel: A History*, was published in 1985 by Harper & Row, New York.

Caidin, Martin. *Vanguard! The Story of the First Man-Made Satellite.* New York: E.P. Dutton, 1957. A layperson's account of the developmental history of Project Vanguard up to, but not including, the launch of Vanguard 1. This well-illustrated but somewhat dated volume details missile development during Project Vanguard and traces the development of payloads, tracking, and telemetry.

Green, Constance M., and Milton Lomask. *Vanguard: A History.* NASA SP-4202. Washington, D.C.: Government Printing Office, 1970. This is a detailed history of Project Vanguard describing the people, agencies, and administrative programs that led to the launchings of Vanguard missiles and satellites. Contains numerous photographs and diagrams. Mission goals and successes are described in detail. The appendices contain flight summaries for the Vanguard and Explorer programs and IGY satellite launches.

Hall, R. Cargill, ed. *Essays on the History of Rocketry and Astronautics.* NASA Conference Publication 2014. 2 vols. Washington, D.C.: Government Printing Office, 1977. A compilation of papers and memoirs written by active participants, this work traces international efforts in rocketry. Volume 2 concentrates on liquid- and solid-propellant rocket research before and after World War II. Accounts of the early phases of the Vanguard and Explorer projects are noteworthy.

Charles Merguerian

THE VIKING PROGRAM

Date: August 20, 1975, to November 13, 1982
Type of program: Unmanned deep space probes

The Viking program, using a pair of heat-sterilized landers, acquired the first data from the surface of Mars. Relay communications equipment on the landers and orbiters enhanced the ability to send high-rate data to Earth and improved the scientific value of these first landings.

PRINCIPAL PERSONAGES
    JAMES S. MARTIN, JR., Project Manager
    HOWARD WRIGHT,
    ISRAEL TABACK, and
    HARPER E. "JACK" VAN NESS, Deputy Project
        Managers
    GERALD A. SOFFEN, Project Scientist

Summary of the Program

Plans to conduct unmanned missions to Mars were initiated shortly after the National Aeronautics and Space Administration (NASA) was established in the fall of 1958. At that time, the Jet Propulsion Laboratory (JPL) was a U.S. Army laboratory operated by personnel from the California Institute of Technology (Caltech). JPL was transferred by executive order from the Army to NASA on December 3, 1958. JPL and NASA had reached agreements that the laboratory would be principally involved in unmanned exploration of the Moon and the planets.

The JPL staff started to plan a series of missions; the early ones involved small spacecraft, with larger, more complicated craft intended for subsequent missions, when more powerful launch vehicles were scheduled to be available. The initial goals were to demonstrate to the country that NASA was an aggressive organization and that it was not necessary to use the military to conduct the civil space program. The lunar and planetary missions included Ranger missions to the Moon and a series of Mariner spacecraft for exploration of Venus and Mars. The Mariners were to be followed by Voyager spacecraft that would require the development of Saturn launch vehicles before they could be flown.

During the early 1960's, developmental problems in the improved launch vehicles and launch failures resulted in the postponement of the earliest missions and deferred the more complicated ones.

The Mariner 2 spacecraft in 1962 and the Mariner 5 spacecraft in 1967 were sent to Venus. The Mariner 4 probe was successfully sent to Mars in 1964. Ambitious plans were supported by NASA for large Mars landers and orbiters to be launched by a single Saturn 5 in 1973. Yet on August 30, 1967, the Voyager project was canceled — because of the lack of adequate support from the Congress, which at that time was concerned about the war in southeast Asia. At the time, a dual flyby mission to Mars for 1969 was under development at JPL. A subsequent deep space mission to Jupiter, Saturn, and Uranus, launched in 1977, was designated by NASA as Voyagers 1 and 2.

For several years prior to the cancellation of the Mars Voyager program, both the Langley Research Center and JPL had been planning unmanned missions to Mars. The research center was interested in the technical challenge of the landing vehicles' passing through the thin Martian atmosphere and in the opportunity to develop and manage an important flight project following Langley's successful role on the Lunar Orbiter project. The design studies that had been performed by the Voyager staff (NASA, JPL, and contractors) were of considerable value in preparing plans for a Mars landing mission at a cost considerably lower than estimates had been for the canceled Voyager project. In December, 1968, NASA supported a soft-lander mission with a ninety-day surface lifetime goal along with a Mariner 1971-class orbiter to be launched in 1973 to continue the exploration of Mars. The project's responsibility was assigned to Langley. NASA assigned the responsibility for the development of the soft-lander system to Langley Research Center, the development of the orbiter system and the tracking and data acquisition system to JPL, and the development of the Titan-Centaur launch vehicle systems to the Lewis Research Center. Further, the control of the flights, following launch from Cape Kennedy, Florida, was to be conducted from JPL facilities in California.

In the spring of 1969, the Viking project office at Langley issued a request for proposals for the development of the Viking lander. The Martin Marietta Corporation's Aerospace Division in Denver, Colorado, won the contract and immediately set about developing the state-of-the-art lander system.

Following the successful Apollo landings on the Moon in 1969, there was an anti-technology furor and a subsequent shortage of federal funds, causing the Bureau of the Budget to

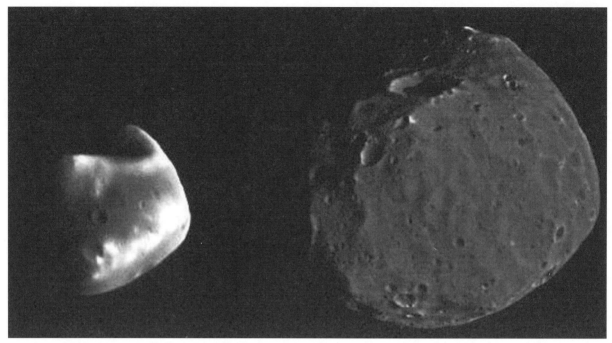

These images of Mars' asteroid-sized satellites Phobos (right) and Deimos (left) were taken by the Viking Orbiter 2 and 1 respectively. (NASA)

make substantial cuts in the NASA budget for fiscal year 1971. To operate with these fiscal restraints, NASA slipped the Viking project from a launch in 1973 to 1975, when the next launch window opened. At Langley, James Martin made the decision to slow the work on the Viking Orbiter substantially, since the risks associated with its development were considered lower than those of the new lander system. The JPL project staff was reduced by almost two hundred people in less than a month to be able to apply the available funds to the new technology. (It was fortunate that JPL was able to reassign these people to the Mariner 8, 9, and 10 projects.)

The Viking lander was considerably more complex than the Surveyor spacecraft that had successfully landed on the Moon. The Viking lander had to use the thin Martian atmosphere to slow it initially so that it could successfully open a large parachute which carried the vehicle to within 1,400 meters of the surface, where the terminal descent engines took over to soft-land the craft. Additionally, all the equipment contained on the lander had had to be sterilized to ensure that the life-detection instruments on the lander did not detect life-forms that it had carried from Earth to Mars. The lander had to survive a sterilization cycle of 40 hours in which the minimum temperature in the lander was 112 degrees Celsius. This included all the scientific and electronic equipment, parachutes, liquid propellants, and a radioactive power source.

During the 1975 launch window, the flight time to Mars was greater than it would have been in 1973, and the spacecraft would be farther from the Sun and Earth. Thus, larger solar panels were necessary, along with a larger high-gain antenna to communicate with Earth. Some additional redundancy of engineering subsystems was included to increase the chance of success during the longer flight.

The Viking project was the most complicated unmanned mission in space conducted by NASA. Activities at four NASA centers — Langley and Lewis research centers, JPL, and Kennedy Space Center— had to be coordinated. The Titan 3E Centaur launcher was used on the two Viking flights. Four separate spacecraft were flown (two orbiters and two landers). Communication links were established between Earth and each of the four spacecraft; in addition, radio links were provided between the orbiters and landers to recover high-rate data from the landers when they were on the surface of Mars.

To ensure that all elements of the project were efficiently coordinated, the project's manager, James Martin, established a management council that met every month at Langley. The managers of each of the systems developed for the project spent roughly two days a month describing the progress and problems of their systems along with plans for activities currently under way. Martin also instituted a "Top Ten Problem List" to highlight the most significant problems that might put the program's success at risk. These difficulties were reviewed in detail at each management council meeting, and Martin had to be satisfied that each problem had been resolved before it could be removed from the list. A total of forty were identified and eventually resolved during the developmental phase of the project.

In a few cases, Martin supported the parallel development of competing designs to ensure that suitable equipment would be available for the mission. The opportunity to fly to Mars occurs roughly every twenty-five months, and if some part of the flight systems were not available, the project would miss its launch opportunity during August-September, 1975. Ensuring that costs were not overrun was a challenge. Viking had been NASA's most expensive unmanned project to date; even so, it was necessary to reduce the number of available subsystems to cut costs.

As a further step to reduce costs, NASA conducted extensive joint tests with structural models of the lander and orbiter and several separate interface tests of the prototype equipment that provided either data or electrical power to the lander or orbiter. The first full physical and operational tests utilizing complete landers and orbiters did not occur until the flight systems had been delivered to the Kennedy Space Center for the final launch preparations. Fortunately, the project manager had included schedule reserves of roughly one month for planned activities at the Cape to ensure that the first Viking would be launched on time. This reserve became vital when a severe thunderstorm activated the electrical power of the initial orbiter on the launchpad, and the flight orbiter and lander had to be switched with those intended for the second launch to replace the discharged batteries.

Knowledge Gained

The Viking mission provided the first and second soft landings on the surface of Mars. Observations and measurements of the surface and atmospheric conditions were obtained for far longer than the three months for which the landers had been designed. The two orbiting spacecraft also operated significantly beyond their design goals; as a consequence, substantially more scientific and visual data were obtained from Mars than had been originally expected.

The flight operations activities were complicated. More than five hundred people were required to conduct these operations during the prelanding and initial postlanding operations. A large team of scientists was present at the JPL operations center, along with personnel from the Viking project office, lander technical specialists from Martin Marietta, and orbiter and tracking and data acquisition personnel from JPL.

The primary mission ended on November 15, 1976, during the solar conjunction (when Mars was on the opposite side of the Sun from Earth and no communications were possible with any of the spacecraft). When communications were again possible in mid-December, 1976, an extended mission was originated. On April 1, 1978, the extended mission was concluded and project management responsibility was transferred from Langley to JPL for a continuation mission that was designed to acquire scientific information as long as the spacecraft continued to operate.

Viking Orbiter 2 developed leaks in its attitude control

gas equipment, and its operations were terminated on July 25, 1978. Viking Lander 2 was shut down on April 12, 1980. Orbiter 1 was silenced on August 7, 1980. Communications with Lander 1 became limited to a once-a-week transmission to Earth at a low data rate, and limited science data were produced. A total of 51,539 orbiter images and more than 4,500 lander images had been returned to Earth.

Context

The Viking missions were the fifth and sixth to visit Mars and the first two soft landings on the surface that provided data. The Soviets had tried earlier, but with minimal success. In 1971 the Soviets launched two identical missions to Mars with orbiters and landers that weighed eight times more than the U.S. orbiter Mariner 9, which was also flown in 1971. The Soviet Mars 2 crashed on Mars on November 27, 1971. The Mars 3 lander landed on December 3, 1971, but communicated with its orbiter for only twenty seconds. The large dust storm on Mars at that time outlasted the Soviet orbiters' lifetimes and little useful data resulted. The Soviets also received some preliminary data from their Mars 6 probe in 1974.

Mariner 4, launched in 1964, had provided the first flyby views of Mars. In 1969, Mariners 6 and 7 conducted flyby missions which supplied roughly two hundred times more imaging data than had been acquired by Mariner 4. In 1971, Mariner 9—the first American Mars orbiter—was successfully flown. When Mariner 9 arrived at Mars on November 13, 1971, a severe dust storm was in progress, which obscured all surface features. After several days, the peaks of the four large volcanic craters on Mars appeared above the pall of dust. In February, 1972, the dust cleared at the lower altitudes, which were of particular interest for Viking landing sites, and useful photographs were obtained. In all, Mariner 9 provided 7,329 pictures covering 85 percent of the Martian surface. As a result, an initial mapping of Mars became possible.

The Mariner missions had showed that Mars has a small magnetic field. Its atmospheric density, altitudinal variations, and surface and atmospheric temperatures had also been determined by the Mariner probes. Before the Viking program, Mars was known to have volcanoes, and water was observed in its polar caps. It was known that the atmosphere consisted principally of carbon dioxide, but the Viking lander determined that a small amount of nitrogen was also present. The 4,000-kilometer-long Valles Marineris had been photographed by Mariner 9.

The scientific instruments on the lander were intended to determine whether life existed on Mars. As a result of the Viking mission, scientists concluded that life does not exist at the two locations where the spacecraft landed. The severe ultraviolet radiation on the planet's surface would have destroyed any organic compounds. It is probable that the severe radiation has highly oxidized chemicals on the surface of Mars. Nevertheless, several scientists still believe that the possibility for life on Mars exists, either farther below the surface of the planet or nearer the polar regions, where it is clear that more water is present than exists at the Viking landers' equatorial locations.

A considerable amount of water must have been present long ago to form the channels that have been observed on Mars. The question is, where did the water go? There is frozen water in the polar caps, but is there also water below the surface? The current atmospheric density is so low that liquid water cannot exist on the surface of Mars. Continued study of the planet's atmosphere and meteorology may be helpful in improving the understanding of processes occurring on Earth.

Bibliography

Baker, Victor R. *The Channels of Mars.* Austin: University of Texas Press, 1982. Includes many illustrations obtained by the Viking orbiters and landers along with a few obtained by Mariners 4, 6, 7, and 9. This text includes a chapter about the existence of water, or water ice, on Mars.

Corliss, William R. *The Viking Mission to Mars.* NASA SP-334. Washington, D.C.: Government Printing Office, 1975. Suitable for high school and college levels. Contains a description of the spacecraft and launch vehicle used, along with information on the scientific exploration accomplished during the mission.

Ezell, Edward Clinton, and Linda Neuman Ezell. *On Mars.* NASA SP-4212. Washington, D.C.: Government Printing Office, 1984. The official history of the Viking missions. Written by historians, this work is suitable for high school and college levels. Contains substantial background information on earlier missions to Mars along with numerous illustrations, photographs of the Viking staff and spacecraft, and pictures returned from the mission.

Kopal, Zdenêk. *The Realm of the Terrestrial Planets.* New York: John Wiley and Sons, 1979. This book deals principally with what has been learned about the inner planets (Mercury, Venus, Earth, and Mars) and the Moon through space exploration. Excellent illustrations are included.

Pollack, James B. "Mars." *Scientific American* 233 (September, 1975): 16, 106-117. Contains an excellent overview of pre-Viking knowledge of the evolution of Mars and its atmospheric and surface characteristics.

Sagan, Carl. "The Solar System." *Scientific American* 233 (September, 1975): 16, 22-31. Suitable for both high school and college audiences. Describes each of the planets of the solar system and briefly discusses the then-current or planned missions for planetary exploration.

Henry W. Norris

VIKING 1 AND 2

Date: August 20, 1975, to November 13, 1982
Type of mission: Unmanned Mars probes

The *Viking mission to Mars was the first long-duration, intensive exploration of the surface of another planet. Vikings 1 and 2 each consisted of a soft-lander, which examined the physical and chemical properties of the Martian surface and atmosphere, and an orbiter, which extensively surveyed Mars from orbit and served as an Earth-Mars transport vehicle and communications relay for its lander.*

PRINCIPAL PERSONAGES

> JAMES S. MARTIN, JR., Project Manager
> A. THOMAS YOUNG, Mission Director
> GERALD A. SOFFEN, Project Scientist
> B. GENTRY LEE, Director of Science Analysis and
>   Mission Planning
> CONWAY W. SNYDER, Orbiter Scientist, primary mis-
>   sion, and Project Scientist, extended mission
> HAROLD MASURSKY, Team Leader, landing site certifi-
>   cation
> THOMAS A. MUTCH, Team Leader, lander imaging
> MICHAEL H. CARR, Team Leader, orbiter imaging
> KLAUS BIEMANN, Team Leader, molecular analysis
> HAROLD P. KLEIN, Team Leader, biology

Summary of the Missions

Vikings 1 and 2 each consisted of an unmanned soft-lander and an orbiter that carried the lander to Mars and served as a communications relay station from Mars orbit. The Viking mission's primary objectives were to investigate the physical characteristics of Mars and to search for Martian life. To this end, the Viking spacecraft carried the experiments of thirteen teams of scientists who, throughout the mission, maintained remote control over their experiments from the mission control center at the Jet Propulsion Laboratory (JPL) in Pasadena, California.

Since Viking was to be the first attempt by the National Aeronautics and Space Administration (NASA) at soft-landing a craft on another planet, the spacecraft designs required tremendous technological advances over previous vehicles. In addition to the usual scientific tasks, the orbiters would carry the landers to Mars, hold them in orbit while certifying the preselected landing sites as safe, position and release the landers for descent to the surface, and serve as relay stations for communications between the landers and Earth. To meet these ends, the highly successful Mariner design— a flat, octagonal body with four protruding rectangular solar panels— was adopted, enlarged, and considerably modified for the task. Compared to their Mariner predecessors, the resulting Viking Orbiters were larger (9.75 meters across the extended solar panels) and heavier (2,328 kilograms launch weight), and they carried vastly superior computer command systems (two redundant 4,096-word computers, either of which could operate the craft independently from pre-programmed instructions or from commands sent from Earth).

The landers were essentially a new design. Hidden beneath a panoply of protruding scientific instruments, cameras, and antennae was a flat (0.457-meter-high), hexagonal body (1.494 meters across) with alternately long (1.1-meter) and short (0.56-meter) sides. A shock-absorbing landing leg extended from each of the three short sides, while inside the body were self-contained, miniaturized laboratories for analyzing the Martian soil and looking for signs of microbial life in it.

To avoid bringing to Mars any terrestrial organisms that might interfere with the Vikings' search for life there (or, worse yet, inadvertently contaminate Mars), each lander was sterilized prior to launch and then sealed inside a contamination-proof bioshield capsule, where it would remain until safely outside Earth's atmosphere. Since the orbiters were not sterilized, the initial launch trajectories were set so the Vikings would miss Mars altogether (rather than crash-land there) if the craft proved uncontrollable after launch. Once in flight and proved directable, they would be redirected toward Mars.

Both Vikings were to be launched on Titan 3E Centaur launch vehicles from Cape Canaveral's Launch Complex 41, the first scheduled for August 11, 1975, and the second ten days later. As the launch dates approached, a faulty thrust vector control valve (one of twenty-four valves that give fine guidance control over the thrust direction), followed by an accidental battery draining, caused the two craft to be switched. The first launch, designated Viking 1, occurred nine days late on August 20 at 2:22 P.M., Pacific standard time. The original craft was then repaired and prepared for launch as Viking 2. After the thrust vector control valve had been repaired, further repair work on its S-band (low-frequency) radio receiver was required. Viking 2 finally lifted off on September 9 at 11:39 A.M., only three minutes before an

Viking Lander 1 scene showing a rocky field, with trenches made by the spacecraft sampling arm visible in the foreground. (NASA)

approaching storm would have canceled the launch. To compensate for these launch delays, minor course connections were made so that Viking 1 would reach Mars only one day behind its original schedule and Viking 2 would arrive on schedule.

On May 1, 1976, Viking 1's sensors first detected Mars—a calibration picture was taken from a distance of eleven million kilometers. Beginning six weeks later, a series of color photographs taken of Mars during approach showed that the dust storms that had hampered Mariner 9 were absent: The 4,500-kilometer-long Valles Marineris (Valley of the Mariners) and the 27-kilometer-high volcano Olympus Mons (Mount Olympus) were clearly visible, as were water-ice fogs and various surface and atmospheric brightenings.

Even as these images were being planned and taken, a leaking pressure regulator was found to be threatening to overpressurize the Viking Orbiter 1 rocket propellant tanks. The pressure was relieved by ordering two unscheduled engine burns (on June 10 and 15), which slowed Viking 1 by 4,000 kilometers per hour and delayed its arrival at Mars by some six hours. Finally, on June 19, a 38-minute engine burn— the longest burn to date in deep space — expelled 1,063 kilograms of propellant and placed Viking 1 into Mars orbit.

An orbit trim maneuver two days later settled Viking 1 into a highly elongated, synchronous orbit (an orbit in which the craft's orbital period is identical to the rotation period of the planet below — in this case 24 hours, 39 minutes, and 36 seconds). With each orbit, Viking 1 dived to a periapsis (closest approach to the planet) of only 1,514 kilometers — directly above its intended landing site — and then ascended to an apoapsis (farthest point from the planet) of 32,800 kilometers. The initial periapsis passage (on June 19) was designated "P0," the next "P1," and the others followed in this sequence; the historic first Mars landing was scheduled for P15, which (not coincidentally) would occur on July 4, 1976— the American Bicentennial.

The primary landing target (designated "A1" for "mission A, site 1") was at 19.5° north latitude and longitude 34° west in Chryse, a now-dry delta region in the outflow pattern of an ancient flood channel. This site was selected because it was at a low elevation (where the higher atmospheric pressure made landing easier), close to the equator (as required by the lander's approach angle), relatively level and devoid of high winds (which made landing safer), and showed evidence of past water (which enhanced the chances of finding life).

The first detailed photographs of A1, taken on P3 and P4, showed spectacular image quality but frightening details. What Mariner 9's cameras had picked up as a smooth plain was revealed to be a confusion of craters, depressions, knobs, and islands — apparently too hazardous a terrain in which to attempt a landing. This assessment was complicated by the fact that, despite their impressive quality, the Viking orbiter photographs could not reveal objects less than 100 meters in size; the greatest hazard to the Viking landers would be from objects 0.1 to 1 meter in size.

In the interest of safety, the planned landing was postponed, and the Viking flight team began feverishly looking for an alternative landing site. Orbiter 1 was directed to photograph areas immediately to the south and northwest of A1 as well as the Viking 2 primary target site B1 (44° north latitude, longitude 10° west) in Cydonia and the alternative site C1 (6.5° south latitude, longitude 42.75° west) in Capri. Viking geologists, assisted by a tireless team of undergraduate interns, worked around the clock assessing the potential landing hazards suggested by these photographs. Leonard Tyler, a member of the Radio Science Team, analyzed data from the Arecibo and Goldstone radio observatories on Earth which, through the pattern of radar signals reflected from Mars, could be used to estimate the average surface roughness resulting from centimeter-sized objects scattered over large areas of the Martian surface. All this information was used by the Landing Site Certification Team, led by astrogeologist Harold Masursky, to form geologic models from which estimates of the landing hazards presented by the unobservable meter-sized objects could be made.

In the end, the leader of the Magnetic Properties Team, Robert B. Hargraves, suggested looking farther "downstream" — that is, northwest — from A1, where the ancient floods might have deposited fine-grained sediments and left a relatively smooth surface. On July 8, Viking 1's orbit was altered to bring its periapsis point over a newly designated A1-NW site, 300 kilometers northwest of A1. After further evaluation of the area, a site 240 kilometers due west of A1-NW was selected as the final landing target. This target lay at 22.4° north latitude and longitude 47.5° west, just within the western edge of Chryse Planitia (the Plains of Gold). The landing was set for July 20, 1976; coincidentally, this would be the seventh anniversary of the first manned lunar landing, a date now known as Space Exploration Day.

On the morning of July 20, Mars was 360 million kilometers from Earth. At that distance, one-way radio communication took 19 minutes — too long for flight controllers on Earth to intervene in the landing. Viking would have to land on its own. At 1:51:15 A.M., Pacific daylight time, Viking Orbiter 1 released Viking Lander 1; a 20-minute burn of the lander's own engines then nudged it out of orbit to begin a long, looping descent toward Mars. At 5:03 A.M., Lander 1 entered the top of the Martian atmosphere at a shallow 16-degree angle; it would need to travel nearly horizontally for a thousand kilometers in that thin atmosphere before its speed decreased sufficiently for landing. Radar monitored the descent while other instruments collected data on the composition and physical characteristics of the atmosphere and radioed them back to Earth. At 5:10 A.M., at an altitude of 5,906 meters, Lander 1 opened its parachute. Forty-five seconds later, at an altitude of 1,462 meters, the parachute was jettisoned, and three terminal descent engines, which exhausted sterile propellants in an outward fan-pattern to avoid contaminating or disturbing the landing site below, immediately burst into life. They slowed the lander to a scant 2 meters per second (5 miles per hour) at which speed it soft-landed at 22.46° north latitude, longitude 47.82° west — within 20 kilometers of the targeted site.

Immediately upon landing, the rate at which Viking Lander 1 was transmitting data back to Earth automatically increased from 4,000 bits per second to 16,000 bits per second. Nineteen minutes later, at 5:12:07 A.M., mission controllers at JPL received that increased bit rate as the first indication of Viking's safe landing. Their immediate shout of "Touchdown. We have touchdown," echoed to the cheers of eight hundred Viking team members at JPL.

On Mars, it was late afternoon (4:13:12 P.M. local lander time upon landing), and Lander 1 was busy. Some 25 seconds after touchdown, it began taking the first photograph — an image of the ground around its right front footpad. During that 4-minute exposure, the lander also erected and aimed a high-gain antenna at Earth and deployed a weather station. Within 15 minutes, a second photograph was completed; both were transmitted to Earth, where they were processed immediately.

By 5:54 A.M. excitement mounted at JPL as the first of these photographs — the first successful picture ever taken from the surface of another planet — appeared on the laboratory monitors. The image was incredibly sharp and clear. Small rocks, up to 10 centimeters in size, littered the area. On the right, the circular footpad was clearly visible; it had barely penetrated the ground, indicating that the surface was solid. On the left, dust still settling from the landing had left dark streaks on the photograph. Undeniably, Viking Lander 1 was on Mars.

The second photograph showed a 300-degree panorama of the Martian landscape. A series of ridges and depressions led to a horizon 3 or 4 kilometers in the distance. Boulders, perhaps meters in size, and smaller rocks were strewn about everywhere. In all, the scene was remarkably Earth-like—reminiscent of a Southwestern desert— except that nowhere was there any visible sign of vegetation or any other form of life.

The first color picture, taken on SOL 1, mistakenly showed a delightfully blue Earth-like sky. (A SOL is a Martian day as defined by one rotation of Mars—equal to 24 hours, 40 minutes of Earth time. The day of the landing was designated SOL 0 and the next SOL 1.) Subsequent corrections to the color balance revealed a salmon-pink Martian sky (made so by dust particles suspended in the atmosphere) overlooking a rusty, reddish-orange landscape. Meteorology team leader Seymour Hess made history by issuing the first weather report from another planet: "Light winds from the east in the late afternoon, changing to light winds from the southwest after midnight…Temperature range from −122 degrees Fahrenheit, just after dawn, to −22 degrees Fahrenheit… Pressure is steady at 7.70 millibars."

On SOL 3, the remaining 60 degrees of landscape was scanned, and a field of large sand dunes with a huge boulder, affectionately dubbed "Big Joe," was found. Viking scientists were struck by the realization that if Lander 1 had come down on Big Joe, it would have never survived the landing. (On SOL 12, a picture of the left front footpad revealed yet another surprise: The footpad was buried several centimeters deep in soft sediment.)

Also on SOL 3 an attempt was made to extend a 3-meter-long surface sampler arm intended to dig up and collect soil samples; the arm stalled upon retraction. Viking engineers, using duplicate Landers on Earth, concluded that a locking pin used to stow the sampler arm during flight had failed to release. They then "repaired" it by sending a sequence of extension and rotation commands that caused the pin to drop free. On SOL 8, the arm successfully dug trenches and delivered soil to biology, inorganic chemistry, and molecular analysis instruments inside the Lander body. Early responses from the biology instruments were suggestive of life having been found. In a climate of utter hopefulness coupled with scientific caution, it gradually became clear that the responses were most likely the result not of biota but of chemical reactions in the Martian soil. The next week and a half was spent attempting to make further repairs to the sampler arm, the molecular analyzer, and a radio transmitter, while obtaining more surface samples, photographing the surroundings in detail, and puzzling over results.

Viking 2 entered Mars orbit flawlessly on August 7, immediately adopting an orbit that would allow its periapsis point to "walk" around the planet, passing over the entire 40- to 50-degree north latitude band every eight days. With assistance from Orbiter 1, Orbiter 2 carefully surveyed Lander 2's prime landing site (B1) in Cydonia, its backup landing site (B2) near Alba Patera, and a hastily chosen additional backup site (B3) in Utopia Planitia. After much analysis, and with some misgivings, the B3 site at 47.9 °north latitude and longitude 225.8 ° west was picked by a Viking crew too exhaust-ed to consider further searching.

On September 3, 1976, Viking Lander 2 headed for a suspenseful landing. When it separated from its orbiter, flight controllers temporarily lost contact with the orbiter. Unable to receive progress reports relayed from the lander, they could only sit and wait as Lander 2 landed automatically and, miraculously, sent home the high-transmission-rate direct signal which indicated a safe landing. Touchdown was recorded at 3:58:20 P.M.

The scene at Utopia was remarkably reminiscent of that at Chryse: A desolate, red, rock-and-boulder-strewn landscape was seen. There were, however, differences in detail: Utopia was generally flatter than Chryse, and it had perhaps twice as many rocks. Most of them were irregularly broken and laced with pits and holes. Apparently, the unphotographable back leg of Lander 2 had, in fact, come down on top of one of those rocks: The Lander was tilted about 8 degrees from the upright position. Lander 2 set about a science analysis sequence similar to that followed by Lander 1. It also set another precedent in planetary exploration: It used its sampler arm to push aside a rock and sample the soil beneath.

Throughout all this, scientific activities with Viking 1 continued unabated. The harried Viking flight team had four spacecraft— two orbiters and two landers — with which to conduct scientific studies simultaneously. Each of the thirteen teams of scientists had cooperated closely in designing their instrumentation to fit and operate in the close confines of the Viking craft. Now they cooperated in sharing the available resources of the craft (for example, data storage and transmission and electrical power capabilities), and in making sure that each of the various investigations complemented the efforts of the others. These goals were all amply met as science investigators followed a hectic routine.

Each day at noon, the scientists met to share the findings of the previous twenty-four hours, discuss their significance, and plan future activities. Every two days, on average, a new set of instructions for future operations was laid out and "uplinked" (sent by radio telemetry) to the craft on Mars. Thus, while the Vikings always had in their on-board computers complete instructions for operating automatically in case the uplink capability was lost, the instructions were continually updated to meet the ongoing needs of the project. This updating capability contributed immeasurably to the mission's success as experimental procedures were altered in response to unexpected findings. Also — and perhaps more important— problems with the craft themselves were continually analyzed and in some cases repaired by remote control.

In an unprecedented acknowledgment of public interest in Viking, Viking scientists reported daily their findings and progress to the more than one hundred members of an international press corps resident at JPL. So close was the cooperation with the press that on one occasion the Viking biology team actually performed an experimental sequence suggested by reporter Jonathan Eberhart of *Science News*. From the seeming chaos of these day-to-day operations, a new picture of Mars gradually began to emerge.

The routine continued until November 15, 1976, when the orbit of Mars was about to take it behind the Sun. Radio communications with the Vikings would then be impossible, so the primary mission was declared ended. In mid-December, Mars reemerged from behind the Sun, and an extended mission began, mostly with replacement personnel.

On February 12, 1977, Orbiter 1's orbit was altered to bring it within 90 kilometers of the Martian moon Phobos for high-resolution photography. On March 11, 1977, its periapsis was lowered to 300 kilometers, allowing photographic identification of features on Mars as small as 20 meters across. On September 25, 1977, Orbiter 2's orbit was altered to bring it within 22 kilometers of the other moon, Deimos; on October 23, 1977, its periapsis was also lowered to 300 kilometers. Both craft continued observing and mapping Mars from orbit until an entire Martian year had elapsed. Then, on July 25, 1978, with nearly sixteen thousand photographs to its credit, Orbiter 2 was powered down when a series of leaks exhausted its steering gas. Orbiter 1 continued observing Mars until July 14, 1980, when it took the last of its more than thirty-four thousand photographs. Almost out of steering propellant, the orbiter was then used in a series of tests to determine exactly how close to empty the tanks could get before control of the craft was lost. This provided information crucial to the design of future space missions. After these tests, Orbiter 1 was finally deactivated on August 7, 1980.

On Mars's surface, Lander 2 had observed ground frost in mid-August of 1977, late in the Martian winter. It continued monitoring its surroundings until April 12, 1980, when, apparently because of battery failure, it stopped transmitting information to Earth. Meanwhile, Lander 1 was programmed to take photographs automatically and to monitor the Martian weather through 1994. On January 7, 1981, it was designated by NASA as the Thomas A. Mutch Memorial Station, dedicated to the Viking Imaging Team leader after his untimely death in a tragic mountain-climbing accident. The Mutch Station continued returning photographs from Mars until November 13, 1982, when its radio transmitter unexpectedly fell silent. A heroic effort to revive the craft was mounted, but after five unsuccessful months the effort was abandoned and the Viking mission was at long last officially terminated. Lander 1's career, however, was not yet finished. On May 18, 1984, ownership of the Mutch Memorial Station was transferred to the National Air and Space Museum of the Smithsonian Institution to begin a new career as the most distant landed historical marker of human civilization.

Knowledge Gained

Each Viking orbiter carried two vidicon cameras (similar to television cameras), with a 475-millimeter telephoto lens and filters allowing color photography. In total, 51,539 pictures were returned, covering the entire planet at a resolution of 200 meters and much of it at resolutions as small as 8 meters. Pictures of the Martian moons, Phobos and Deimos, were also taken; those of Deimos were the highest-resolution pictures ever taken of a planetary body from a flyby or orbiting spacecraft.

Mars was revealed as a planet with a tremendous variety of land features. Of greatest interest were those that gave clear evidence of past water flow: Broad, dry flood channels, apparently formed by episodes of catastrophic flooding two or three billion years ago, and smaller, dry networks of runoff channels generally more than three and a half billion years old were commonly seen. Such features were surprising because the Martian atmosphere is too thin and the climate too cold for liquid water to exist. Indeed, no signs of liquid water, nor any past or present ocean or lake basins, were seen. The water was apparently underground, frozen in a 1- to 3-kilometer-thick layer of permafrost. As evidence, some meteorite impact craters resembled giant mud splats, apparently created when heat from the impact had melted subsurface ice. Other areas showed large-scale polygonal features or fretted terrain (smooth, flat lowlands bounded by abrupt cliffs) perhaps caused by the activity of ground ice. The polar regions consisted of alternating layers of water ice (as verified below) and entrapped dust; apparently, these were the regions where the underlying ice breached the surface.

Volcanism was also apparent. Large shield volcanoes (in which the lava flows out through cracks), including Olympus Mons and others in the Tharsis region, were seen, as were composite volcanoes (which sometimes erupt explosively). Unique to Mars were the low, broad volcanic vents known as pateras. Pedestal craters (craters sitting atop raised plateaus) were revealed by Viking to be similar to Icelandic table mountains (shield volcanoes erupting beneath thick ice sheets). Some volcanoes showed evidence of having been active within the last few hundred million years.

Valles Marineris, a 4,500-kilometer-long system of steep-walled canyons up to 9 kilometers deep, was apparently formed by large-scale faults. Subsequent slumping (in-falling or landsliding) of the canyon walls was easily visible, as were horizontal sedimentary rock layers in the canyon walls.

Other features seen included numerous meteorite impact craters and basins, bright streaks indicative of wind-blown dust, dark streaks interpreted as erosion scars, atmospheric haze caused by airborne dust, and carbon-dioxide condensate clouds and fogs. Spectacular clouds routinely formed around Olympus Mons and other large volcanoes in the late morning. Two global dust storms and several dozen localized dust storms were observed during the extended mission.

High-resolution pictures of the Martian moon Phobos revealed sharp, fresh-looking impact craters and a peculiar system of parallel linear grooves, apparently formed during the impact which created Phobos' large crater Stickney. The other moon, Deimos, was seen to be saturated with impact craters and covered, apparently, with a layer of dust.

An infrared spectrometer on board each orbiter measured the concentration of water vapor suspended in the Martian atmosphere. The amounts found were highly variable, ranging from zero parts per million during the Martian winter to eighty-five parts per million near the poles during the Martian summer. Over the northern polar cap in mid-Martian summer, the atmosphere was saturated (held as much water vapor as possible without condensation occurring), strongly suggesting that the polar cap itself contained water ice. In total, the equivalent of 1.3 cubic kilometers of water ice was found in vapor form in the atmosphere.

A radiometer on each orbiter mapped the surface of Mars in infrared light (heat rays); from the amount of infrared radiation being given off, the temperature of the surface could be determined. The most significant finding was that the temperature of the permanent northern polar cap was −73 to −58 degrees Celsius—too warm to be frozen carbon dioxide (dry ice); it therefore had to be made of water ice. A layer of carbon-dioxide ice apparently settled over the water ice in the winter but evaporated in the summer. The southern polar cap was much colder, suggesting that it was a mix of water and carbon-dioxide ice with a year-round cover of carbon-dioxide frost. The global temperature extremes ranged from −140 degrees Celsius at the winter poles to 20 degrees Celsius at the noonday equator.

The radio communications and radar instruments on board Viking were used for a number of experiments auxiliary to their primary purposes. When Mars went behind the Sun, the Sun's gravity was observed to slow the transit time of radio signals passing near it. This was in accord with predictions made by Albert Einstein's general theory of relativity, and it provided new confirmation of that theory. Also at that time, radio signals passing through the solar corona yielded information on the small-scale structure of the corona. Other radio data led to improved measurements of the planet's size, orientation, spin rate, orbital characteristics, and distance from Earth, finding the exact location of the landers on Mars (to 11-kilometer accuracy), and information on the electrical properties of the Martian surface.

During the descent to the surface, instruments on board the Viking landers examined the properties of the atmosphere. The upper atmosphere was found to be mostly carbon dioxide, with small amounts of nitrogen, argon, carbon monoxide, oxygen, and nitrogen monoxide. The concentration of nitrogen 14 (normal nitrogen) relative to that of nitrogen 15 (a nitrogen atom with one extra neutron) was found to be lower than it is expected to have been early in the planet's history. Since nitrogen 14 escapes from the Martian atmosphere more rapidly than does nitrogen 15, there must have been much more nitrogen 14 present in the past. This suggests that Mars had a much denser atmosphere sometime in the past—perhaps one dense enough to have allowed the flow of liquid water.

Each lander carried two facsimile cameras — cameras in which the field of view is divided into a grid of small squares, which are scanned one at a time to produce the whole picture. The horizon around Lander 1 showed meter-sized boulders and a crater 500 kilometers in diameter at a distance of 2.5 kilometers fronting a ridge 10 kilometers distant. The near surface was duricrust (a coinage for the hard, cemented Martian soil) littered with small rocks. Throughout the field were numerous ventifacts (angular, multifaced blocks shaped by winds), a ubiquitous litter of blocks thought to be either debris from impact craters or deposition from past floods, sediment drifts (probably of silt and clay), and outcrops of bedrock. The drifts were stable; in fact, the only sign of movement observed was a small slumping of soil around Big Joe (significant, nevertheless, as the first observation of geologic change ever made on another planet). During the extended mission, darkening of the sky and softening of shadows were observed during two global dust storms.

The horizon around Lander 2 showed gently undulating plains with much less relief than at the Lander 1 site; the terrain was almost undoubtedly formed by ejecta (debris) from the impact crater Mie, 160 kilometers to the east. Like the terrain around Lander 1, there were a duricrust surface, ventifacts, and a litter of blocks and boulders. Most of the Lander 2 blocks, however, were vesicular (pitted and porous-looking), probably the result of gas bubbles trapped in cooling volcanic lava, but some were smoother, typical of finer-grained volcanic rock. Signs of wind erosion included the ventifacts, wind-scalloped blocks, and wind-sculpted pedestals under some boulders. A linear trough, 10 to 15 centimeters deep and more than 10 meters long, was seen as part of a polygonal network near the lander; it may have been formed by seasonal freezing and thawing of groundwater. The frost observed was probably water ice or clathrate (a mixture of carbon dioxide and water); the ambient temperatures were too high for the existence of frozen carbon dioxide alone.

Other activities of lander imaging included monitoring the sampler arm, the soil sample collection, and the magnetic properties experiment. Lander imagery also proved invaluable as an aid to diagnosing problems with other parts of the spacecraft. Some astronomy—photographs of Phobos, the Sun, and the shadow of Phobos during a solar eclipse—was also done. A few pictures, particularly those of the American flags on the landers and of the spectacular sunsets and sunrises, were taken primarily for their aesthetic value. In total, more than 4,500 lander photographs were taken.

Data compiled from photographs, the forces exerted by the ground on the landing legs and sampler arm, and other such clues were used to understand properties of the Martian surface material. The surface was generally firm (with the exception of the area under Lander 1's left footpad) and adhesive (meaning that it stuck to things, such as parts of the lander), but only weakly cohesive (meaning that it did not stick well to itself; instead, it crumbled like a clod of dirt).

An array of magnets attached to the collector head of each sampler arm picked up any magnetic materials in the soil. Photographs of the magnets then revealed the material and allowed determination of its abundance and magnetic properties. Roughly 10 percent of the soil was found to be magnetic, most likely maghemite (a compound of iron and oxygen similar to rust). If so, then the planet's surface is red because it is, quite literally, rusty.

Each lander carried a seismometer designed to detect "Marsquakes." During the first five months at Utopia Planitia, no major disturbances were detected, although data believed to indicate a small quake (2.8 on the Richter scale) were recorded on SOL 80. Analysis of the underground reflections of that disturbance indicated that the crust in the region near Lander 2 was 15 kilometers thick. An analysis of the structure of the planet's deep interior, the primary goal of this investigation, was not possible because the seismometer on Lander 1 failed to unlock from its stowed position after landing. The Lander 2 seismometer routinely picked up spacecraft vibrations resulting from winds and the activities of other instruments on board the craft. It was unexpectedly pressed into use as a wind monitor to supplement the meteorology instruments.

A complete weather station, including air temperature, pressure, and wind condition sensors, occupied the end of a boom extending 1 meter from each lander. Subsequent reports collected during six years on Mars provided data important for understanding both global and local atmospheric circulation patterns. Lander 1 recorded temperatures ranging from −88 degrees Celsius to −12 degrees Celsius. Winds were mild — generally only a few meters per second — although they did exceed 50 meters per second during violent storms. The atmospheric pressure was observed to vary by some 30 percent as carbon dioxide from the atmosphere froze onto or sublimated from the polar caps with the seasons.

The chemical composition of Martian soil samples was analyzed by an X-ray fluorescence spectrometer, a device that identifies various chemicals through their absorption and reemission of X rays. The soil was basically an iron-rich clay. Its composition was 5 percent magnesium, 3 percent aluminum, 20.9 percent silicon, 3.1 percent sulfur, 0.7 percent chlorine, less than 0.25 percent potassium, 4 percent calcium, 0.51 percent titanium, and 12.7 percent iron, with traces of rubidium, strontium, ytterbium, and zirconium; the remainder consisted of elements which could not be identified by the spectrometer.

A gas chromatograph mass spectrometer (essentially a sophisticated, miniaturized version of a breath alcohol content analyzer) on board each lander identified organic molecules by breaking them down into their constituent elements. Their primary finding was that the Martian soil contained no organic compounds in quantities that could be detected — making the probability of finding life there extremely unlikely. This instrument was also used to analyze the chemical composition of the Martian atmosphere. It corroborated the atmospheric composition found by the entry science experiments (95 percent carbon dioxide, 2.7 percent nitrogen, 1.6 percent argon, 0.13 percent oxygen, and smaller amounts of other constituents), and it revealed an underabundance of argon 36 (argon with four fewer neutrons than normal) relative to argon 40 (normal argon) of 10 percent, compared with Earth's atmosphere. This observation provided vital clues to models suggesting that early in its history Mars had a substantially warmer and denser atmosphere, which might have allowed liquid water to exist on its surface.

Three experiments designed to detect microorganisms in the Martian soil were on each lander. Controversy over their results continued throughout the mission.

In the gas exchange experiment, a soil sample was placed in an artificial atmosphere of helium, krypton, and carbon dioxide humidified with a liquid nutrient. The presumption was that if microorganisms were present, they would reveal themselves by metabolizing the nutrient and then emitting hydrogen, nitrogen, oxygen, methane, or carbon dioxide. Surprisingly, the first run on Lander 1 produced copious amounts of oxygen, suggesting the presence of life. Subsequent experiments — particularly the demonstration that heat sterilization did not prevent the release of oxygen —indicated that the reaction was not indicative of life but of a peculiar physical chemistry process in the Martian soil. Apparently, ultraviolet light from the Sun had bound oxygen to material in the soil, creating compounds known as peroxides (or perhaps variants known as superoxides or ozonides). Water vapor from the nutrient apparently released that bound oxygen, mimicking a biological response. Similar results were obtained by Lander 2.

In the labeled release experiment, a soil sample was moistened with a liquid nutrient seeded with radioactive carbon 14 (which served as an easily monitored "label" or tracer of the nutrient). The presumption was that if any microorganisms were present, they would take in the nutrient and expel carbon dioxide or other gases containing the carbon 14 labels. In fact, labeled carbon dioxide was released, again mimicking a biological reaction. When more nutrient was added to the sample, however, the carbon dioxide release did not increase (as it would if organisms were releasing it); instead, it decreased, consistent with the premise that the nutrient had been decomposing peroxides to create the carbon dioxide. Heat sterilization here prevented carbon-dioxide production completely, a result consistent with either biological or physical chemistry explanations.

In the pyrolytic release experiment, a soil sample was exposed to normal Martian air that had been labeled with radioactive carbon monoxide 14 and carbon dioxide 14. The sample was then incubated under simulated sunlight, with the presumption that if any plantlike microorganisms were present, they would take in the radioactive gases. After prolonged incubation, the sample was vaporized (by heating, hence the term "pyrolytic") and the vapors examined for radioactive carbon. In the first run on Lander 1, a substantial amount of radioactive carbon was found — again suggesting life. Subsequent runs, however, failed to confirm that result.

Experiment designer Norman H. Horowitz concluded that the chance of the results being caused by biological activity was negligible in that the first high carbon reading could not be reproduced, adding water vapor had no effect on the reactions, turning off the light had no substantial effect on the reactions, and laboratory simulations on Earth could reproduce the results through the interaction of the labeled gases with iron oxide compounds such as maghemite (which the magnetic properties experiment had suggested was prevalent on Mars).

In the final analysis, virtually all the Viking scientists agreed that although a biological basis for the results could not be completely ruled out, it was extremely unlikely. The lack of organic compounds in the soil argued strongly against a biological explanation, and scientists had shown that all the results could be explained by purely physical chemistry processes. Labeled release experiment designer Gilbert V. Levin continued to argue for the minute possibility that life might nevertheless have been found.

Context

Before the space age, Mars was a fantasy place. From Edgar Rice Burroughs' tales of John Carter on Mars to Percival Lowell's scientific speculations about Mars as an "abode of life," the Western world equated "Martians" with "extraterrestrial life." The Mariner missions replaced those fantasies with orbital photographs of Moon-like craters and Earth-like volcanoes, valleys, and flood channels; the surface landscape itself, however, remained a matter of speculation. When Viking 1 touched down on Mars, all that changed: The entire world saw the barren, red, rock-strewn desert landscape of Chryse Planitia. Instantly, Mars became real.

Project Scientist Gerald Soffen has often said that in Viking's first month on Mars, more was learned about Mars than had been previously learned in the entire history of humanity. The complete absence of detectable organic compounds in the soil and the failure to find confirmable signs of life there was disappointing, even as the copious evidence for a history of flowing water was tantalizing. What allowed that water to flow in the past? Was there once a warm, dense atmosphere on Mars? Where has that water gone? Theorists analyzing the Viking data now suspect that there indeed was once such an atmosphere, and that the water is now locked in a kilometers-thick layer of ice-laden permafrost, which undergirds the entire surface of the planet and emerges at the polar caps. This planetary picture — a once-hospitable, now dry, cold, and lifeless world — contrasts with the sulfuric acid-laden hothouse of Venus and leaves scientists astounded at the temperate, water-rich world called Earth.

It is easy to argue that Viking was the pinnacle of unmanned exploration of the solar system. Certainly, the task of simultaneously operating four craft in the vicinity of Mars while landing two of them successfully on the surface was an unprecedented challenge for mission controllers. At the time, the ability of the Vikings to operate either completely automatically or upon continually updated instructions from Earth made them the most versatile and sophisticated deep space craft ever flown. The spectacular operational lifetime of Viking Lander 1 of 6 years, 3 months, and 24 days on Mars set a long-lasting record for continual scientific operations on the surface of another planet.

Subsequent exploration of Mars did not resume until July, 1988, when the Soviet Union launched Project Phobos, two craft intended to examine Phobos at close range and land probes upon the Martian moon for which they are named. The American Mars Observer, a long-duration mission to continue mapping and monitoring Mars from orbit, was scheduled for launch by the space shuttle sometime after resumption of flights following the *Challenger* accident. Beyond that, there remained the desire for further Mars landings, including a Mars rover and a Mars sample return mission, and the firm belief that someday human beings would themselves walk on Mars.

Bibliography

Baker, Victor R. *The Channels of Mars*. Austin: University of Texas Press, 1982. A detailed analysis and summary of Martian geomorphology as revealed by Viking, focusing on water-cut channels and other evidence of water on Mars. Extensive references and numerous illustrations. Advanced college level.

Burgess, Eric. *To the Red Planet*. New York: Columbia University Press, 1978. A good, chronological summary of the Viking missions, beginning with historical perceptions of Mars and continuing through the early Viking results. Some illustrations; written for general audiences.

Carr, Michael H. *The Surface of Mars*. New Haven, Conn.: Yale University Press, 1981. A comprehensive survey by the leader of the Orbiter Imaging Team of physical processes affecting the surface of Mars. Also includes illustrations, a chapter on the search for life on Mars, and a chapter on the Martian moons. Advanced college level.

Cooper, Henry S. F., Jr. *The Search for Life on Mars: Evolution of an Idea*. New York: Holt, Rinehart and Winston, 1980. A masterful account by *The New Yorker's* premier science writer. Covers the history and chronology of the Viking biology experiments in exquisite detail, with much attention to the human side of the experiments. Accurate and easily accessible for general audiences.

Eberhart, Jonathan. "Operation Red Planet." *Science News* 109 (June 5/12, 1976): 362. This article and its accompanying special reports began an extensive series of articles on Viking that continued in subsequent issues throughout the mission. Eberhart, the undisputed dean of space science reporters, is lucid, insightful, and accessible to all audiences.

Ezell, Edward Clinton, and Linda Neuman Ezell. *On Mars: Exploration of the Red Planet, 1958–1978*. NASA SP-4212. Washington, D.C.: Government Printing Office, 1984. NASA's official history of the Viking program. Includes extensive administrative, political, financial, and scientific background on the development of the U.S. Martian exploration program. Also contains an in-depth review of Viking landing-site selection and certification procedures and an overview of Viking science results. Detailed appendices and references; suitable for general audiences.

Horowitz, Norman H. *To Utopia and Back: The Search for Life in the Solar System*. New York: W. H. Freeman and Co., 1986. An authoritative, firsthand analysis of the search for life in the solar system, written by one of the principal Viking biologists. Covers biological background, development of Viking biology experiments, and the Viking mission itself. Clear and accurate; college level.

Moore, Patrick. *Guide to Mars*. New York: W. W. Norton and Co., 1977. Great Britain's celebrated astronomer and writer presents a tourist's guide to Mars, beginning with basic introductions to the solar system and telescopes, continuing with a history of Mars exploration, and culminating with Viking and a look toward future exploration. Highly accessible and informative for all audiences.

National Aeronautics and Space Administration, Viking Lander Imaging Team. *The Martian Landscape*. NASA SP-425. Washington, D.C.: Government Printing Office, 1978. Viking Lander Imaging Team leader Thomas Mutch's anecdotal account of the conception, design, building, and operation on Mars of the lander cameras, followed by more than two hundred of the best and most representative photographs taken by the Viking landers—some in color, some in stereo, all with explanatory text. Stereo viewer included. For all audiences.

Science 193/194 (August 27, October 1, and December 17, 1976). These three issues are dedicated to the Viking missions; the overview was written by Gerald Soffen and Conway Snyder, two members of the various Viking teams (see entry below).

Soffen, Gerald A., and Conway W. Snyder. "The First Viking Mission to Mars." *Science* 193 (August 27, 1976): 759. This overview begins the first of three special issues of *Science* dedicated to the Viking missions (see entry above). Advanced college level.

Spitzer, Cary R., ed. *Viking Orbiter Views of Mars*. NASA SP-441. Washington, D.C.: Government Printing Office, 1980. The Viking Orbiter Imaging Team's public report on their findings. Contains brief introductions to Viking and Mars, followed by spectacular photographs— some in stereo (viewer provided)—illustrating all the major landforms and atmospheric phenomena observed by the Viking orbiters. Includes detailed photographs of the Viking landing sites. The extensive captions are college level; the photographs are enthralling to all audiences.

Washburn, Mark. *Mars at Last!* New York: G. P. Putnam's Sons, 1977. An engaging, popularized (though sometimes scientifically misleading) account of Viking. The first half focuses on the cultural and mythological motivations for going to Mars; the second half covers the Mariner and Viking missions. Suitable for general audiences.

Philip J. Sakimoto

THE VOYAGER PROGRAM

Date: Beginning August 20, 1977
Type of program: Unmanned deep space probes

The Voyager probes executed the first Grand Tour in planetary exploration by successively encountering Jupiter, Saturn, Uranus, and Neptune. Such a tour, using the "planetary-gravity-assist" technique to travel from planet to planet, is possible only once every 175 years.

PRINCIPAL PERSONAGES

MICHAEL A. MINOVITCH, the inventor of the gravity-assist concept

GARY A. FLANDRO, the discoverer of the 175-year period between Grand Tour alignments of the outer planets

EDWARD C. STONE, Voyager Project Scientist

CHARLES E. KOHLHASE, Principal Mission Designer

ELLIS D. MINER, Assistant Project Scientist

ANDREI B. SERGEYEVSKY, Principal Trajectory Designer for the Neptune encounter

BRADFORD A. SMITH, Principal Investigator, imaging science

G. LEONARD TYLER, Principal Investigator, radio science

HARRIS SCHURMEIER, Voyager Project Manager from project inception through development

JOHN R. CASANI, Voyager Project Manager from before launch through Jupiter-encounter preparations

RAYMOND L. HEACOCK, Voyager Project Manager for the Jupiter and Saturn encounters

RICHARD P. LAESER, Voyager Project Manager for the Uranus encounter

NORMAN R. HAYNES, Voyager Project Manager for the Neptune encounter

Summary of the Program

Voyager conducted the first planetary Grand Tour in history. Two Voyager spacecraft were launched from Earth in 1977; Voyager 1 encountered Jupiter in 1979 and Saturn in 1980, and Voyager 2 encountered Jupiter in 1979, Saturn in 1981, Uranus in 1986, and Neptune in 1989. It is this latter sequence of planetary encounters that is called the Grand Tour.

The Voyager spacecraft used the "planetary-gravity-assist" technique to move from one planet to the next. Concepts of using gravity to propel a spacecraft from one body to another

have existed since the 1920's. The actual technique of executing a gravity assist was not well understood, however, until Michael Minovitch developed it in the early 1960's.

With twentieth century technology, it is not possible to accomplish a Grand Tour unless the gravity-assist technique is used. The fuel requirements of a nongravity-assist tour are vastly beyond the existing technology. To illustrate, Voyager saved 1.5 million kilograms of fuel by using Jupiter's gravity to propel it toward Saturn. Similar amounts of fuel were saved at Saturn (using gravity assist to reach Uranus) and at Uranus (using gravity assist to proceed to Neptune).

The outer planets must be properly aligned or the tour is not possible. In 1966, Gary Flandro discovered that this alignment occurs only every 175 years, that the last alignment had occurred in 1802, and that the next time the planets would be properly aligned for a Grand Tour launch would be in 1977. The Voyager mission was designed to take advantage of that opportunity.

The National Aeronautics and Space Administration (NASA) authorized the Jet Propulsion Laboratory (JPL) in 1972 to start the Voyager project. Initially, only a four-year mission to Jupiter and Saturn, with a launch date in 1977, was funded. This circumstance led to the original name of the project: Mariner Jupiter Saturn 77 (MJS77). In 1977, the name of the project was changed to Voyager.

Two spacecraft were to visit Jupiter, then Saturn. If the Voyager 1 encounter with Saturn was successful, then Voyager 2 would be permitted to go on to Uranus and then Neptune. The Uranus option was authorized in 1981, the Neptune option in 1986.

The Voyager spacecraft, which were identical, used the Mariner spacecraft's decagonal shape. The ten bays housed electronics boxes. In the middle was a fuel tank for the propulsion system. Attached to one end was a large communication antenna for receiving spacecraft commands from Earth and for transmitting data back to it. Also attached were two deployable booms. At the end of one lay the three nuclear power plants that provided the electrical power to run the spacecraft. Along the sides and at the end of the other boom lay various scientific instruments.

The Voyager Grand Tour mission was one of scientific exploration. Each spacecraft carried the same eleven instruments; the complete package included sensors that point at an object (target body instruments) and ones that make *in situ* measurements (field, particle, and wave instruments). The target body sensors included wide-angle and telephoto cameras, an infrared telescope, an ultraviolet telescope, an instrument to measure certain characteristics of light, and a radio transmitter to send radio signals through planetary atmospheres and rings back to Earth. The *in situ* sensors included a magnetic field sensor, a radio receiver, a plasma wave sensor, and three particle detectors.

The two spacecraft were to be launched from Titan-Centaur launch vehicles, approximately three weeks apart. The first spacecraft would be on a slower path to Jupiter and Saturn; the second would be on a faster path to the planets and thus would arrive first. As the order in which the spacecraft would encounter each planet was far more important than the order of launch, Voyager 2 was launched first, at 10:29:45 A.M. eastern daylight time on August 20, 1977, and Voyager 1 was launched second, at 8:56:01 A.M. on September 5, 1977.

Each spacecraft orbited Earth several times and then was injected onto a Jupiter-encounter path by a solid-fueled rocket. After burnout, all deployable booms were extended, and the spacecraft were thoroughly checked. Thirteen days after launch, Voyager 1 turned its telephoto camera toward Earth and shuttered the first image in history to contain both Earth and the Moon.

The Voyagers' first planetary encounter was with Jupiter. Two spacecraft had preceded the Voyagers there: Pioneers 10 and 11. Pioneer 10 had encountered the planet in 1973, and Pioneer 11 a year later. Neither spacecraft had an imaging camera, although each had an imaging photopolarimeter which could (and did) take rather crude photographs. It was left to Voyager to provide the first high-quality images of the solar system's largest planet.

Voyager 1 took exactly eighteen months to reach Jupiter, making its closest approach on March 5, 1979. Voyager 2 made its closest approach on July 9, 1979. Jupiter was known to have four huge moons (discovered by Galileo in 1610 and named "the Galilean satellites" in his honor) and nine smaller ones. The planet itself and the four large Galilean satellites were the main targets of interest; thus, the Voyager trajectories were designed to permit a close encounter with each of these five main bodies.

The planetary-gravity-assist technique was used by both Voyagers at Jupiter to propel the spacecraft on to Saturn. Before the Voyagers executed the gravity assist, they had been in an elliptical orbit about the Sun and Jupiter. After the maneuver, the spacecraft had enough energy to escape the gravitational pull of the Sun permanently. It was the gravity assist at Jupiter that started Voyager 2 on the Grand Tour.

The Voyagers' second planetary encounter was with Saturn. One spacecraft had preceded Voyager to Saturn: Pioneer 11, which had encountered Saturn on September 1, 1979. After a little more than twenty months of interplanetary cruising from Jupiter, Voyager 1 made its closest approach

to Saturn on November 12, 1980. Voyager 2 followed nine months later, making its closest approach on August 25, 1981.

Saturn was known to have nine moons of varying sizes. Between the two spacecraft, close encounters with seven of the nine were made. One of the moons, Titan, which is much larger than the others, was known to have an atmosphere. Thus, there was the possibility of an environment capable of supporting life. For these reasons, Voyager 1 was targeted to pass within 4,000 kilometers of Titan's surface. Unfortunately, this close passage flung the spacecraft out of the plane of the ecliptic (the plane in which the planets orbit the Sun) at about a 35-degree angle, costing Voyager 1 the opportunity to encounter Uranus, Neptune, and Pluto.

Voyager 2's third planetary encounter was with Uranus, which had not yet been encountered by any spacecraft. After a four-year, four-month interplanetary cruise, Voyager 2 made its closest approach to Uranus on January 24, 1986. This planet is tilted on its side. As it orbits the Sun, first its northern pole, then its equator, then its southern pole, and then its equator points toward the Sun. When Voyager 2 encountered Uranus the planet's southern pole was pointed at the Sun.

Five moons were known to orbit Uranus about its equator. As Voyager 2 approached the southern pole, the five moons appeared to orbit the planet in concentric circles, creating a bull's-eye effect. Because of this geometric pattern, Voyager 2 could pass close to only one moon. Fortunately, the gravity assist required at Uranus for the trip to Neptune allowed the spacecraft to pass close to Miranda, the moon closest to Uranus.

The final Voyager 2 planetary encounter was with Neptune, and the spacecraft approached the planet in 1989. Neptune is known to have one large moon, Triton, and one small one, Nereid. Of the two, Triton is known to have an atmosphere, and thus scientific interest in it is high. Voyager 2 came close to both bodies, receiving a final gravity assist at Neptune to bend the spacecraft's path by almost 45 degrees so that it could pass within 40,000 kilometers of the surface of Triton. This dual encounter marked the end of the Grand Tour and began Voyager's interstellar mission.

Knowledge Gained

The Voyager mission provided the first high-quality visual study and the most comprehensive scientific investigation of the Jovian system until that time. The outer atmosphere of Jupiter is now known to be made up of about 89 percent hydrogen, about 11 percent helium, and trace amounts of many elements. Cloud-top lightning was observed at all latitudes on the dark side of the planet. At the place in the atmosphere where the pressure is the same as that at Earth's surface, the temperature is about −108 degrees Celsius. Voyager discovered a set of rings, no more than thirty kilometers thick, composed of fine particles.

For the first time, humanity knows what the surfaces of Jupiter's four major moons look like. On Io, nine volcanoes in the process of eruption were observed. Europa was observed to have the smoothest known surface in the solar system, with a difference in altitude between the highest peak and the lowest valley of two hundred meters or less. Ganymede proved to be the largest known moon in the solar system. Three new Jovian moons were discovered.

Jupiter's magnetic field is the largest of all the planets in the solar system. (The magnetic field of the Sun is larger.) At times Jupiter's magnetic field stretches beyond the orbit of Saturn. An electric current of more than five million amperes flows between Io and Jupiter.

Voyager also provided the first high-quality visual images and a highly comprehensive scientific study of the Saturnian system. The outer atmosphere of Saturn is now known to be made up of about 94 percent hydrogen, about 6 percent helium, and trace amounts of many other elements. Saturn radiates about 80 percent more energy than it receives from the Sun. Lightning was observed at low latitudes on the dark side of the planet. At the place in the atmosphere where the pressure is the same as that at Earth's surface, the temperature is about −139 degrees Celsius. Saturn's day is 10 hours, 39 minutes, and 15 seconds long.

Each of Saturn's three great rings (the A-, B-, and C-rings) actually contains many hundreds of thousands of ringlets. Pioneer 11's discovery of the thin F-ring was confirmed, and Voyager discovered two new rings: the diffuse D- and G-rings. Two small moons, one on either side of the thin F-ring, were discovered; the existence of these moons had been predicted before Voyager 1's encounter with Saturn.

An enormous crater, one-third Mimas' diameter, was discovered on Saturn's innermost major moon, Mimas. An even larger crater, 400 kilometers in diameter, was observed on Saturn's moon Tethys. The moon Enceladus has the second smoothest surface in the solar system. The moon Iapetus is half light and half dark, giving it the largest light-to-dark ratio of any body in the solar system. Voyager 1 established that Saturn's largest moon, Titan, is the second largest in the solar system. Titan's atmosphere has a near-surface pressure 1.5 times that of Earth, has a temperature of −179 degrees Celsius, and is composed of 90 percent nitrogen, with methane, argon, and trace carbon and hydrocarbon compounds making up the remaining 10 percent. Three new moons were discovered: the two moons on either side of the F-ring and a moon just outside the A-ring.

Saturn's magnetic field is aligned with its north pole to within one degree. A torus of hydrogen and oxygen ions that orbits Saturn is probably provided by the moons Tethys and Dione. Titan provides its own orbiting torus of neutral hydrogen atoms.

As another first, Voyager provided the first encounter of any kind with the Uranian system. The quality of the visual imagery obtained and the completeness of its scientific investigation there were comparable to those attained at Jupiter and Saturn. For the first time, the appearance of the planet Uranus, its eleven rings, and its five major moons became

known. The outer atmosphere of Uranus is now known to be made up of 85 percent hydrogen, about 15 percent helium, and trace amounts of many hydrocarbon compounds. Unlike Saturn, Uranus radiates only about one-third as much energy as it receives from the Sun. Lightning was also detected in Uranus' atmosphere. The Uranian day is 17 hours, 14 minutes, and 40 seconds long.

Voyager 2 discovered two new rings, both of them very thin. All eleven of Uranus' rings are thin (no more than 100 kilometers wide). Two moons, one on either side of the outermost (and thickest) Uranian ring, the epsilon ring, were discovered. These moons help confine the epsilon ring particles. Ten new moons (including the two discussed above) were discovered. The surface of Uranus' innermost major moon, Miranda, revealed nearly every type of geological process observed anywhere else in the solar system.

Voyager 2 discovered that Uranus has a magnetic field. The field is tilted by 58.6 degrees with respect to Uranus' north pole. In contrast to any other planet that has been explored, the magnetic field's center is offset from the center of Uranus by nearly one-third of the planet's radius.

The color of Neptune has already been determined to be almost the same as that of Uranus: aqua. Neptune's largest moon, Triton, appears to have an orange color.

Context

Voyagers 1 and 2 were the third and fourth spacecraft to encounter Jupiter, and the second and third to encounter Saturn. Voyager 2 was the first spacecraft to encounter Uranus and was to be the first spacecraft to encounter Neptune. It was also the first spacecraft to go on the Grand Tour of the outer solar system.

The solar system contains one star, nine planets, more than fifty moons, thousands of asteroids, and hundreds of comets. Together, the two Voyager spacecraft used the planetary gravity assist a total of six times to encounter five of the nine planets and fifty-one moons. Voyager provided the first high-quality visual imagery of Jupiter and Saturn and the first imagery of any kind of the moons of Jupiter, the rings and moons of Saturn, the planet Uranus and its rings and moons, and the planet Neptune and its rings and moons. An extensive scientific investigation of the four gas giants' systems was conducted.

Voyager conducted the first scientific reconnaissance of the entire outer solar system, contributing greatly to the understanding of the characteristics of the solar system's parts that is necessary before a full-fledged theory of the solar system can be developed. Such a theory would explain the creation and evolution of the entire system and each of its parts. The theory might then permit accurate predictions regarding the system's future evolution. The inhabitants of Earth have a vested interest in knowing what will happen to their planet. A well-supported theory of the solar system might also allow accurate predictions to be made of the density and characteristics of solar systems in the Galaxy and beyond.

Bibliography

Beatty, J. Kelly, et al., eds. *The New Solar System*. Cambridge, Mass.: Sky Publishing, 1982. Gives a comprehensive description of the solar system, using the results of planetary exploration missions from all countries. Each of the twenty-one chapters is written by a pioneer in planetary exploration. Contains many illustrations and reproductions of images returned by planetary spacecraft. Suitable for general audiences.

Fimmel, Richard O., William Swindell, and Eric Burgess. *Pioneer Odyssey*. NASA SP-396. Washington, D.C.: Government Printing Office, 1977. Discusses the state of knowledge of the Jovian system before the Pioneer 10 and 11 encounters, the history of the Pioneer 10 and 11 missions, and the knowledge gained from the two spacecraft. Contains many illustrations, a list of project participants, and recommendations for further reading.

Fimmel, Richard O., James Van Allen, and Eric Burgess. *Pioneer: First to Jupiter, Saturn, and Beyond*. NASA SP-446. Washington, D.C.: Government Printing Office, 1980. This book is an update to *Pioneer Odyssey*, covering the Pioneer 11 encounter with Saturn and the mission after the planetary encounters. A more mature discussion of the knowledge gained from the Jupiter encounters is also provided. Contains many illustrations, images from both Jupiter and Saturn, a listing of project participants, and a bibliography.

Frazier, Kendrick. *Solar System*. Alexandria, Va.: Time-Life Books, 1985. Considers the state of knowledge about the solar system, starting with the Sun, working out to the inner terrestrial planets and the outer gas giants, and ending with the comets and asteroids. Also discusses what little is known about the beginning and evolution of the solar system and the galaxy. Contains many color illustrations, photographs, and charts.

Gallant, Roy A. *Our Universe*. Washington, D.C.: National Geographic Society, 1980. Treats the state of knowledge of the solar system, the Galaxy, and the rest of the universe. Also discusses the future of manned spacecraft. Contains many color illustrations and reproductions of images returned from various exploratory spacecraft.

Morrison, David. *Voyages to Saturn*. NASA SP-451. Washington, D.C.: Government Printing Office, 1982. Examines the state of knowledge of the Saturnian system before Pioneer 11 and gives an account of the Pioneer 11 encounter, the history of the Voyager mission, and the Voyager encounters. Contains many images `returned by the Pioneer 11 and Voyager spacecraft, a list of Voyager project personnel, and suggestions for additional reading.

Morrison, David, and Tobias Owen. *The Planetary System*. Reading, Mass.: Addison-Wesley Publishing Co., 1988. This is one of the most helpful books on planetary science; it was designed to be used in an introductory college course for nonscience undergraduates. Discusses the entire solar system, making reference to fundamental concepts from physics, astronomy, geology, and atmospheric science when necessary.

Morrison, David, and Jane Samz. *Voyage to Jupiter*. NASA SP-439. Washington, D.C.: Government Printing Office, 1980. Discusses the state of knowledge of the Jovian system before any of the spacecraft encounters; proceeds to give an account of the Pioneer encounters, the history of the Voyager project, the Voyager encounters, and the prospects for a return to Jupiter. Reproduces many images returned by the Pioneer and Voyager spacecraft and includes a list of the Voyager project personnel and suggestions for additional reading.

Trefil, James S. *Space, Time, Infinity: The Smithsonian Views the Universe*. New York: Pantheon Books, 1985. Explores the history of astronomy and the state of knowledge of the solar system, of the Galaxy, and of the universe. Predictions regarding the future of astronomy and the evolution of the solar system, the Galaxy, and the universe are made. Illustrated. Suitable for a general audience.

Yeates, C. M., et al. *Galileo: Exploration of Jupiter's System*. NASA SP-479. Washington, D.C.: Government Printing Office, 1985. Discusses the state of knowledge of the Jovian system and the design of the Galileo spacecraft (which was projected to orbit Jupiter), scientific instruments, and trajectory. Includes many photographs of parts of the Jovian system, a list of Galileo project personnel, and a list of references. Suitable for general audiences.

William J. Kosmann

VOYAGER 1
JUPITER

Date: March 5 to April 15, 1979
Type of mission: Unmanned deep space probe

Voyager 1 collected detailed information on the planet Jupiter, its rings, satellites, and surrounding environment, including detailed photographs of the four Galilean satellites: Io, Europa, Ganymede, and Callisto. Voyager 1 demonstrated the viability of building a complex, semiautonomous spacecraft capable of lasting more than a decade.

PRINCIPAL PERSONAGES

HARRIS SCHURMEIER, the first Voyager
    Project Manager
RAY HEACOCK, Voyager Project Manager for the
    Jupiter encounter
EDWARD C. STONE, Voyager Project Scientist
ROCHUS E. VOGT, Principal Investigator,
    cosmic ray experiment
RUDOLF A. HANEL, Principal Investigator,
    infrared experiment
BRADFORD A. SMITH, Principal Investigator,
    imaging experiment
S. M. (TOM) KRIMIGIS, Principal Investigator,
    low-energy charged particle experiment
NORMAN F. NESS, Principal Investigator,
    magnetometer experiment
HERBERT S. BRIDGE, Principal Investigator,
    plasma science experiment
CHARLES W. HORD, Principal Investigator,
    photopolarimetry experiment
JAMES W. WARWICK, Principal Investigator,
    planetary radio astronomy experiment
FREDERICK L. SCARF, Principal Investigator,
    plasma wave experiment
G. LEONARD TYLER, Principal Investigator,
    radio science experiment
A. LYLE BROADFOOT, Principal Investigator,
    ultraviolet spectrometer experiment

Summary of the Mission

The story of Voyager 1 began in 1966 at the Jet Propulsion Laboratory (JPL) in Pasadena, California, where a team of scientists and engineers conceived the idea of a Grand Tour of the outer planets. A spacecraft can use the gravity of one planet to speed up and deflect its trajectory toward another planet— a technique called gravity assist. In the late 1970's, Jupiter, Saturn, Uranus, and Neptune were all positioned in an arc on the same side of the Sun, making possible a gravity-assist trajectory from one planet to the next. This special alignment of planets occurs only once every 175 years.

To study the feasibility of a mission to the outer planets, the thermoelectric outer planet spacecraft (TOPS) group was formed. The TOPS group considered many problems posed in designing a spacecraft for the outer solar system. For example, the spacecraft would have to function at greater distances from the Sun and Earth than had any other spacecraft. Since a mission to the outer planets would take about ten years to complete, all parts in the spacecraft had to be designed to last that number of years or have a failproof backup system. The spacecraft needed to be more automatic and more independent than any previous spacecraft. Because the outer planets are so distant from Earth, it would take hours for engineers to correct a spacecraft malfunction.

When some of the more significant questions were answered, the Space Science Board, a group of appointed scientists, carefully studied the recommendations of the TOPS group. In 1969, a series of five separate outer planet missions to visit Jupiter, Saturn, Uranus, Neptune, and Pluto were recommended. At the same time, many other missions were competing for funds within the National Aeronautics and Space Administration (NASA) in Washington, D.C. Because of budget constraints in 1972, Congress approved a revised plan to build only three spacecraft. Voyager 1 would fly by both Jupiter and Saturn. If the Voyager 1 mission was a success, then Voyager 2 would be targeted to fly by not only Jupiter and Saturn but Uranus and Neptune as well. The third spacecraft would be built as a ground spare. To provide advance knowledge about the environments around both Jupiter and Saturn, two additional, less complex spacecraft, Pioneer 10 and Pioneer 11, were separately funded and launched in 1972.

JPL was selected to implement the mission. The Mariner Jupiter Saturn 77 (MJS77) project, later renamed Voyager, began on July 1, 1972, under the management of Harris Schurmeier. When funds were authorized for the mission, NASA issued an "announcement of flight opportunity" to select the scientific instruments for the craft. Eventually, eleven instruments were built for Voyager 1. Edward C. Stone was selected as Project Scientist and charged with coordinating scientific activity. For Voyager 1's Jupiter flyby, Ray Heacock was Project Manager.

The Voyager 1 spacecraft was modeled on the Mariner spacecraft series, which had flown earlier to Venus and Mars. The spacecraft was about the size of a compact car. It weighed 825 kilograms, including 117 kilograms of scientific instruments. Voyager 1 was not a spinning spacecraft; it was stabilized on all three axes, using one sensor locked on the Sun and a second sensor locked on a star. Voyager 1 was a ten-sided aluminum structure, containing its key electronic elements inside its inner walls. The center of the structure contained a spherical propellant tank filled with hydrazine fuel. The fuel was used for trajectory corrections and to control the orientation of the spacecraft so that the high-gain antenna, 3.66 meters in diameter, pointed toward Earth.

The spacecraft was powered by three nuclear power sources, radioisotope thermal generators (RTGs) that produced about 400 watts of electrical power. A digital tape recorder could store about 500 million bits of information — equivalent to about one hundred images. The spacecraft was controlled by six on-board computers (two of each kind): the attitude and articulation control subsystem, the flight data subsystem, and the computer command subsystem. The attitude control subsystem controlled the stability and orientation of the spacecraft and the scan platform. The flight data subsystem provided instrument control for the scientific instruments and digital tape recorder and formatted the scientific and engineering data before they were sent to the ground. The computer command subsystem provided primary control of the spacecraft. These Voyager 1 computers could accept precoded sets of instructions that could provide autonomous operation for days or even weeks. These systems also included detailed instructions to detect and correct problems without human intervention.

On Labor Day, September 5, 1977, Voyager 1 was launched from Cape Canaveral, Florida, at 8:56 A.M., eastern daylight time, five days after the launch window had opened. The launch vehicle was a Titan 3E-Centaur rocket. Unexpectedly, the Titan main engine shut down early during the launch, and the Centaur stage had to make up the difference during the trajectory-insertion burn. After completing the insertion burn, the Centaur stage shut down with less than five seconds' worth of fuel left in its tank. If the launch had proceeded five days before, as scheduled, the thrust from the remaining fuel would not have been enough to allow Voyager 1 to reach Jupiter. With a little bit of luck, Voyager 1 was on its way to Jupiter.

During the autumn of 1977, a series of small problems challenged the JPL engineers. Attitude control thrusters fired at the wrong times, and sometimes the computer control systems overrode the commands from the ground. The onboard computers had been programmed to be too sensitive to slight changes on the spacecraft, and some reprogramming of the computers was necessary.

On February 23, 1978, the scan platform malfunctioned and prematurely stopped. This platform contained important remote-sensing instrument, and full mobility of the platform during the planetary flybys was essential. At JPL, tests were run on an exact copy of the scan platform. Slowly and carefully, the spacecraft platform was commanded to move, and normal operation was resumed. Engineers suspected that some material caught in the platform gears had been moved out of the way or crushed.

On January 4, 1979, the science-intensive Jupiter encounter began. Voyager 1 was now transmitting information not obtainable from Earth. For the next three months, Voyager 1 carried out a scientific survey of Jupiter, its satellites, and its magnetosphere. More than 30,000 images were transmitted to Earth during the encounter with Jupiter. Throughout the month of January, the Voyager 1 cameras sent back a series of images every two hours. The images were then turned into a color "motion picture" of Jupiter's weather patterns.

On February 28, Voyager 1 crossed Jupiter's bow shock— the boundary between the solar plasma that flows from the Sun (solar wind) and the planet's magnetosphere. Six hours later, the solar wind had pushed the magnetosphere back toward Jupiter, and Voyager 1 was once again in the solar wind. Over the next several days, variations in the solar wind pressure allowed Voyager 1 to cross the bow shock five times in all, as the Jovian magnetosphere repeatedly expanded and contracted.

A single eleven-minute imaging exposure of space, just above the equatorial cloud tops, was taken as the spacecraft passed through the plane of Jupiter's equator on March 4. Faint rings circling the planet were discovered. Close-range observations of the four Galilean satellites —Io, Europa, Ganymede, and Callisto — were made between March 4 and March 6. The closest flyby distances from each satellite were Io, 22,000 kilometers; Europa, 734,000 kilometers; Ganymede, 115,000 kilometers; and Callisto, 126,000 kilometers.

Voyager 1's closest approach to Jupiter took place on March 5, 1979, at 4:05 A.M., Pacific standard time. Thirty-seven minutes later, at 4:42, the signals from the spacecraft reached Earth. At 8:14, the spacecraft passed out of sight behind Jupiter, and the radio occultation of the Jovian atmosphere began. As the spacecraft flew behind Jupiter, the varying strength of the radio signal was used to probe the cloud structure in the atmosphere. Two hours later, Voyager 1 safely reappeared.

On March 5, during the period when it was closest to Jupiter, harsh radiation from the planet caused problems on the spacecraft. The main spacecraft clock slowed a total of eight seconds, and two computers were out of synchronization with the flight data subsystem computer. Some of the best images of Io and Ganymede were out of focus because the spacecraft started to move before the camera's shutter closed. Once the spacecraft moved farther away from Jupiter, this problem was corrected. On March 20, Voyager 1 crossed the Jovian bow shock, leaving Jupiter's magnetosphere and once again entering the solar wind. After a spectacular encounter with Jupiter, Voyager 1 was on its way for an encounter with Saturn on November 12, 1980.

Knowledge Gained

Voyager 1's pictures of Jupiter provide details of a turbulent, colorful atmosphere unlike any seen before. Images of Jupiter's Great Red Spot, a feature whose diameter is three times that of Earth and which is more than three hundred years old, reveal a huge hurricane-like storm towering above the surrounding clouds. White oval-shaped features about the size of Earth are other storms similar to the Great Red Spot.

Images of Jupiter reveal a stable zonal pattern of east-west winds. This planet-wide flow is more fundamental than the shifting cloud patterns within the east-west alternating belts (dark, deeper atmospheric regions) and zones (light, higher atmospheric regions). Within the belts and zones reside dark, brownish regions known as hot spots, holes in the uppermost cloud tops. These regions are warmer than the surrounding atmosphere, and both water vapor and germanium were discovered there. The minimum temperature on Jupiter was 110 Kelvins at 0.1 bar (on Earth, the surface pressure is typically 1.0 bar).

Cloud-top lighting bolts, similar to superbolts on Earth, were photographed, and radio-frequency emissions associated with the lightning were observed by the instrument for the planetary radio astronomy experiment. Auroral emissions in the polar region were seen in both the ultraviolet and the visible spectra.

With the imaging system, eight active sulfur volcanoes were discovered on the surface of Io, with plumes extending 250 kilometers above the surface. Tidal heating as a result of interactions with the other satellites and with Jupiter's powerful gravity melts the interior of Io, producing spectacular volcanoes. The moon's surface is uncratered and young because the volcanoes bring about continual resurfacing. The infrared instrument discovered hot spots on the surface of the satellite, and infrared scientists independently concluded that volcanoes existed there. Infrared measurements identified sulfur dioxide gas over the volcano named Loki.

Imaging observations showed numerous intersecting, linear features on the surface of Europa. Two distinct types of terrain, craters and grooves, characterize the surface of Ganymede. Ganymede was the first body other than Earth to display evidence of tectonic activity. Callisto displays a heavily cratered surface.

Voyager discovered rings of material orbiting Jupiter, with an outer edge about 128,000 kilometers from the center of the planet. The rings consist of small, dusty particles. Two newly discovered satellites, Metis and Adrastea, are embedded in the rings and are probably the source of the tiny ring parti-

cles.

An electric current of more than one million amperes flows in a magnetic flux tube linking Jupiter and Io. An Io torus — an invisible "doughnut" containing ionized sulfur and oxygen circling Jupiter at Io's orbit — was discovered. Jupiter has unusual radio emissions at the kilometer wavelength, which may be generated by plasma interactions with the Io torus.

Pioneer had shown that the magnetic field of Jupiter was dipolar, with a tilt of 11 degrees, and Voyager verified it. Both Pioneer and Voyager measurements showed that the magnetosphere of Jupiter is large. If viewed from Earth, it would be twice the size of the Moon at its fullest.

Context

Voyager 1 was the third spacecraft to fly through the Jovian system and the first to take high-resolution pictures of all four Galilean satellites. Voyager 2 followed four months later. The flybys five years earlier had both been U.S. missions as well, Pioneer 10 and Pioneer 11.

In the late 1960's, Earth observations had established that Jupiter had an internal heat source. Pioneer investigations had confirmed this finding: Jupiter is still cooling from its initial collapse and formation. Infrared instruments aboard Pioneer had provided the first measurement of the ratio of hydrogen to helium on Jupiter, and Voyager was able to refine this value further. This ratio, roughly ten hydrogen atoms for every helium atom, is comparable to the value for the Sun, supporting the idea that Jupiter and the Sun have similar compositions.

Observations of Jupiter are important for the understanding of the origin and evolution of the solar system. Jupiter has an extremely large mass. Had it been roughly one hundred times more massive it would have formed a star. With the same basic composition as the Sun, Jupiter constitutes a sample of the original material from which the solar system formed.

The four large Galilean satellites form a miniature planetary system. Ranging in size from just larger than the planet Mercury (Ganymede and Callisto) to just smaller than the Moon (Io and Europa), these satellites decrease in density with increasing distance from Jupiter, as do the planets in the solar system with increasing distance from the Sun. With the exception of Callisto's, the surfaces on the other Galilean satellites are in general much younger than expected. Cratering records showed multiple periods of bombardment, interspersed with resurfacing.

Plans for a return to Jupiter focused on the Galileo spacecraft, a combined orbiter and probe. Galileo's projected launch date was spring or summer of 1989; it was to reach Jupiter about six years later. The probe would be sent into the atmosphere of Jupiter, and the orbiter was to circle the planet, recording data, during the two years of its prime mission. One final opportunity for a Jupiter flyby was to come with the Cassini mission to Saturn. The planned year of launch was

1997, with a Jupiter flyby in the year 2004. The gravity of Jupiter would be used to speed up and deflect the Cassini spacecraft toward Saturn.

Voyager established the viability of building a spacecraft to last a decade or more. Scientists expected that this technology would play a major role in the design of spacecraft for extended missions to the outer planets, craft that would relay information to Earth for many years.

Both Voyager and Pioneer expanded the frontier of the outer solar system. With the pictures and information sent back by Voyager, the Galilean satellites were transformed from pinpoints of light into tiny worlds. Jupiter's turbulent atmosphere provides a model against which Earth's circulation patterns can be compared. Still, Voyager 1 only began to address fundamental questions about the formation and evolution of Jupiter and planets in the outer solar system.

Bibliography

Beatty, J. Kelly, et al., eds. *The New Solar System*. Cambridge, Mass.: Sky Publishing, 1982. Twenty chapters by distinguished researchers synthesize knowledge of the solar system. Findings on the Sun, the planets, the satellites, and the medium between are clearly discussed, with an emphasis on the discoveries of space probes such as Voyager.

Eberhart, Jonathan. "Jupiter and Family." *Science News* 115 (March 17, 1979): 164-165, 172-173. Highlights of Voyager 1's Jupiter encounter, particularly the discovery of volcanoes on Jupiter's moon Io, are described in this article published less than two weeks after the closest approach to Jupiter. Color photographs of Jupiter and its moons are included.

Gore, Rick. "Voyager Views Jupiter's Dazzling Realm." *National Geographic* 157 (January, 1980): 2-29. This readable article describes the findings of the Voyager Jupiter flybys. Includes quotes from key individuals involved in the mission. Beautiful color images of Jupiter and the Galilean satellites are included.

Lauber, Patricia. *Journey to the Planets*. Rev. ed. New York: Crown Publishers, 1987. The history and physical character of the nine planets and their moons are described, including their differences and similarities. Photographs and observations from space probes, including Voyager, are discussed. Suitable for general audiences, including junior high school students.

Poynter, Margaret, and Arthur L. Lane. *Voyager: Story of a Space Mission*. New York: Macmillan, 1981. Relating stories of the people behind the scenes during the Voyager encounters with Jupiter, this book describes the process leading to mission selection and funding. The interactions between the scientists and engineers in building and flying the Voyager spacecraft are discussed, culminating in an inside look at the excitement and wonder of new discoveries about Jupiter. Suitable for junior high school and high school levels.

Simon, Seymour. *Jupiter*. New York: William Morrow and Co., 1985. This well-written book summarizes knowledge of the planet Jupiter in a format suitable for elementary levels and older. Twenty color photographs taken during the Voyager encounters highlight this introduction to Jupiter.

Soderblom, L. "The Galilean Moons of Jupiter." *Scientific American* 242 (January, 1980): 88-100. The four Galilean satellites, Io, Europa, Ganymede, and Callisto, are discussed in detail in this journal article. Physical characteristics of the satellites and the parameters of the Voyager flybys of each satellite are detailed. Possible evolution scenarios and cratering rates are outlined. Suitable for high school and college students.

Linda J. Horn

VOYAGER 1
SATURN

Date: November 12 to December 15, 1980
Type of mission: Unmanned deep space probe

On its second planetary flyby, Voyager 1 encountered Saturn and sent back to Earth information on the planet's rings, satellites, and atmosphere. In the process, the probe helped demonstrate that a complex spacecraft could last more than a decade while operating in space semiautonomously.

PRINCIPAL PERSONAGES

HARRIS SCHURMEIER, the first Voyager Project
Manager
RAY HEACOCK, Voyager Project Manager for the
Saturn encounter
EDWARD C. STONE, Voyager Project Scientist
ROCHUS E. VOGT, Principal Investigator,
cosmic ray experiment
RUDOLF A. HANEL, Principal Investigator,
infrared experiment
BRADFORD A. SMITH, Principal Investigator,
imaging experiment
STAMATIOS "TOM" KRIMIGIS, Principal Investigator,
low-energy charged particle experiment
NORMAN F. NESS, Principal Investigator,
magnetic fields experiment
HERBERT S. BRIDGE, Principal Investigator, plasma
science experiment
ARTHUR L. LANE, Principal Investigator,
photopolarimetry experiment
JAMES W. WARWICK, Principal Investigator, planetary
radio astronomy experiment
FREDERICK L. SCARF, Principal Investigator, plasma
wave experiment
G. LEONARD TYLER, Principal Investigator,
radio science experiment
A. LYLE BROADFOOT, Principal Investigator, ultraviolet
spectrometer experiment

Summary of the Mission

The Grand Tour of the outer planets in the solar system was conceived by a team of engineers and scientists working at the Jet Propulsion Laboratory (JPL) in Pasadena, California. A plan to build three spacecraft for a mission to the outer planets was approved by Congress in 1972. One of those spacecraft was Voyager 1; its trajectory took it past both

Jupiter and Saturn. Jupiter's gravity was used to increase the spacecraft's velocity and divert its trajectory toward Saturn, shortening the travel time to Saturn from 6 years to 3.3 years.

JPL was chosen to implement the mission. The Mariner Jupiter Saturn 77 (MJS77) project, later renamed Voyager, began on July 1, 1972, under the management of Harris Schurmeier. When funds were authorized for the mission, the National Aeronautics and Space Administration (NASA) began the process of selecting the scientific instruments that would be installed on the craft. Eventually, eleven instruments were built for Voyager 1. To coordinate the scientific activity, Edward C. Stone was selected as Project Scientist, and Ray Heacock was Project Manager for the Saturn flyby.

There were two groups of Voyager instruments. The first group comprised the remote-sensing instruments. Most of these were mounted on a movable scan platform and obtained data on remote targets. The second group included *in situ* instruments, those that made direct measurements of the surrounding charged particles, magnetic field, and plasma waves.

The remote-sensing instruments were mounted on a movable scan platform at the end of a 2.3-meter science boom. The scan platform could move in two axes, scanning all the sky except for the region blocked by the spacecraft itself. The instruments mounted on the scan platform included the infrared interferometer spectrometer and radiometer, two imaging cameras, a photopolarimeter, and an ultraviolet spectrometer. Together, these instruments measured the properties of the objects they detected in wavelengths from the infrared through the visible to the ultraviolet.

Attached at the midpoint of the science boom were two *in situ* instruments, the cosmic ray experiment and the low-energy charged particle experiment. The plasma experiment was an *in situ* instrument farther out on the science boom, near the scan platform. These instruments measured the distribution of energetic charged particles such as electrons, protons, and ions.

The planetary radio astronomy and plasma wave experiments shared an antenna. The antenna consisted of two thin metal rods, each 10 meters long, set at right angles to each other. These rods were extended from the spacecraft after launch. The planetary radio astronomy experiment was a remote-sensing instrument, and the plasma wave experiment took samples *in situ*. The *in situ* magnetic field experiment consisted of four three-axis magnetometers, two high-field sensors attached to the spacecraft, and two low-field sensors mounted at the end of a 13-meter boom. This boom was packed tightly in a canister during launch; once in space, the canister opened, and the boom extended automatically.

The final Voyager instrument, the radio science experiment, used the dish-shaped high-gain antenna. The high-gain antenna was the radio communications link between Earth and Voyager, relaying data from the outer science instruments as well as making remote-sensing measurements of various atmospheres as the spacecraft passed behind planetary bodies.

On Labor Day, September 5, 1977, Voyager 1 was launched from Cape Canaveral, Florida, at 8:56 A.M., eastern daylight time, five days after the launch window opened. The launch vehicle was a Titan 3E-Centaur rocket. The Titan main engine cut out early, forcing the Centaur stage to make up the difference during the trajectory insertion burn. After completing its burn, the Centaur stage shut down with very little fuel left in its tank. If the launch had not been delayed, the thrust from the remaining fuel would have been insufficient to allow Voyager 1 to reach first Jupiter and then Saturn.

During the one-and-one-half-year cruise period between the Jupiter and Saturn encounters, Voyager 1 continued to measure the solar wind. On January 1, ten months before the Voyager 1 encounter, the planetary radio astronomy instrument discovered very long wavelength radio bursts from Saturn. These bursts were highly regular, allowing calculation of the rotation rate for Saturn's interior, which proved to be 10 hours and 39.4 minutes.

In the fall of 1980, the science-intensive Saturn encounter began. Voyager 1 performed a detailed survey of Saturn, its ring system, and its satellites — including a close flyby of the large moon Titan. More than 18,000 images were transmitted to Earth during the encounter with Saturn.

On October 6, an extensive set of images of Saturn's rings revealed unexpected and detailed structure in the rings. As a result of this observed structure, the spacecraft was reprogrammed, and on October 25 the spacecraft cameras pointed toward one ansae (end) of the rings and imaged the rings every five minutes for ten hours. These images were used to produce a "film" of ring activity that highlighted the "spokes," dark fingers of material extending radially outward over a portion of the rings. In images from this film, two tiny satellites, Prometheus and Pandora, were first discovered. They orbit on each side of the narrow F-ring.

On November 11, Voyager 1 crossed the Saturn bow shock, the boundary between the solar plasma, which flows from the Sun (solar wind), and Saturn's magnetosphere. Just inside the bow shock, a boundary called the magnetopause separates the turbulent area between the bow shock and the actual magnetosphere. The location of the magnetopause changes dynamically with variations in solar wind pressure. Five magnetopause crossings occurred during the period of one hour before

Voyager 1 entered the magnetosphere for the final time.

The closest approach to Titan took place on November 11, within 4,000 kilometers. Infrared measurements of varying atmospheric levels were performed from the edge of Titan's north pole. As Voyager 1 passed behind Titan, the varying strength of its radio signal probed Titan's atmosphere (a technique known as radio occultation), measuring its pressure and temperature as a function of distance above the surface. The ultraviolet instrument on board the spacecraft observed the sunset on Titan (solar occultation), and another measurement of Titan's atmosphere was made. About fifteen minutes later, the spacecraft safely reappeared from the shadow of Titan. As Voyager 1 was passing close to Titan, it simultaneously passed through the Saturn ring plane, which passes through Saturn's equator and through the known rings. No damaging particles hit the spacecraft during this passage.

November 12, 1980, was encounter day. Voyager 1's closest approach to Saturn took place at 6:45 P.M., eastern standard time, only 124,000 kilometers above the cloud tops. At 7:08 P.M., the spacecraft passed behind Saturn, and the structure of the clouds in the planet's atmosphere was probed by measuring the varying strength of the radio signals sent to Earth. A solar occultation by the atmosphere was simultaneously observed. The spacecraft reappeared briefly as it reached the other side of Saturn, and then it passed behind Saturn's rings, where it measured the distribution of ring material in the main rings by using the variation in the strength of the radio signal. Almost forty-five minutes after the occultations ended, Voyager 1 crossed the ring plane again, at the orbit of the satellite Dione, called the Dione clear zone, where the risk of collision with a ring particle was less likely. Observing the sunlight shining through them, Voyager 1 spent twenty-three hours underneath the rings photographing views never before seen from Earth.

Close-up views of five of the major Saturnian satellites were obtained on November 12. The closest flyby distances to each satellite were Mimas, 88,000 kilometers; Enceladus, 202,000 kilometers; Tethys, 416,000 kilometers; Dione, 161,000 kilometers; and Rhea, 74,000 kilometers. The images with the highest resolution were of Rhea. To obtain sharp images, the entire spacecraft was turned to keep Rhea motionless in the cameras.

By the end of November, Voyager 1 had crossed the Saturn bow shock to leave Saturn's magnetosphere and enter the solar wind for the final time. No further planetary encounters were possible for Voyager 1. In order to bring it close to the satellite Titan, the spacecraft trajectory, using Saturn's gravity, was bent out of the plane containing the orbits of the major planets. Voyager 1 is outward bound on a path leaving the solar system. Perhaps in several hundred thousand years Voyager 1 will fly close to another star. Mounted on the spacecraft is a gold-plated recording of the sights and sounds of planet Earth. Should a distant civilization find Voyager 1 and decode the recording, greetings from

Earth spoken in fifty-three languages and various sounds of the world will be heard. One hundred fifteen images will display the diversity of life and culture on Earth. With its departure from Saturn, Voyager 1 was only beginning a journey of epic proportions.

Knowledge Gained

Images of Saturn reveal alternating east-west belts (darker bands) and zones (lighter bands) similar to those in the Jovian atmosphere, although on Saturn they are considerably more muted. Saturn's equatorial jet stream blows four times harder (about 500 meters per second) around Saturn's equator than Jupiter's winds blow around Jupiter. Measured temperatures in Saturn's atmosphere include a minimum temperature of 80 Kelvins at 0.1 bar (on the surface of Earth the average atmospheric pressure is 1.0 bar).

Voyager 1's data revealed that Titan is hidden by an atmosphere thicker than Earth's, and that it may possess hydrocarbon oceans. A thick, smoglike haze covers Titan and creates a greenhouse effect, warming the surface. Voyager 1 provided the first measurements of the near-surface atmospheric pressure (1.6 bars) and temperature (95 Kelvins) on Titan. The main constituent of Titan's atmosphere was found to be nitrogen. Infrared measurements detected trace amounts of various hydrocarbons in the atmosphere. Nitriles—molecules composed of hydrogen, nitrogen, and carbon—were discovered also.

Images from the Voyager 1 cameras revealed three new, tiny satellites. Atlas orbits just outside the outer edge of the main ring system; Prometheus and Pandora fall on either side of the narrow F-ring, located 3,000 kilometers outside the main ring system. The F-ring appears braided and clumpy as a result of gravitational interactions with these satellites.

Using the imaging system, the sizes of the seventeen known satellites were determined for the first time, thus permitting a more accurate estimate of their densities. On Mimas, a huge crater, 130 kilometers in diameter and one-third the diameter of Mimas itself, was discovered. Enceladus is the brightest satellite in the solar system; some regions of its bright surface are almost devoid of craters, indicating a young surface. An enormous canyon, covering nearly three-quarters of its circumference, engulfs Tethys. Both Dione and Rhea display a dark surface overlaid with bright, wispy terrain, possibly a product of internal processing. Iapetus has an unusual distribution of light and dark material on its surface: It is bright on one side and completely dark on the other. The light material is approximately ten times brighter than the dark material.

Voyager 1 revealed a ring system of structural complexity and variety. The rings are not bland sheets of material, as originally thought prior to the Voyager 1 flyby, but possess detailed structures on scales smaller than a kilometer. The rings are composed of particles in a wide range of sizes, from tiny pebbles to giant boulders, including a sprinkling of fine dust. Spokes and elliptical as well as discontinuous ringlets were discovered in the main ring system by Voyager 1. A new ring, the tenuous G-ring, was discovered between the orbits of Mimas and two coorbiting satellites.

Voyager 1's instruments confirmed that, unlike Earth, Saturn has a magnetic dipole axis that is closely aligned with the planet's spin axis. The relative tilt between the magnetic dipole axis and the spin axis is only 0.7 degree for Saturn, compared to a tilt of 11.5 degrees for Earth. The number of charged particles in Saturn's magnetosphere is much smaller than the number measured at Jupiter. The rings are effective particle absorbers in the inner magnetosphere.

Context

Voyager 1 was the second spacecraft to fly through the Saturn system and the first to take high-resolution images of Saturn, Titan, the icy satellites, and the rings. Voyager 2 followed nine months later. One year earlier, Pioneer 11, another U.S. mission, had also conducted a flyby.

Saturn, the sixth planet from the Sun, is second in size only to Jupiter. Studies of the Saturn system have contributed to the understanding of the origin and evolution of the solar system. Earth-based observations established that Saturn has an internal heat source, and infrared measurements by both Pioneer 11 and Voyager 1 confirmed it. Primordial cooling from Saturn's initial formation and collapse should be complete; thus, an excess of heat was unexpected. Voyager 1 also measured the hydrogen-to-helium ratio of Saturn's upper atmosphere: roughly thirty hydrogen molecules for every helium molecule. Saturn has suffered a threefold depletion of helium relative to the solar abundance value. The mechanism responsible for this depletion in the upper atmosphere may also generate the excess heat.

With the exception of the Jovian satellite Ganymede, Titan is the largest satellite in the solar system, and it is the only one known to possess a substantial atmosphere. At a greater distance from the Sun than Ganymede, it is much colder and richer in ices. Some of the chemical reactions occurring in Titan's atmosphere provide possible analogues to some of the prebiotic chemistry that took place on primitive Earth to form the nucleic acids found in living organisms.

An interesting puzzle is the apparent youth of the ring system, only 10^7 to 10^8 years, much shorter than the solar system age of 4.5×10^9 years. Understanding the evolution of planetary rings may also lead to a better understanding of planetary accretion from the disk of material originally surrounding the Sun.

Plans for a return to Saturn focused on the Cassini mission, a combined U.S. and European project consisting of an orbiter and a probe. The planned launch date was in 1997, with arrival at Saturn in 2004. The probe was to be released into the Titan atmosphere, and measurements were to be taken during its descent to the surface. The orbiter was to circle Saturn, recording data for four years.

Voyager 1 established the possibility of building a spacecraft to last for more than a decade. Sophisticated engineering for Voyager 1's Saturn flyby utilized complex maneuvers to track the limb of the planet during the radio occultation and to compensate for the rapid motion of Rhea during the close flyby. Both Voyager 1 and Pioneer 11 extended the frontier of the outer solar system. With images and information relayed to Earth, a wealth of new knowledge was provided about the Saturnian system.

Bibliography

Beatty, J. Kelly, et al., eds. *The New Solar System*. Cambridge, Mass.: Sky Publishing, 1982. Well-known scientists have provided a useful overview of the solar system, including discussions of the Sun, the planets and their satellites, and the medium between the various bodies.

Branley, F. *Saturn, the Spectacular Planet*. New York: Thomas Y. Crowell, 1983. This clear, straightforward book presents current information on Saturn and its satellites and rings. Simple diagrams illustrate basic concepts about the Saturn system, enhanced with photographs from the Voyager flybys. Suitable for general audiences, including junior high school students.

Cooper, H. *Imaging Saturn*. New York: Holt, Rinehart and Winston, 1981. The Voyager mission is chronicled in a day-by-day description of events that occurred during the Saturn encounters. Accounts of key individuals, in particular the Voyager imaging team members, are detailed. Suitable for general audiences.

Eberhart, Jonathan. "Secrets of Saturn: Anything but Elementary." *Science News* 120 (September 5, 1981): 148-158. Highlights of the Voyager Saturn encounters are described in this article with a focus on Saturn's rings. Several photographs from the encounter are included. Suitable for general audiences.

Gore, Rick. "Voyager 1 at Saturn." *National Geographic* 160 (July, 1981): 2-31. This readable article describes the results of Voyager 1's Saturn flyby and includes statements from key individuals involved in the mission. Beautiful color images of Saturn, Titan, the icy satellites, and the rings are included.

Lauber, Patricia. *Journey to the Planets*. Rev. ed. New York: Crown Publishers, 1987. Lauber describes the history and nature of the planets, noting their resemblances and differences. Includes photographs and information from the Voyager missions.

Morrison, David, and Jane Samz. *Voyages to Saturn*. NASA SP-451. Washington, D.C.: Government Printing Office, 1982. The official account of the Voyager encounters with Saturn. Well illustrated. Includes appendices of information about the personnel involved in the Voyager missions as well as a list of suggested reading material.

Poynter, Margaret, and Arthur L. Lane. *Voyager: The Story of a Space Mission*. New York: Macmillan, 1981. Provides a behind-the-scenes account of the missions during all phases, from planning to the planetary encounters.

Simon, Seymour. *Saturn*. New York: William Morrow and Co., 1985. This easy-to-read book surveys knowledge of the planet Saturn. Includes twenty large color photographs taken during the Voyager encounters.

Linda J. Horn

VOYAGER 2
JUPITER

Date: August 20, 1977, to July 11, 1979
Type of mission: Unmanned deep space probe

The Voyager 2 flyby of Jupiter provided vital information about the Jovian system, in spite of a number of technical problems and equipment failures. Complementing the Voyager 1 flyby, this mission helped map Jupiter's moons, collected valuable data on the magnetic and radiation fields surrounding Jupiter, and monitored atmospheric phenomena.

PRINCIPAL PERSONAGES

RAY HEACOCK, Project Manager
EDWARD C. STONE, Project Scientist
BRADFORD A. SMITH, imaging science team leader
RUDOLF A. HANEL, Principal Investigator,
    infrared interferometry
A. LYLE BROADFOOT, Principal Investigator,
    ultraviolet spectroscopy
C. R. LILLIE and
CHARLES W. HORD, principal investigators, photopolarimetry
JAMES W. WARWICK, Principal Investigator, planetary
    radio astronomy
NORMAN F. NESS, Principal Investigator,
    magnetic fields
HERBERT S. BRIDGE, Principal Investigator,
    plasma particles
FREDERICK L. SCARF, Principal Investigator,
    plasma waves
STAMATIOS "TOM" KRIMIGIS, Principal Investigator,
    low-energy charged particles
ROCHUS E. VOGT, Principal Investigator,
    cosmic ray particles
VON R. ESHLEMAN, radio science team leader

Summary of the Mission

Voyager 2 was launched on August 20, 1977, sixteen days before the launch date of its twin, Voyager 1. The flight paths of the two spacecraft were such that between the orbits of Mars and Jupiter, in the asteroid belt, Voyager 1 overtook Voyager 2 and arrived at Jupiter four months and four days before Voyager 2. The differences in these two flight paths allowed the two spacecraft to complement each other, so that Voyager 2 obtained data on features of the Jovian (and Saturnian) system that Voyager 1 was unable to probe. Voyager 2's flight path was designed to carry it past Uranus and Neptune, while Voyager 1 would leave the plane of the solar system after its encounter with Saturn.

The data and pictures sent back to Earth by Voyager 2 are remarkable not only for their quality and uniqueness but also because of the number of problems that were overcome to obtain them. Even before launch, failures in two of the computer subsystems of the VGR77-2 spacecraft (which later became Voyager 1) delayed its launch and forced the substitution of the identical VGR77-3 spacecraft, now known as Voyager 2. Also just before the August 20 launch, the low-energy charged particle instrument had to be replaced. During launch, Voyager 2 behaved as if it had been jolted or bumped, switching to backup systems and losing telemetry signals. Flight engineers later determined that the Attitude and Articulation Control Subsystem (AACS) had experienced some electronic gyrations. Just after launch, the lock on the scientific instrument boom failed to signal that the boom was extended, even though it was. During the fall of 1977, Voyager 2 continued its erratic behavior. Finally, it was determined that the spacecraft's systems had been programmed to be too sensitive to environmental changes; after reprogramming, the erratic behavior subsided.

A more serious problem soon developed. The Voyager spacecraft are equipped with two receivers, a primary and a secondary or backup, through which commands are received from Earth. In late November, 1977, Voyager 2's primary receiver began losing power; it failed in late March, 1978. The failure would have caused no major difficulty if the backup receiver had been working properly, but when the spacecraft's Computer Command Subsystem (CCS) switched to the backup receiver seven days after it received its last communication from Earth, Voyager 2 still did not respond properly. Twelve hours later, on April 5, 1978, the CCS switched back to the primary receiver, which worked for half an hour before a power surge blew its fuses and permanently disabled it. During the next seven days, flight engineers devised a way of communicating through the faulty secondary receiver. The problem with the backup receiver was that it had lost its ability to compensate for slight frequency changes in signals from Earth, so that now only the most accurate signals could be recognized. Slight changes in frequency and receiver response result from Earth's rotation (Doppler shift), temperature fluctuations caused by electronic components switching on and off, and environmental factors, such as the magnetic and electric fields near planets and the associated radiation to which the spacecraft is subjected. Now all these fluctuations had to be accounted for when a signal was beamed from Earth, which made programming Voyager 2 from Earth much more difficult. On June 23 and October 12, 1978, Voyager 2 was programmed for backup automatic missions at Saturn and Jupiter in the event that the faulty backup receiver should fail completely. In August, 1978, it was again reprogrammed to ensure better scientific results during its Jupiter encounter. In particular, Voyager 2 was instructed to compensate for motions caused by its tape recorder. These motions would cause time-exposed television images to blur and lose details. Throughout the Voyager 2 mission, the faulty backup receiver was used repeatedly to reprogram the spacecraft. This reprogramming was crucial to the amount and quality of information sent back to Earth.

A serious fuel shortage also threatened the mission. Course corrections took 15 to 20 percent more fuel than had been expected because the maneuvering jets were partially blocked by the struts that had connected the spacecraft to its last booster rocket. To reduce the number and duration of course corrections, Voyager 2 was tipped upside down. In this attitude the spacecraft was not as easily blown off course by the solar wind (charged particles streaming from the Sun), but its attitude control system had to be reprogrammed to steer by different guide stars. Another major fuel-saving maneuver was accomplished only two hours after Voyager 2's closest approach to Jupiter. Flight engineers determined that about 10 percent of the original fuel load could be saved by rescheduling a major course correction to that time. Executing this course correction so close to Jupiter made it much more difficult to monitor. In fact, just as the spacecraft began the 76-minute thruster firing which would send it on to Saturn, communication with Earth was lost. Afterward, once the interference from the Jovian magnetosphere (the region dominated by Jupiter's magnetic field instead of the Sun's) was penetrated, the flight controllers found that Voyager 2 had executed its new programming perfectly and was headed toward Saturn, with enough fuel to redirect it toward Uranus and beyond.

In spite of all these problems, Voyager 2 began sending back information about interplanetary space, days after launch. During its cruise to Jupiter, the spacecraft's instruments were being calibrated and tested. Not only did this provide ground controllers and investigators with a better understanding of the instruments' behavior, but it also provided a chance to study infrared and ultraviolet radiation, magnetic fields, solar flares, and the solar wind far from Earth's influences. Unlike the Pioneer missions, Voyager 2 was not equipped to analyze the particulates of matter in the asteroid belt. Fortunately, the passage through this "shooting gallery" was uneventful, and by October, 1978, Voyager 2 had passed through the asteroid belt

and was slightly more than halfway to Jupiter.

The July 9, 1979, encounter with Jupiter was planned so that Voyager 2 could photograph the unseen sides of Jupiter's largest moons. (Because of Jupiter's tremendous gravitational and tidal pull on its satellites, one side of these bodies always faces Jupiter and the other side is turned out toward space.) When Voyager 2 passed through the Jovian system, the moons were encountered first, so that their spaceward sides were facing the Sun and Voyager 2's cameras. The flight path followed by Voyager 2 brought it near enough to Europa to resolve features as small as four kilometers across. In contrast, the best Voyager 1 pictures of Europa revealed only features 33 kilometers across or larger. Even though Io was always more than one million kilometers away from Voyager 2, the volcanoes discovered with the help of Voyager 1 data had so fascinated project scientists that a ten-hour "Io volcano watch" was planned. Just after its closest approach to Jupiter, Voyager 2 turned its cameras and instruments toward Io and the glowing gases which surround its orbit about Jupiter.

Because Voyager 2 was scheduled to go to Uranus and Neptune, it could not come as close to Jupiter as Voyager 1 had. In spite of this, Voyager 2 data revealed a moon (later dubbed Adrastea), closer to Jupiter than Amalthea, and took spectacular pictures of Jupiter's rings. Voyager 2 data also showed that the magnetosphere and radiation belts surrounding Jupiter had enlarged and intensified since Voyager 1's flyby. Jupiter's radiation belts resemble Earth's Van Allen belts, except that they can be ten thousand times stronger. Their radiation is generated by energetic charged particles that are trapped by Jupiter's magnetic field. Voyager 2's closest distance from Jupiter was nearly twice Voyager 1's closest distance, yet Voyager 2 instruments detected radiation levels that were three times stronger than those detected by Voyager 1 (a circumstance that may explain the communication interruption at closest approach). Furthermore, Voyager 2 first encountered Jupiter's bow shock (where the solar wind collides with a planet's magnetosphere) at a greater distance than Voyager 1 had four months earlier. Jupiter's bow shock flutters in the solar wind, however, so that in three days Voyager 2 passed through the bow shock ten times.

Another important aspect of the Voyager 2 flyby of Jupiter was the information it returned on Jupiter's meteorology. Six weeks after the first Voyager flyby, Voyager 2 turned its cameras toward Jupiter and began to document the cloud movements of Jupiter. A series of pictures were taken and were combined into a motion picture of the dynamics of Jupiter's atmosphere. Throughout the encounter, the spacecraft took pictures of Jupiter itself; scientists hoped thereby to detect changes in cloud patterns, lightning flashes, or auroras.

Knowledge Gained

The information returned by Voyager 2 dealt with three basic aspects of the Jovian system: Jupiter's atmosphere, satellites and rings, and magnetosphere. This information permit-

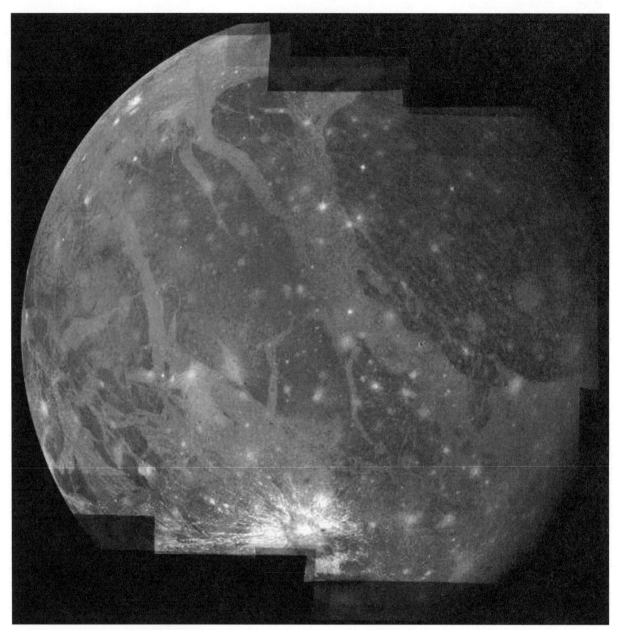

In this Voyager 2 mosaic, photographed at a range of 300,000 kilometers, the ancient dark area of Regio Galileo lies at the upper left. (NASA/JPL)

ted important comparisons with Voyager 1's data or provided detailed images and measurements of phenomena documented only briefly by Voyager 1. Not only did Voyager 2 confirm many of the sightings of Voyager 1, but it also supplied significant new information.

The motions of Jupiter's upper atmosphere were revealed in a series of photographs taken by Voyager 2. High-altitude jet streams were shown to form bands at constant latitudes and alternate with Jupiter's belts (dark bands) and zones (light bands). The jet stream velocities varied slightly from those measured by Voyager 1. At the boundaries of Jupiter's belts and zones can be found turbulence and storms. Some of these storms are very persistent; the Great Red Spot, for example, has lasted more than four centuries, and three white ovals have been studied since 1939. Voyager 2 revealed that these ovals, like the Great Red Spot, are all anticyclonic (rotating counterclockwise in the southern hemisphere and clockwise in the north). This finding indicates that these ovals, and four other white spots identified by Voyager 2, are sites of upsurging material. In the four months between the Voyager flybys, a protrusion formed to the east of the Great Red Spot, blocking the circulation of small structures about it. During the Voyager 2 flyby, a white region covered a brown oval in the north, showing that the brown ovals are actually breaks in the higher-altitude clouds of white ammonia crystals. Ultraviolet studies of Jupiter by Voyager 2 revealed an absorbent layer of haze above the cloudtops, precipitation of charged particles from the magnetosphere, and auroras near the poles. The high-latitude auroras are induced by charged particles from Io and play a part in the atmospheric chemistry of Jupiter, as does lightning, which produced eight flashes detected by Voyager 2.

Jupiter's moons and main ring were also scrutinized by Voyager 2's cameras and instruments. The smooth surface of Europa was photographed with unprecedented resolution, so that the moon was shown to have uniform, bright terrain crisscrossed by dark lines and ridges and almost no features resembling craters. In contrast, Callisto's outward face, like its inward face (which is always turned toward Jupiter), is heavily cratered and very ancient. Ganymede, between Callisto and Europa, exhibits a variety of surface features such as the old and cratered surface of Regio Galileo, discovered by Voyager 2, and ancient parallel mountain ridges, nearly the size of Earth's Appalachian Mountains. Six of Io's volcanoes were still active at the time of the Io volcano watch. The Pele volcano, which had been the most violent, was now quiet, but its surrounding terrain had been visibly altered since the time of the Voyager 1 images; the plume of the volcano Loki was much larger than it had been. High-resolution images of Jupiter's ring were obtained, revealing a bright, narrow segment with a slightly brighter center surrounding a broader, dimmer disk— all surrounded by a halo of very fine particles. The outer edge of the ring is sharp, and two moons orbit just outside the edge. The inner disk of the ring probably extends all the way down to the cloud tops of Jupiter.

The outer edge of Jupiter's magnetosphere was observed to fluctuate considerably as Voyager 2 crossed it several times. In spite of the boundary's instability, it was clear that the magnetosphere had changed shape since the first Voyager probed it, protruding more toward the Sun and narrowing in its long tail. Inside the magnetosphere, a hot plasma (a gas so hot that its atoms cannot keep their electrons) of hydrogen, oxygen, and sulfur was slowly spiraling outward. The amount of oxygen and sulfur had decreased, though, since the Voyager 1 passage, suggesting that these heavier atoms were slowly settling back toward Jupiter and that Io was producing less of them. These heavier atoms of sulfur and oxygen are most likely injected into the magnetosphere by the volcanoes on Io and first collect in a plasma cloud that surrounds Io's orbit. During the Voyager 2 encounter, this plasma cloud was glowing twice as bright with ultraviolet radiation as it had four months before, yet its temperature had decreased. It was also determined that the auroras observed on Jupiter were caused by Io's plasma cloud. Data from Voyager 2 also revealed that Ganymede swept up some of the charged particles from the magnetosphere's plasma, producing a plasma wake similar to the wake of a speedboat on a calm lake. A similar effect had been detected near Io by Voyager 1.

Context

Voyager 2 was the fourth spacecraft to encounter Jupiter. The first two, Pioneers 10 and 11, principally showed that the later Voyager missions were possible. Voyager 2's design and flight path were altered after these two Pioneer missions supplied measurements of Jupiter and its moons, permitted identification of certain problems — such as the intense radiation surrounding the planet— and suggested interesting phenomena for study. Some Voyager discoveries had been hinted at by Earth-based observations. Radio signals from Jupiter suggested that electrical current was flowing between Io and Jupiter, but the current proved much stronger than had been imagined.

Data from Voyager 1 provided the most guidance for the second Voyager flyby of Jupiter. For example, theories had predicted that Jupiter could not have a ring, because gravity would pull the material into Jupiter's atmosphere; yet Voyager 1 found a ring. The ring could not be studied carefully at the time, because the discovery was unexpected. The presence of the ring could be explained by Io's volcanoes, which were subsequently detected by Voyager 1. Some of the material ejected by Io may find its way down toward Jupiter, where it could replace precipitating material. Another possible explanation is that the rings are renewed by collisions between high-velocity particles and Jupiter's nearest moons, Adrastea and Metis, which undoubtedly shape the ring even if they do not regenerate it. In any event, much of the information Voyager 2 collected about the rings, Io's volcanoes, and Adrastea would have been undiscovered if data from Voyager 1 had not pointed the way.

Another surprise of the Voyager mission to Jupiter was the smooth surface of Europa. The lack of large impact craters suggests a relatively new surface, and the smooth regions between intersecting lines resemble an aerial photograph of Arctic ice crossed by pressure ridges. If this interpretation is correct, Europa may be covered by a frozen ocean. How thick the ice is and whether liquid water lies underneath is still speculative; more information is needed. Without the Voyager 2 images, however, the idea of a recently frozen surface would seem very unlikely.

The information from Voyager 2's flyby of Jupiter has taken years to analyze, and some of the mysteries cannot be solved without another mission. The Galileo mission has had to wait for a number of difficulties to be resolved. In this next step, a spacecraft would not merely pass by Jupiter but would become another satellite of that planet, providing long-term observations of phenomena that Voyager instruments could detect only briefly.

Bibliography

Beatty, J. Kelly, et al., eds. *The New Solar System*. Cambridge, Mass.: Sky Publishing, 1982. A collection of articles by noted experts. Chapters 11, 12, 13, 14, and 19 are particularly relevant for readers who wish to learn more about Voyager 2's flyby of Jupiter.

Couper, Heather, and Nigel Henbest. *New Worlds: In Search of the Planets*. Reading, Mass.: Addison-Wesley Publishing Co., 1986. Includes a summary of the explorations of the Jovian system. Accessible to the general reader.

Editors of Time-Life Books. *The Far Planets*. Alexandria, Va.: Time-Life Books, 1988. This volume is notable not only for its photographs and informative illustrations but also for its lively and complete account of the Voyager 2 mission. This is a volume in the Voyage Through the Universe series.

Hunt, Garry E., and Patrick Moore. *Jupiter*. New York: Rand McNally and Co., 1981. A succinct yet complete treatment of almost all aspects of Jupiter and the Voyager missions to Jupiter. Its maps of the Galilean satellites are notable.

Morrison, David, and Tobias Owen. *The Planetary System*. Reading, Mass.: Addison-Wesley Publishing Co., 1987. A very complete summary of all the Sun's satellites by two members of the Voyager imaging team. Contains photographs and information never before published.

Morrison, David, and Jane Samz. *Voyage to Jupiter*. NASA SP-439. Washington, D.C.: Government Printing Office, 1980. The official summary of the Voyager missions to Jupiter. This volume is notable for its chronological approach to both Voyager flybys of Jupiter, its summaries, its photographs and tables, and its maps of Jupiter's moons.

Murray, Bruce C., ed. *The Planets*. New York: W. H. Freeman and Co., 1983. This collection of reprints of *Scientific American* articles includes articles on Jupiter, its Galilean moons, and its planetary rings.

Smoluchowski, Roman. *The Solar System*. New York: W. H. Freeman and Co., 1983. A well-written and well-illustrated summary of man's understanding of the solar system. The discussion of Io is particularly fascinating.

Snow, Theodore P. *Essentials of the Dynamic Universe: An Introduction*. 2d ed. St. Paul, Minn.: West Publishing Co., 1987. A well-written introduction to astronomy and astrophysics. Intended for nonscientists. The chapter on Jupiter contains a summary of the Voyager missions.

Larry M. Browning

VOYAGER 2
Saturn

Date: June 5 to September 4, 1981
Type of mission: Unmanned deep space probe

The Voyager 2 flyby of Saturn produced high-resolution images of Saturn's ring system and of the satellites Iapetus, Hyperion, Enceladus, and Tethys. In addition, this second Voyager mission to Saturn collected valuable data on the magnetic and radiation fields surrounding that planet and observed atmospheric phenomena on Saturn.

Principal personages

ESKER DAVIS, Project Manager
EDWARD C. STONE, Project Scientist
BRADFORD A. SMITH, imaging science team leader
RUDOLF A. HANEL, Principal Investigator, infrared
   interferometry
A. LYLE BROADFOOT, Principal Investigator, ultraviolet
   spectroscopy
ARTHUR L. LANE, Principal Investigator, photopo-
   larimetry
JAMES W. WARWICK, Principal Investigator, planetary
   radio astronomy
NORMAN F. NESS, Principal Investigator, magnetic
   fields
HERBERT S. BRIDGE, Principal Investigator, plasma
   particles
FREDERICK L. SCARF, Principal Investigator, plasma
   waves
STAMATIOS "TOM" KRIMIGIS, Principal Investigator,
   low-energy charged particles
ROCHUS E. VOGT, Principal Investigator,
   cosmic ray particles
G. LEONARD TYLER, radio science team leader

Summary of the Mission

The Voyager 2 mission to Saturn was the last formal objective for the Voyager program, but the flight controllers knew that with careful planning and a little luck, Voyager 2 could be sent to Uranus and Neptune on a "grand tour" of the outer solar system. Voyager 2's flight path took the probe only 32,000 kilometers from the edge of Saturn's F-ring, through the region where Pioneer 11 nearly collided with the satellite now known as Janus. This trajectory was chosen so that Saturn's gravitational pull would assist in sending Voyager 2 toward a 1986 encounter with Uranus. The timing of Voyager 2's passage through the Saturnian system was also planned to optimize measurements of the moons Iapetus, Enceladus, and Tethys, as well as of the rings, which were better illuminated by the Sun than they had been during Voyager 1's flyby. Other satellites would also be photographed and compared with data returned by Voyager 1, which had achieved very close encounters with the moons Mimas, Rhea, and Titan.

During its approach to Saturn, Voyager 2 monitored the solar wind, the charged particles that stream from holes in the Sun's corona. This information warned Voyager 1 flight controllers and scientists of gusts or other changes in the solar wind that would affect the boundary of Saturn's magnetosphere (the region surrounding a planet that is dominated by that planet's magnetic field). The data were useful in interpreting information sent from Voyager 1 as it crossed Saturn's bow shock. (A planet's bow shock is the region where the solar wind first encounters the planet's magnetic field and loses most of its energy.)

In February of 1981, Voyager 2 passed through the tail of Jupiter's magnetosphere. It was expected that in August of 1981, during Voyager 2's Saturn encounter, Jupiter's magnetosphere would extend over Saturn, shading Saturn from the solar wind. That did not happen, however, and Saturn's magnetosphere was subjected to the solar wind's full fury, robbing Voyager 2 of the opportunity to study the collision of Jupiter and Saturn's magnetospheres.

Ten weeks before periapsis (closest approach), Voyager 2 began taking a series of pictures that were later combined into films showing the motion of Saturn's atmosphere and rings. The atmospheric films showed banding and turbulence similar to those on Jupiter; wind speeds, however, were an amazing 400 to 500 meters per second, and cloud layers were deeper, giving Saturn a more uniform appearance than Jupiter. In the nine months since Voyager 1 had visited Saturn, atmospheric activity had increased so that more storms, spots, and waves could be seen in the cloudtops. The ring images showed the three major rings easily visible from Earth, which are named A, B, and C from the outermost inward. In the B-ring, radial "spokes" which rotate with the planet were clearly visible. The detection of these spokes by Voyager 1 had been a surprise, because such structures cannot orbit Saturn as a result of the gravitational forces that shape the concentric, nearly circular rings. The Voyager 2 pictures helped to establish that the spokes are very tiny dust or haze layers suspended over larger ring particles by Saturn's magnetic field and, consequently, rotate with the field as the planet rotates.

Between the orbits of the moons Iapetus and Titan, Voyager 2 encountered Saturn's bow shock. As on its flyby of Jupiter, Voyager 2 crossed the bow shock many times, because the solar wind would alternately gust, compressing the magnetosphere, and relax, allowing the magnetosphere to expand. Because the solar wind's strength had increased since Voyager 1's encounter, Voyager 2 finally passed into Saturn's magnetopause, or the magnetosphere's boundary, just inside Titan's orbit. The spacecraft then proceeded to pass through three distinct regions of Saturn's magnetosphere. The first is dominated by Titan, whose orbit is surrounded by a cloud of hydrogen gas emanating from the moon's atmosphere and extending toward Saturn up to Rhea's orbit. Between Rhea and Mimas are intense radiation belts composed of charged particles that rotate with Saturn's magnetic field. Closer to the planet, the rings almost completely neutralize the charged particles, making Saturn's rings one of the most radiation-free regions in the solar system.

The rings themselves were also very carefully studied by Voyager 2 scientists. Almost two and a half hours on the day of periapsis were devoted to a very careful photopolarimeter scan of the rings. This scan resolved objects as small as one hundred meters across and was done by measuring the change in intensity of starlight as the ring passed between the probe and the star Delta Scropii.

On August 22, 1981, Voyager 2 began its survey of Saturn's moons as it flew by Iapetus. The moons, except for Phoebe and possibly Hyperion, have synchronous orbits around Saturn, so that one side of the moon always faces Saturn. Put another way, one side of the moon always faces ahead, leading the moon in its orbit, and the other side always faces behind, trailing the moon as it circles. In the case of Iapetus, the leading edge is darker than the trailing edge; the moons Dione and Rhea, however, have brighter leading faces and darker trailing faces crossed by bright streaks. After passing Iapetus, Voyager 2 flew by Hyperion. The probe returned pictures that revealed this moon to be irregularly shaped, with its long axis pointing out toward space instead of toward Saturn, as had been expected. Just before the closest approach to Saturn, Voyager 2 turned its cameras toward the small satellites Telesto, Calypso, and Helene, and toward the larger moons Enceladus and Tethys, even though the closest approach to the larger moons would occur after periapsis.

The preliminary studies of Enceladus and Tethys proved fortunate, because fifty-five minutes after Voyager 2 crossed the plane of Saturn's rings, its camera platform's azimuth control became stuck. When the scan platform's back-and-forth motion stopped, Voyager 2 was behind Saturn and out of contact with Earth. The spacecraft was on automatic control

and was recording its cameras' images for later relaying. Reviewing the images hours later, flight controllers and scientists watched as the cameras moved progressively off target so that first the high-resolution pictures were lost and then even the wide-angle pictures were blank.

As soon as contact was reestablished with Voyager 2 as it moved out of Saturn's shadow, flight controllers realized that the cameras and other sensitive instruments were pointed toward the Sun. If left in this position, the instruments would be destroyed, effectively blinding the spacecraft and making the rest of the mission practically useless. Quickly, commands were sent to rotate the entire spacecraft. Voyager 2 was so far from Earth, however, that an hour and a half would pass before these commands would be received. Adding to the danger was the possibility that Voyager 2 would not be able to decode the commands, as its primary receiver had failed soon after launch, and its backup receiver was faulty. If the flight engineers had not taken into account the fact that the spacecraft had cooled in the shadow of Saturn, and adjusted the commands accordingly, Voyager 2's last images would have recorded its cameras burning out in the Sun's glare.

Three days passed before the instrument platform was partially freed and a few final images of Saturn and the moon Phoebe were sent back to Earth. Unfortunately, the highest-resolution images of Enceladus and Tethys had been lost, along with three-dimensional pictures of the F-ring, a photopolarimeter scan of the F- and A-rings, and images of the night side of Saturn with backlit rings. It was also clear that for future flybys, the entire spacecraft would have to be slowly rotated to keep time-exposed pictures from blurring.

Knowledge Gained

During its flyby of Saturn, Voyager 2 conducted the first high-resolution reconnaissance of the moons Iapetus and Hyperion and of the two previously scanned moons, Tethys and Enceladus. The images of Iapetus revealed a dark leading edge surrounded by a concentric, dark circle. The trailing edge was much brighter, but the deepest craters showed dark bottoms. Voyager 2 flew so close to Iapetus that it was able to make the first direct measurement of this moon's mass as its gravity bent the spacecraft's trajectory.

The Hyperion pictures revealed an elongated moon, pockmarked by meteoric impacts, with its longest axis pointing away from Saturn. This dark, icy moon's orientation was seen to be so unusual that mission scientists suspected that Hyperion is not synchronously rotating about Saturn, as are all the other moons but Phoebe. Unfortunately, Voyager 2 could not observe Hyperion long enough to determine its exact orientation and rotation period.

Voyager 2's images of Tethys revealed a huge and ancient impact crater 400 kilometers wide—bigger than the moon Mimas. This crater, named Odysseus, is relatively shallow, because its floor rebounded after the initial meteoric impact; it now has nearly the same curvature as the rest of the moon.

It was also discovered that the huge trench Ithaca Chasma, first photographed by Voyager 1, extends three-fourths of the way around Tethys.

The surface of Enceladus is incredibly bright, reflecting nearly all the light that reaches it. This extreme brightness suggested to the Voyager mission scientists that Enceladus' surface is very new and perhaps frequently restored by ice volcanoes. Voyager 2 detected no such ice volcanoes, but it did find a variety of terrain, which indicates recent geological activity. Such activity was unexpected in such a small moon and is unknown among Enceladus' neighbors.

Voyager 2's high-resolution images of Saturn's rings showed much more detail than had ever been seen before. When various sections of the rings were compared, it was realized that the rings are very dynamic; their fine structure is constantly shifting and changing. The Cassini Division and the Keeler Gap (also called the Encke Division), which from Earth had appeared devoid of material, were shown to have small, dark ringlets. Scientists had expected to find small moons in the divisions sweeping away material, but none were detected. Voyager 2 also measured the rings' thickness and found it to be less than 300 meters; corrugations in the rings make them appear ten times thicker from Earth.

Weather patterns on Saturn proved to be remarkably stable. Storms identified by Voyager 2 as it approached Saturn were seen six weeks later at the same latitude and traveling with the same speed. Even more remarkable was that many storms appeared just where Voyager 1 data had predicted they would.

Context

The third probe to fly by Saturn, Voyager 2 provided much detailed information that Pioneer 11 and Voyager 1 could not. Pioneer 11 had made several important discoveries, including Saturn's F-ring, but its imaging systems and other instruments could not match the resolution capabilities of the Voyager spacecraft. Voyager 1 had observed most of the previously unknown and in many cases unexpected aspects of the Saturnian system nine months before Voyager 2's passage, but a single flight through the Saturnian system could not capture every aspect for scrutiny. Also, to have a close look at Titan, which was the only moon in the solar system known

to have a substantial atmosphere, Voyager 1 mission scientists had to give up the chance to send the probe on to Uranus and Neptune. The second Voyager craft's trajectory was calculated to take it past those two planets. Voyager 2 had slightly better instruments, as well; its imaging system had about 50 percent more sensitivity and produced sharper pictures. Finally, Voyager 1's photopolarimeter was destroyed by the intense radiation near Jupiter, and important details about Saturn's rings and the polarization of light scattered from Titan were left for Voyager 2 to record.

Voyager 1 provided information that greatly influenced the planning of Voyager 2's Saturn flyby. For example, Voyager 1's data on the complexity of Saturn's rings made scientists and flight controllers realize how important a high-resolution scan of the rings would be. Nevertheless, not all the guidance for the Voyager 2 mission was provided by Voyager 1. During 1966 and 1980, Saturn's rings' edges were facing Earth. Without the brighter rings to obscure them, a number of new satellites and rings were observed in the Saturnian system by several teams of astronomers. Voyagers 1 and 2 confirmed their existence, and these objects are now known as the two coorbital satellites Epimetheus and Janus; the satellite Helene, which shares an orbit with Dione; and the E-ring, near Enceladus. Another, much older, Earth observation was confirmed by Voyager 2 when it sent back images of Iapetus showing a dark leading surface and a bright trailing surface. In 1671, the astronomer Gian Domenico Cassini, who discovered Iapetus and the gap between the A- and B-rings which now bears his name, observed that Iapetus was easy to see when it was west of Saturn but could barely be seen when it was east of Saturn. These observations are consistent with a synchronously rotating satellite with a dark leading edge, which the Voyager 2 images revealed Iapetus to be.

Despite Voyager 2's successes, there was some sadness at the press conference held after the Saturn flyby. Everyone there knew that it would be the last such conference for quite some time. Even though Voyager 2 would eventually encounter Uranus, in 1986, that was five years in the future, and, more important, no new probes had been launched. The Voyager 2 flyby of Saturn not only marked the end of the Voyager program's formal objectives but also marked the end of a period of intense planetary exploration.

Bibliography

Beatty, J. Kelly, et al., eds. *The New Solar System.* Cambridge, Mass.: Sky Publishing, 1982. A collection of articles by noted experts and authors. Several chapters are particularly relevant to Voyager 2's flyby of Saturn.

Cooper, Henry S. F., Jr. *Imaging Saturn: The Voyager Flights to Saturn.* New York: H. Holt and Co., 1985. A very readable, chronological account of the Voyager missions to Saturn.

Couper, Heather, and Nigel Henbest. *New Worlds: In Search of the Planets.* Reading, Mass.: Addison-Wesley Publishing Co., 1986. A summary of planetary exploration, including the exploration of Saturn. Also contains information about how to find and observe Saturn. Accessible to all readers.

Editors of Time-Life Books. *The Far Planets.* Alexandria, Va.: Time-Life Books, 1988. A volume in the Voyage Through the Universe series, this source is notable not only for its pictures and informative illustrations but also for its lively and complete account of the Voyager 2 mission.

Frazier, Kendrick. *Solar Systems*. Alexandria, Va.: Time-Life Books, 1985. A volume in the series Planet Earth, this work is attractively illustrated and understandable to the layman.

Hunt, Garry, and Patrick Moore. *Saturn*. New York: Rand McNally and Co., 1981. A succinct and complete treatment of almost all aspects of Saturn and the Voyager missions to that planet. Maps of some of the satellites and diagrams of the Voyager spacecraft and its instruments are included.

Morrison, David. *Voyages to Saturn*. NASA SP-451. Washington, D.C.: Government Printing Office, 1980. The official summary of the Voyager missions to Saturn, this work is notable for its chronological approach to both Voyager fly-bys of Saturn, its summaries, its pictures and tables, and its maps of Saturn's moons. The Voyager flybys of Jupiter are also summarized.

Morrison, David, and Tobias Owen. *The Planetary System*. Reading, Mass.: Addison-Wesley Publishing Co., 1987. A complete summary of all the Sun's known satellites by two members of the Voyager imaging team. Contains pictures and information never before published.

Murray, Bruce, ed. *The Planets*. New York: W. H. Freeman and Co., 1983. A collection of reprints of *Scientific American* articles. Includes discussions of Saturn, Saturn's moons, and planetary rings.

Smoluchowski, Roman. *The Solar System*. New York: W. H. Freeman and Co., 1983. A well-written and well-illustrated summary of mankind's understanding of the solar system.

Snow, Theodore P. *The Dynamic Universe: An Introduction to Astronomy*. 3d ed. St. Paul, Minn.: West Publishing Co., 1988. A well-written introduction to astronomy and astrophysics intended for nonscientists. The chapter on Jupiter contains a summary of the Voyager missions, and the chapter on Saturn discusses the use of Voyager 2's photopolarimeter to resolve Saturn's rings.

Larry M. Browning

VOYAGER 2
URANUS

Date: November 4, 1985, to February 25, 1986
Type of mission: Unmanned deep space probe

Voyager 2 was the first spacecraft to collect and return data from the planet Uranus. This encounter was the third of four potential encounters made possible by a planetary alignment that occurs only once every 175 years.

PRINCIPAL PERSONAGES

RICHARD P. LAESER, Voyager Project Manager
for the Uranus encounter
EDWARD C. STONE, Voyager Project Scientist and
Principal Investigator, cosmic ray experiment
ELLIS D. MINER, Assistant Project Scientist
CHARLES E. KOHLHASE, Principal Mission Designer
BRADFORD A. SMITH, Principal Investigator,
imaging experiment
G. LEONARD TYLER, Principal Investigator,
radio science experiment
RUDOLF A. HANEL, Principal Investigator,
infrared experiment
STAMATIOS "TOM" KRIMIGIS, Principal Investigator,
low-energy charged particle experiment
NORMAN F. NESS, Principal Investigator,
magnetometer experiment
HERBERT S. BRIDGE, Principal Investigator,
plasma science experiment
ARTHUR L. LANE, Principal Investigator,
photopolarimeter experiment
JAMES W. WARWICK, Principal Investigator,
planetary radio astronomy experiment
FREDERICK L. SCARF, Principal Investigator,
plasma wave experiment
A. LYLE BROADFOOT, Principal Investigator,
ultraviolet spectrometer experiment

Summary of the Mission

The Voyager 2 spacecraft first encountered the planet Uranus while the probe was on its way out of the solar system. This swing-by of Uranus was the third of four planetary encounters made possible by an alignment of the outer planets — Jupiter, Saturn, Uranus, and Neptune— which occurs only once every 175 years. This alignment allowed Voyager 2 to arrive at Uranus in nine years instead of sixteen by using the gravity of each planet to boost it on to the next.

In 1981, prior to Voyager 2's encounter with Saturn (but after Voyager 1's Saturn encounter), the National Aeronautics and Space Administration (NASA) approved the Voyager 2 Uranus mission. The gravity of Saturn would be used to direct the spacecraft's trajectory toward Uranus. The journey between these worlds would take more than four years and 1.5 billion kilometers, for a spacecraft that had already spent that long and traveled that distance in space.

The time between the Saturn and Uranus encounters was used by Voyager engineers to modify the craft's onboard computer programs. These modifications were needed to enable the probe to overcome the problems associated with visiting a planet that is twice as far from the Sun as is Saturn. One such problem was the decreasing light levels. At a distance of 3.2 billion kilometers from the Sun, the light at Uranus would be four hundred times dimmer than it is at Earth. This level of light made necessary longer photographic exposures, which would lead to blurred pictures if the spacecraft could not keep the target in the camera's field of view during the exposure. To address this need, a computer algorithm known as target motion compensation was written and sent to the spacecraft. This algorithm allowed Voyager to drift in such a way as to keep the target in the camera's sights. In essence, this routine allowed the spacecraft to "pan" the cameras, like a human photographer does to photograph a moving object.

The greater distance also forced ground engineers to reduce the rate at which the spacecraft transmitted data. This reduction compensated for the ever-decreasing signal strength as the spacecraft moved away from Earth. It also meant that as the spacecraft got farther away, more time would be required to send the same amount of information. At Jupiter, for example, one picture could be transmitted every 96 seconds; at Saturn, the rate had fallen to one picture every 3.2 minutes, and at Uranus, it would be one every 8.8 minutes. To reduce the impact of this problem during the cruise toward Uranus, project managers made a bold decision. Instead of using one flight data subsystem computer to format the data to be sent and the other as a backup, the two would be used together with no backup. The primary subsystem would still format the data; the secondary subsystem, however, would be used to combine the data more efficiently. This combination routine, known as image data compression, could send a complete photograph using less than 40 percent of the informa-

Uranus and satellites mosaic. (NASA)

tion bits normally required. Thus, even with the slower transmission rates from Uranus, one picture could be transmitted every 4.8 minutes.

Many more changes were made to Voyager 2 to compensate for the greater distance from the Sun and Earth and the lack of knowledge of the outer solar system. These changes involved not only the spacecraft program but also procedures used to operate the spacecraft. The spacecraft that finally arrived at Uranus was far superior to the one that had passed by Jupiter and Saturn.

On November 4, 1985, the first phase of the Uranus encounter began. Known as the observatory phase, it started when the quality of data from the spacecraft instruments surpassed the quality of data from ground-based instruments. During this phase, systematic searches were made of the Uranian system for new satellites and rings. In addition, atmospheric measurements began to be taken to gain information regarding atmospheric structure and composition and wind patterns.

Twenty-five days prior to the spacecraft's closest approach to Uranus, on December 31, 1985, Voyager 2 discovered its first Uranian satellite, which was given the name Puck. Puck is 170 kilometers in diameter and is located between the outer Uranian ring, known as the epsilon ring, and the innermost satellite, known as Miranda. Very little information about Puck's surface was gained from the discovery pictures because of its small size and great distance from the spacecraft. Because of the importance of surface geology, however, the various Voyager teams worked quickly to modify the spacecraft's program to photograph Puck immediately before the busy near encounter phase.

To modify the program, ground controllers decided to eliminate a planned observation of Miranda and replace it with an observation of Puck, then known as 1985U1. The spacecraft photographed Puck successfully, but the receiving ground antenna station started to drift, producing a poor alignment of the ground antenna with the spacecraft and resulting in the loss of the spacecraft signal that contained the Puck photograph. Fortunately, the photograph had been recorded on the spacecraft's tape recorder and could be played again. The commands to replay the data had to be sent quickly, however, before new information was recorded over the Puck data. The second playback was successful, revealing the surface of the moon discovered only days earlier.

The next phase of the encounter was known as the far encounter. This phase started on January 10, 1986, when Uranus and its rings were too large to fit comfortably into one narrow-angle photograph. To capture the entire planet and its rings required that the observation be designed as a mosaic of four pictures. Each picture in the mosaic would contain one quarter view of the planet and its rings.

During the far encounter, the spacecraft experienced its first hardware failure since its encounter with Saturn. On January 18, 1986, six days prior to Voyager 2's closest approach

to Uranus, unexpected gaps started appearing in the photographs. Engineers quickly reviewed both the ground data system and the multimission image processing laboratory for hardware failures or software problems. Two days later, convinced that the problem was on the spacecraft and not on the ground, spacecraft controllers sent commands to the spacecraft instructing it to transmit the contents of its flight data subsystem memory. After reviewing the information, analysts found that one memory location had an incorrect value. The following day, on January 21, a command was sent to the spacecraft to use one of the few remaining spare memory locations instead of the one that appeared to have failed. After the craft received the command, the gaps in the spacecraft photographs disappeared, proving that a memory location had indeed failed.

On January 22, 1986, the near encounter phase began. During this phase, which lasted only four days, the spacecraft would be closest to the planet and would be gathering most of the important scientific data of this part of its mission. Timing was critical during this phase. The spacecraft had to be in the correct place at the correct time if its commands were to execute properly. The Voyager project teams worked throughout the night of January 23 and early into the next morning, adjusting the spacecraft commands. The final commands were transmitted to the spacecraft less than eight hours prior to their execution.

On January 24, at 1759 Greenwich mean time, Voyager 2 passed 107,000 kilometers from the center of Uranus. The navigation team had done a superb job of directing the spacecraft to its destination. The spacecraft was off its schedule by only sixty-one seconds, and its placement was such that Voyager 2 would pass by Neptune several years later.

The last phase of the encounter, which began on January 26, 1986, was the post-encounter. During this time, measurements resembled those that had been executed during the observation phase. Pictures showed Uranus as a crescent as Voyager 2 headed toward Neptune.

Knowledge Gained

The visible atmosphere of Uranus was found to be composed predominantly of hydrogen and helium, with concentrations very similar to those found in the Sun. Carbon, however, was found to be twenty times more abundant than in the Sun. This relatively large amount of carbon, which is combined with hydrogen to form methane, gives Uranus its blue-green color. Methane absorbs mainly in the red wavelengths, reflecting the blue light to the observer.

The infrared experiment found that almost the same amount of energy is emitted by the poles of Uranus as by the equator. This was surprising, for the poles are exposed to more sunlight during the eighty-four-year orbit of Uranus and thus should radiate more energy. In addition, the spin axis of Uranus was found to be tilted 98 degrees relative to its orbital plane. Thus, even though Uranus spins on its axis once every 17.24 hours, the south pole faced the Sun during the entire Voyager 2 encounter. It takes forty-two years (one-half of a Uranian orbit) for the Sun to rise and set at the planet's poles.

Few discrete clouds were observed in the atmosphere of Uranus. The motion of those few clouds seen, however, indicated that winds rotated with the planet at a maximum speed of about 200 meters per second at a southern latitude of 60 °. The winds slowed on either side of this region until they became almost nonexistent at the pole and at 20 ° south latitude. Farther north at the equator, however, the radio science experiment found the winds to flow in a direction opposite the rotation, with speeds up to 100 meters per second.

The upper atmosphere of Uranus was found to extend far above the planet. This part of the atmosphere has an extremely high temperature of 500 degrees Celsius and may produce drag forces on the particles located in the rings. These forces may be responsible for removing dust-sized particles from the Uranian ring system. The extended atmosphere also interacts with sunlight, giving off emissions that were detected by the ultraviolet experiment.

The nine previously known rings were photographed by Voyager's cameras. They are very dark and contain few particles in the 1-to-10-centimeter range. The darkness may be the result of the bombardment of the methane-water ice particles by protons trapped in the Uranian magnetosphere. The result of this bombardment is a carbon residue on the ice, which makes the ring particles as dark as coal. Two additional rings were found as Voyager 2 approached Uranus. As the probe left, one single long-exposure photograph indicated that micrometer-sized dust exists throughout the entire ring system.

Voyager 2 discovered ten new satellites in all, increasing the total number of known Uranian satellites from five to fifteen. The new satellites range from 40 to 170 kilometers in diameter and are located between the outer part of the ring system and the orbit of Miranda. Compared to Saturn's icy moons, the five previously known moons of Uranus were found to be relatively dark. Ariel, the second moon out from Uranus, has a fractured surface, which may indicate that ice once flowed across it. It also possesses one of the most geologically active surfaces in the Uranian system. Yet the most distinctive surface was found on Miranda. This satellite may have broken apart and reformed during its early history.

The magnetic field experiment found that the axis of the Uranian magnetosphere was tilted at 59 degrees to the plan-

et's rotational axis. In addition, instead of being generated at the planet's center, as on Jupiter and Saturn, the magnetic field was found to be offset one-third of a Uranian radius.

Context

On March 13, 1781, Sir William Herschel discovered Uranus, the seventh planet in the solar system, from his home in Bath, England. At that time, Uranus was the first planet to be observed by modern astronomers that had not been known to the ancients; its discovery showed that the solar system's outer boundary was much farther than had been previously believed.

Since its discovery, many telescopes have been pointed at Uranus. Its extreme distance from the Sun, however, has made unraveling its secrets very difficult. If one counted all the photons of light collected from Uranus by all the telescopes in the years from its discovery up to the Voyager 2 encounter, the amount of light would equal that given off by a flashlight in one second. Thus, the Voyager encounter with Uranus greatly increased knowledge of the Uranian system.

As a result of the encounter, planetary scientists now have the ability to compare the system of Uranus with those of Jupiter and Saturn. This study, known as comparative planetology, will allow scientists to understand the physical characteristics of these worlds better. This knowledge can then be applied to Earth. For example, atmospheric studies of the outer planets can improve meteorologists' understanding of terrestrial atmospheric dynamics.

Study of the Uranian ring system showed that it had many similarities to the ring systems of both Jupiter and Saturn; nevertheless, models of ring dynamics cannot yet completely explain the thinness of the nine Uranian rings. In addition, there is much to learn from the interaction of the Uranian magnetosphere with the ring system. Better models need to be developed to improve the understanding of the Uranian magnetosphere as well as of the magnetic fields of Mercury, Earth, Mars, Jupiter, and Saturn.

The Voyager 2 mission, sometimes referred to as the Grand Tour mission (a grand tour past Jupiter, Saturn, Uranus, and Neptune), is truly one of the most remarkable missions of all time. At its completion, it will have observed all the gaseous giant outer planets and most of the major moons in the solar system. The design, engineering, and operation of such a spacecraft will remain a monument to the creativity and curiosity of humanity in its quest to understand the universe.

Bibliography

Davis, Joel. *Flyby: The Interplanetary Odyssey of Voyager 2.* New York: Atheneum Publishers, 1987. This book, intended for general audiences, gives a behind-the-scenes look at the Voyager 2 encounter with Uranus. It describes the individuals responsible for making the encounter as successful as it was and the events in which they took part.

Gore, Rick. "Uranus: Voyager Visits a Dark Planet." *National Geographic* 170 (August, 1986): 178-195. This article, intended for general audiences, gives an overview of the Voyager 2 encounter of Uranus. It contains many photographs and drawings that are helpful in elucidating theories and physical characteristics of the Uranian system.

Hunt, Garry E., ed. *Uranus and the Outer Planets.* Cambridge, England: Cambridge University Press, 1982. This collection of papers, intended for college students, describes the history of the discovery of Uranus and the knowledge of the Uranian system prior to Voyager 2's swing-by.

Laeser, Richard P., et al. "Engineering Voyager 2's Encounter with Uranus." *Scientific American* 225 (November, 1986): 36- 45. This article, written by the Voyager project manager, the manager, and the deputy manager for the Flight Engineering Office, is intended for high school and college students. It explains how Voyager 2 had to be modified in preparation for the Uranus encounter. Contains photographs and illustrations.

Morrison, David, and Tobias Owen. *The Planetary System.* Reading, Mass.: Addison-Wesley Publishing Co., 1987. This book, intended for first-year undergraduates, gives an overview of each planetary system in the solar system as well as a description of asteroids and comets. It contains many photographs, illustrations, graphs, and tables.

Radlauer, Ruth, and Carolyn Young. *Voyager 1 and 2: Robots in Space.* Chicago: Children's Press, 1987. This book, intended for elementary and junior high school students, describes the entire Voyager 1 and 2 program. It contains many photographs and includes Voyager 2 data from Uranus.

Randii R. Wessen

VOYAGER 2
Neptune

Date: Beginning August 20, 1977
Type of mission: Unmanned deep space probe

Voyager 2 essentially completed the mission of the originally proposed Grand Tour of the outer solar system, flying by Jupiter, Saturn, Uranus, and Neptune, on a trajectory that will ultimately take it beyond the solar system.

PRINCIPAL PERSONAGES

> LENNARD A. FISK, Associate Administrator,
> Office of Space Science and Applications,
> National Aeronautics and Space Administration
> (NASA)
> LEW ALLEN, Director, Jet Propulsion Laboratory (JPL)
> EDWARD C. STONE, project scientist for Voyager,
> California Institute of Technology
> NORMAN HAYNES, Voyager project manager
> BRADFORD SMITH, Voyager imaging team leader
> LAURENCE SODERBLOM, Voyager imaging team
> member
> GARY FLANDRO, suggested the possibility of a
> Grand Tour of the outer solar system using
> gravitational assists

Summary of the Mission

Voyager 2 departed from Cape Canaveral's Pad 41 on August 20, 1977, atop a Titan IIIE-Centaur D1 booster. The spacecraft was inserted into a coasting Earth orbit. At the proper point in the orbit, the Centaur upper stage ignited to boost Voyager 2's speed by 12,000 meters per second above orbital speed. Its fuel exhausted, the Centaur was separated from Voyager 2. A solid rocket motor attached to Voyager 2 then fired to provide an additional 2,000 meters per second to send the spacecraft out of Earth's gravitational influence on the proper heading to send it through the asteroid belt and fly past Jupiter.

Because of the trajectories chosen for the two spacecraft, Voyager 2 was launched sixteen days earlier than its sistership, Voyager 1. The latter overtook Voyager 2 on a faster trajectory and encountered Jupiter and then Saturn before Voyager 2's flybys of those same two planets.

With launch and Earth escape behind them, Voyager spacecraft controllers began configuring the spacecraft for the long cruise ahead. Several booms and the science scan platform had to be properly deployed. A few hours into the mis-

sion, hope for a successful Grand Tour of the outer solar system began to look bleak. Voyager 2's science scan platform apparently failed to lock into the proper position. Without that lock, it appeared that there would be no way to adequately steer television cameras and other scientific equipment to desired targets during closest approach to the outer planets. After detailed analysis, it was determined that the platform had indeed extended to within less than one degree of the proper position and would support movements required for scientific observations.

With the scan platform problem overcome, technicians began determining the health of Voyager 2's science instruments. All were turned on and found to be in working order by September 2.

A much more serious problem developed in April, 1978. Sidetracked by problems with Voyager 1, controllers forget to check in with the spacecraft. As part of its computer code, Voyager 2 switched from its primary radio receiver, believing it to be malfunctioning since it had not heard from Earth within a pre-determined period, to the backup receiver. Unfortunately, the backup receiver had suffered a malfunction in its tracking-loop capacitor while not in use. This left Voyager 2 unable to communicate with Earth. Also, as part of its computer code, the spacecraft switched back to the primary receiver after a pre-determined period, but this time the primary receiver suffered a power-supply short circuit and was permanently lost. One week later, the spacecraft, as required by computer programming, switched again to the backup receiver. As a result of the tracking-loop capacitor problem, the frequency with which the backup receiver could acquire commands drifted. Controllers on the ground sent commands to Voyager 2 at a variety of frequencies, one of which reestablished contact with the spacecraft. For the remainder of the mission, controllers had to monitor the frequency drift, eventually developing a technique that allowed them to properly determine the drift behavior and maintain command of Voyager 2. The spacecraft passed through the asteroid belt without any significant damage.

Voyager 2 flew by Jupiter on July 9, 1979, coming within 645,000 kilometers of the giant planet. It provided enough high-resolution images of Jupiter to produce time-lapse motion pictures of Jupiter's atmospheric circulation patterns, including the interaction of high speed winds with the planet's Great Red Spot, a feature known through telescopic observations from the time of Galileo. In traversing the Jovian system, Voyager 2 made a close pass by Callisto, the outermost Galilean moon; Ganymede (at only 62,000 kilometers); and Europa. Voyager 2 did not pass as close to Io, the innermost Galilean moon, as had Voyager 1, but the spacecraft monitored its volcanic activity. Only two hours after closest approach to Jupiter, Voyager 2 fired its propulsion system for 76 minutes, putting the spacecraft on course for its next target— the ringed planet Saturn.

Voyager 2 flew by Saturn on August 26, 1981, coming within 101,000 kilometers of the mysterious ringed planet. Although Saturn's numerous and diverse moons were objects of study on the inward and outbound portions of the closest approach, the complex ring system was scrutinized with tremendous intensity. Unlike Voyager 1, Voyager 2 did not make a close pass by Saturn's largest moon Titan, one of the few moons in the solar system to hold its own atmosphere. A Titan encounter was sacrificed on Voyager 2 in order to target the spacecraft for a gravitational assist that would send it on its way to Uranus.

It took Voyager 2 four and one-half years to blaze the previously untraveled trail to Uranus, a ringed gas giant that rotates on its side. Whereas Voyager 2's Jupiter and Saturn encounters provided sensational photographs that received significant attention in the press, the spacecraft's Uranian encounter preceded the *Challenger* accident (January 28, 1986). The death of seven astronauts overshadowed the results being sent from Voyager 2 at Uranus.

The spacecraft passed within 71,000 kilometers of Uranus on January 24, 1986. Although Uranus was a rather bland looking world, high resolution photographs of its five larger moons provided striking evidence of unusual surface features, particularly on Miranda. Because of the spacecraft's tremendous speed, special motion compensation techniques were developed to prevent images from blurring. Since the scan platform could not be skewed enough to accomplish that, the entire spacecraft was carefully moved using short, precise thruster firings. Ten new moons and several more rings were discovered during the Uranian encounter.

On February 14, 1986, Voyager 2 burned 12 kilograms of propellant during a 2.5-hour-long course adjustment that provided 21.1 meters per second more to the spacecraft's speed, placing Voyager 2 on a trajectory aimed to take it to within only 1300 kilometers of Neptune's outer atmosphere and allow it to pass by Neptune's largest moon, Triton, at a distance of 6,000 kilometers. Although the spacecraft was in reasonably good health and flight controllers had an excellent record of maintaining contact with Voyager 2's pesky radio receiver, in September, 1986, a backup mission load was transmitted to the spacecraft's computer, programming it to carry out the entire Neptune encounter autonomously and send its data back to Earth even if contact was lost during the 3.5-year cruise to Neptune.

Voyager 2 provided the first close-up pictures of the blue planet Neptune, twelve years after launch and at a distance of 4.4 billion kilometers from Earth, generating a wealth of information about Neptune's atmospheric circulation patterns, its frigid, light-pink and white moon Triton, and asymmetrical ring systems.

A deep blue storm system, almost the size of Earth, referred to as the Great Dark Spot, located 22 ° south of the planet's equator, completed a rotation in 18 hours, 18 minutes, while spinning in a counterclockwise sense. The Great Dark Spot changed shape as it rotated, interacting with white cirrus clouds passing around its boundaries.

Another, smaller dark spot was found at 54° south latitude. This one contained bright white clouds near its center, indicative of atmospheric upwelling. Between this and the Great Dark Spot was a feature referred to as the Scooter, a rapidly moving bright cirrus cloud, possibly composed of frozen methane.

Prior to Voyager 2, segments referred to as ring arcs were known to orbit Neptune, which was verified by the first ring photographs imaged prior to close encounter. Long exposures, some up to ten minutes in duration and back-lit by sunlight, clearly revealed a total of four complete diffuse rings of non-uniform density and a broad sheet of dust. Brighter clumps were the sources of ring arcs detected from Earth. To determine particle density and size, Voyager 2's photopolarimeter recorded the brightness of the star Sigma Sagitarii as it was occulted by the rings and Neptune itself.

Voyager 2 recorded fewer charged particles (hydrogen, helium, and nitrogen) in the magnetosphere of Neptune than in that of the other gas giants. Not only was Neptune's magnetic field tilted from the rotation axis, its structure was quite complex, with a repetition rate of 16 hours and 3 minutes. Triton moves through the magnetic field equator, contorting the field's shape. (Triton was the source of the nitrogen ions observed in Neptune's magnetosphere.) The field was displaced from Neptune's center, suggesting an electric dynamo effect occurring in conductive layers surrounding Neptune's core rather than in the core itself (as on Earth).

Voyager 2 approached Triton from the south, providing high resolution views of the polar cap, a white cover of frozen nitrogen, and much of the southern hemisphere then illuminated by solar conditions experienced during the moon's summer. It appeared that the polar cap was in the process of sublimating as the height of season approached. Spacecraft images revealed several distinct surface features, of various ages, but no heavily cratered areas indicative of an age greater than three billion years. Near the equator was a relatively young region marked with linear regions reminiscent of a cantaloupe's skin. In the northern hemisphere were calderas and frozen lakes suggesting a volcanic flow of water, ammonia, methane, and nitrogen mixtures. Voyager 2 discovered evidence of active volcanism — nearly a dozen dark plumes (blowing northeasterly) depositing radiation-darkened

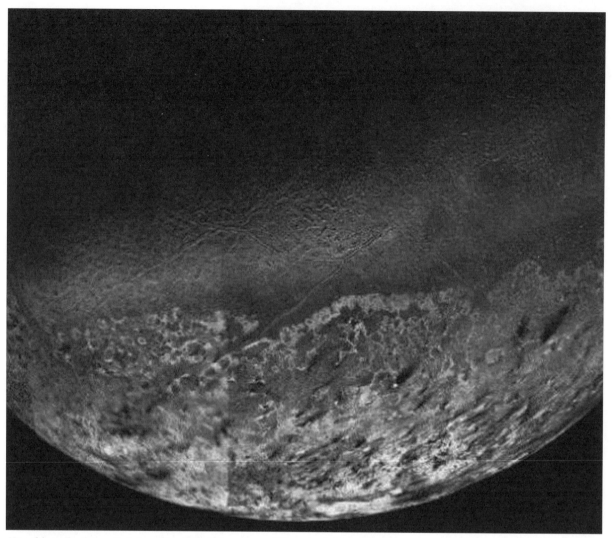

View of the south pole (near the bottom of the photo) of Neptune's moon Triton, taken by Voyager 2. (NASA)

methane on the surface. The spacecraft probed Triton's tenuous atmosphere, using an occultation of the star Beta Majoris.

After the spacecraft headed out of the Neptunian system, it looked back and transmitted images of both a crescent Neptune and Triton. Voyager 2's trajectory was bent by Neptune's gravity in such a way that it headed 48 ° below the plane of the solar system. This portion of the spacecraft's mission was referred to as the Voyager Interstellar Mission (VIM).

Knowledge Gained

Voyager 2 refined measures of Neptune's mass (17.135 times that of Earth), diameter (24,764 kilometers), density (1.64 grams per cubic centimers), rotation rate (16.05 hours), atmospheric wind speed (up to 1,000 kilometers per hour from the west), and magnetic field (inclined 50 ° to the planet's rotation axis).

Prior to Voyager 2's encounter, Neptune was known to have only two moons: Triton and Nereid. Voyager 2 discovered six additional relatively small moons, initially given designations 1989N1 through 1989N6. Although Voyager 2 came no closer than 4.7 million kilometers to Nereid, it was determined that Nereid, in actuality, was smaller than 1989N1 (400 kilometers in diameter). All the other new moons were less than 210 kilometers in diameter and quite irregular in shape. With the exception of 1989N6, which was inclined 4.5°, the new moons orbited in the plane of Neptune's rings. Two of the small moons exercised shepherding roles close to rings.

Triton was a curious world, almost certainly formed independent of Neptune, and later captured by the gas giant when it strayed too close to Neptune. This would account for Triton's unusual orbital inclination: 20 ° relative to the ring plane and the plane of most of the other small Neptunian moons. Voyager 2 refined measures of Triton's mass (2.13 x 1022 kilograms), diameter (2,720 kilometers), and density (2.03 grams per cubic centimeter). Triton's atmosphere consisted of nitrogen, with traces of methane. Surface pressure was a mere 10 microbars. Nitrogen extended as far as 800 kilometers from the surface, with a haze layer existing in the closest 25 kilometers. Methane, a heavier molecule, remained much closer to the surface.

Although Voyager 2 could not be diverted to intercept Pluto, the spacecraft may have provided clues to the nature of the Pluto/Charon system through its examination of Triton, revealing the presence of a combination of rock and frozen nitrogen, methane, and water. It is expected that Triton undergoes seasonal changes, and its atmosphere of nitrogen and methane may well freeze out during winter. Triton's average temperature, 37 Kelvins, is typical of Pluto as well, so Pluto may also have a similar behavior and composition.

Context

In 1966, Gary Flandro, then an aeronautics graduate student at JPL, published a paper describing a unique alignment of the outer planets that would permit spacecraft to use energy provided by the planets themselves to alter the spacecraft's trajectory and allow a "Grand Tour" of the outer solar system in an amount of time far less than would be required by least-energy-transfer-orbits. In 1969, NASA began planning for an ambitious program that it entitled "The Grand Tour," in which four spacecraft would be dispatched from Earth to encounter Jupiter, Saturn, Uranus, Neptune, and Pluto. Two would be sent to Jupiter, Saturn, and Pluto, and the other two would fly by Jupiter, Uranus, and Neptune. Budget cutbacks forced the cancellation of this ambitious program, but NASA was able to obtain funding for what it initially called the Mariner-Jupiter-Saturn (MJS) project. The name was later changed to Voyager, and although the scope of the scientific investigation was scaled back, the opportunity still remained for at least one of the Voyager spacecraft to use a gravitational assist at Saturn to then send it on to Uranus and possibly also Neptune.

Without gravitational assists, the Voyager 2 mission to visit four planets would have been impossible. That navigational technique was first demonstrated on the Mariner 10 mission (1973-1975) to Venus and Mercury. Part of Voyager 2's journey was flown in the wake of previous explorers. The pathway to Jupiter and Saturn had been blazed by the successful Pioneers 10 (1973 Jupiter flyby) and 11 (1974 Jupiter flyby and 1979 Saturn flyby), and Voyager 1 spacecraft. But Voyager 2 entered virgin territory when it investigated the Uranian and Neptunian systems in 1986 and 1989, respectively. It then joined the Pioneers 10 and 11, and its fellow Voyager 1 spacecraft as a quartet of interplanetary travelers, leaving our solar system, heading away in different directions, each carrying a message from the people of Earth, and investigating the outer solar system boundaries before eventually running out of energy some time between 2005 and 2010.

After about 8,500 years in flight, Voyager 2 should pass within four light-years of Barnard's Star. Its closest approach to a stellar system would occur in 40,457 years, when it should breeze within 1.5 light-years of Ross 248. Those last two encounters presuppose that, in the intervening time, humanity would fail to develop interstellar travel and seek out the intrepid Voyager 2, returning it to Earth for an honored resting place in an appropriate museum.

Bibliography

Beatty, J. Kelly, and Andrew Chaikin, eds. *The New Solar System*. Cambridge, Mass.: Sky Publishing Corporation, 1990. Pages 107 through 206 contain detailed information about all major solar system objects, including photographs and data returned by unmanned spacecraft from Mariner 2 through Voyager 2. Many chapters were written by principal investigators from the Voyager program.

Berry, Richard. "Triumph at Neptune." *Astronomy* 17 (November, 1989): 20-28. Contains excellent color images returned by Voyager 2 from Neptune, its rings, and moons. Thorough presentation of preliminary results, primarily Neptune and Triton.

Berry, Richard. "Neptune Revealed." *Astronomy* 17 (December, 1989): 22-34. Continues the presentation begun in the November, 1989 issue. Results are presented after further reflection. Makes comparisons of Neptune data with that from Voyager 2's Uranus encounter.

Burgess, Eric. *Far Encounter: The Neptune System*. New York: Columbia University, 1991. Provides a full account of the Voyager 2 journey from launch through the Neptune encounter.

U.S. Congress, House. Committee on Science, Space, and Technology. *Voyager 2 Flyby of Neptune*. Washington, D.C.: U.S. Government Printing Office, 1990. A report to the Congressional subcommittee on Space Sciences and Applications by Drs. Lennard Fisk, Lew Allen, and Edward Stone (October 4, 1989). Includes black and white Voyager 2 Neptune images as well as descriptions of the scientific results of the mission, its importance, and implications for future planetary science programs.

Davis, Joel. *Flyby: The Interplanetary Odyssey of Voyager 2*. New York: Atheneum, 1987. Provides a readable account of the journey of Voyager 2 from launch through the Uranus encounter. Provides numerous discussions with scientists involved with the program, and details important discoveries made at Jupiter, Saturn, and Uranus. Profiles the expected Neptune encounter.

Kohlhase, Charles, ed. *The Voyager Neptune Travel Guide.* (JPL Publication 89-24) Washington, D.C.: Superintendent of Documents, June 1, 1989. An excellent pre-encounter publication describing the objectives and maneuvers of Voyager 2 at its Neptune encounter. Provides numerous detailed charts and diagrams. Some parts are rather technical, but most are accessible to the general reader.

Moore, Patrick. *The Planet Neptune.* Chichester, West Sussex, England: Ellis Horwood, 1988. Provides a historical context of the investigation of the planet Neptune, from early searches for its existence up to the advent of spacecraft. Does not include data from Voyager 2.

"Voyager's Last Picture Show." *Sky & Telescope* 78 (November, 1989): 463-470. A pictorial review of the images returned by Voyager 2 at Neptune. Includes close-ups of atmospheric features such as the Great Dark Spot and Scooter, and surface features on Triton.

Wilson, Andrew. *Solar System Log.* London: Jane's Publishing Company, Limited, 1987. A comprehensive encyclopedic log of both American and Soviet/Russian interplanetary probes, including Voyagers 1 and 2. Does not include information about the Voyager 2 Neptune encounter, however.

David G. Fisher

GLOSSARY

Ablative heatshield: A heatshield that is composed of material that ablates (that is, melts and eventually vaporizes) as a spacecraft reenters Earth's atmosphere. The consequent removal of excess heat prevents the spacecraft from burning up.

ABM. See Antiballistic missile.

Abort: To terminate a launch or a mission, usually as the result of equipment failure. As a noun, the action of aborting.

Absolute magnitude: The brightness of a star or other celestial body measured at a standard distance of 10 parsecs, (See also Apparent magnitude, Luminosity, Parsec.)

Absolute temperature scale: A temperature scale that sets the lowest possible temperature (absolute zero, or the temperature at which molecular and atomic motion stops) at zero. (See also Kelvin.)

Absorption spectrum: An electromagnetic spectrum that shows dark lines which result from the passage of the electromagnetic radiation through an absorbing medium, such as the gases found in a star's atmosphere. The resulting absorption lines are characteristic of certain chemical elements and reveal much about the composition of the star's atmosphere. (See also Electromagnetic spectrum, Emission spectrum, Spectrum.)

Acquisition: The detection and tracking of an object, signal, satellite, or probe to obtain data or control the path of a spacecraft. (See also Star tracker)

Active experiment: An experiment package carried by a satellite, usually in a canister, which typically has a control circuit, a battery-driven power system, data recording instruments, and environmental control systems. (See also Passive experiment.)

Active satellite: A satellite equipped with on-board electrical power that enables it to transmit signals to Earth or another spacecraft. Most artificial satellites fit this description. (See also Passive relay satellite.)

Advanced vidcon camera system (AVCS): Spaceborne imaging systems made up of two 800-line cameras with nearly twice the resolution of a normal television camera. Capable of photographing a 3,000-kilometer-wide area with a resolution of 3 kilometers.

Aerodynamics: The study of the behavior of solid bodies, such as an airplane, moving through gases, such as Earth's atmosphere.

Aerography: The study of land features on Mars.

Aeronautics: The study of aircraft and the flight of these man-made objects in the atmosphere.

Aeronomy: The study of the physics and chemistry of the atmospheres of Earth and other planets.

Aerospace: The space extending from Earth's surface outward to the farthest reaches of the universe.

Air lock: A small enclosed area (especially in the space shuttle and space stations) which is located between the interior of a spacecraft and outer space or another spacecraft that an astronaut or cargo can pass without depressurizing the spacecraft.

Airbus (bus): A spacecraft used to deliver payloads into orbit or to another spacecraft, such as a space station. The space shuttle performs airbus functions.

Airglow: A faint glow emitted by Earth that results from interaction between solar radiation and gases in the ionosphere, perceived from space as a halo around the planet. Airglow is known for interfering with Earth-based astronomical observations, making space-based telescopes desirable.

Albedo: The amount of electromagnetic radiation reflected from a nonluminous body, measured from 0 (perfectly black) to 1 (perfectly reflective).

Altitude: The distance of an object directly above a surface. Also, the arc or angular distance of a celestial object above or below the horizon.

Angstrom: One ten-thousand-millionth of a meter; a unit used to measure electromagnetic wavelengths.

Angular momentum: A property of a rotating body (or a system of rotating bodies) which is defined as the product of the distribution of the body's mass around the rotational axis (the moment of inertia) and the speed of the body around the axis (the angular velocity). Angular momentum remains constant; that is, an increase in one of the two factors is compensated by a decrease in the other. Hence, a spinning ice skater will rotate faster as he pulls his arms toward his body; a planet or artificial satellite will move faster in its elliptical orbit as it approaches the point closest to the object around which it is orbiting (periapsis).

Anemometer: An instrument for measuring the force of wind.

Antiballistic missile (ABM): A missile designed to destroy a ballistic missile in flight.

Antimatter: Matter in which atoms are composed of antiparticles; positrons in place of electrons, antiprotons in place of protons, and so forth. The existence of such antiparticles is accepted, although their configuration as antimatter has yet to be discovered. Theoretically, the result of a meeting between matter and corresponding antimatter is mutual obliteration in a release of energy.

Antisatellite (ASAT) system: A weapons system used to destroy potentially hostile orbiting satellites.

Aphelion: The point in an object's orbit around the Sun at which it is farthest away from the Sun.

Apoapsis: The point in one object's orbit around another at which the orbiting object is farthest away from the object being orbited.

Apocynthion: The point in an object's orbit around the Moon at which it is farthest away from the Moon.

Apogee: The point in an object's orbit around Earth at which it is farthest away from Earth.

Apolune: Apocynthion of an artificial satellite.

Apparent magnitude: The brightness of a star or other celestial body as seen from a single point, such as Earth. The brightness is "apparent" only, because stars vary in their distance from Earth. (See also Absolute magnitude, Luminosity.)

Apparent motion: The path of movement of a body relative to a fixed point of observation.

Applications technology satellite: A satellite designed for developing applications (meteorology, navigation, communication, Earth resources, and the like). Also used as a relay between other satellites and Earth stations.

APU. See Auxiliary power unit.

Arm. See Remote manipulator System, Robot arm.

Array: A system of multiple devices (such as radio aerials or optical telescopes) situated to increase the strength of the data received.

Artificial satellite: A man-made satellite or object sent into orbit around a celestial body. Generally referred to simply as "satellites." These spacecraft are usually unmanned.

ASAT. See Antisatellite system.

Ascent stage: One of two stages of the Apollo lunar module; it carried the crew, their gear and samples, the life-support system, and fuel for lift-off from the Moon, orbit, and docking with the command and service module. (See also Descent stage.)

Asteroid: A small solid body (also known as a planetoid), ranging in size from about 200 meters to 1,000 kilometers in diameter, which orbits the Sun. The solar system is home to thousands of these bodies, most of which occur in a region between the orbits of Mars and Jupiter.

Asteroid belt: The region between Mars and Jupiter (between 2.15 and 3.3 astronomical units from the Sun) where most of the solar system's asteroids have been found.

Asthenosphere: The layer of Earth below the lithosphere

Astronaut: A space traveler, usually from the United States but also from other nations. (See also Cosmonaut, Spacenaut.)

Astronautics: The science and technology of spaceflight, including all aspects of aerodynamics, ballistics, celestial mechanics, physics, and other disciplines as they affect or relate to spaceflight. (See also Aerodynamics, Aeronautics, Celestial mechanics.)

Astronomical unit (AU): The mean distance between the centers of Earth and the Sun: 92,955,630 miles or 149,597,870 kilometers. Used for measuring distance within the solar system.

Astronomy: The study of all celestial bodies and phenomena within the universe.

Astrophysics: The branch of astronomy dealing with the chemical and physical properties and behaviors of celestial matter and their interactions.

Atmosphere: Any gaseous envelope surrounding a planet or star. Earth's atmosphere consists of five layers: the troposphere, stratosphere, mesosphere, themosphere (which roughly coincides with the ionosphere), and exosphere.

Atmospheric pressure: The force exerted by Earth's atmosphere, which at sea level is approximately 14.7 pounds per square inch, 101.325 newtons per square meter, or 1 bar. One "atmosphere" refers to any of these sea-level measures and can be used to refer to atmospheric pressure on other planets: Venus' surface pressure, for example is 90 atmospheres.

Atom: The smallest particle of an element that can exist alone or in combination with other atoms. Most atoms consist of one or more electrons (negatively charged particles) and neutrons (particles with no charge). The combinations of these particles determine the identity of the atom as a particular chemical element or isotope of that element.

Attitude: The orientation of a spacecraft or other body in space relative to a point of reference.

Attitude control system: The combined mechanisms working together to maintain or alter a spacecraft's position relative to its point of reference, including on-board computers, gyroscopes, star trackers, or a combination of these.

AU. See Astronomical unit.

Aurora: The colored lights appearing in the sky near the poles when charged particles issuing from the Sun become trapped in Earth's magnetic field. The arching, spiraling glows result from these particles interacting with atmospheric gases as they follow Earth's magnetic force lines.

Aurora borealis: The aurora occurring near Earth's North Pole.

Aurora australis: The aurora occurring near Earth's South Pole.

Auxiliary power unit (APU): A backup system on board a space shuttle, space station, or other spacecraft for generating non-propulsion electrical power in the event of a main-system failure.

AVCS. See Advanced vidcon camera system.

Avionics: The electronic devices used on board a spacecraft, or the development, production, or study of those devices.

Axis: The imaginary line around which a celestial body or man-made satellite rotates.

Azimuth: The arc, or angular distance, measured horizontally and moving clockwise between a fixed point (usually true north) and a celestial object. (See also Altitude.)

Backup crew: A group of astronauts trained to perform the same functions as the crew members of a particular space mission (such as the commander, pilot, and flight engineer), in the event that an emergency requires replacement of the original crew.

Ballistic missile: A missile that is not self-guided but rather is aimed and propelled at the point of launch only, and hence follows the trajectory that is determined at launch.

Ballistics: The study of the motion of projectiles in flight, including their trajectories; it is especially important in the launching and course-correction maneuvers of spacecraft.

Band. See Frequency, Hertz.

Bar. See Atmospheric pressure.

Barbecue maneuver: The deliberately maintained slow roll of a spacecraft in orbit so that all exterior surfaces will be evenly heated by the Sun.

Barycenter: The center of mass of a system of two or more bodies.

Bhangmeter: An optical-flash detector used to detect nuclear explosions from satellites.

Big bang theory: The cosmological theory, accepted by many scientists, that the universe has evolved from a gigantic explosion of a compressed ball of hot gas many billions of years ago. The theory holds that matter is still flying outward uniformly from the center of this explosion, and is therefore also referred to as the expanding universe theory. (See also Steady state theory.)

Binary star: A star system formed of two stars orbiting their combined center of mass. Three types of binaries include visual binaries, which emit radiation in the visible wavelength range (the most common binaries); spectroscopic binaries, detected by their Doppler shifts; and eclipsing binaries, in which one star periodically blocks light from the other as they rotate around each other. It is estimated that more than half of the stars in the Galaxy are binaries.

Biotelemetry: The remote measurement and monitoring of the life functions (such as heart rate) of living beings in space, and the transmissions of such data to the monitoring location, such as Earth.

Biosatellite: Formed from "biological" and "satellite," a biosatellite is an artificial satellite carrying life forms for the purpose of discovering their reaction to conditions imposed in the space environment.

Black dwarf: A star that has cooled to the point that it no longer emits visible radiation; the end state of a white dwarf star. (See also White dwarf.)

Black hole: A hypothetical celestial body whose existence is predicted by Albert Einstein's general theory of relativity and is accepted by many scientists. In a black hole, matter is so condensed and gravitational forces are so strong that not even light can escape. Black holes are thought to be either the product of a collapsed star (stellar black holes) or the result of the original big bang (primordial black holes).

Blueshift: An apparent shortening of electromagnetic wavelengths emitted from a star or other celestial object, indicating movement toward the observer. (See also Doppler effect.)

Boom: A long arm extending outward from a satellite or other spacecraft to hold an instrument or device such as a camera.

Booster. See Rocket booster.

Bow shock: The "wave" created by a planet's magnetic field when it forms an obstacle to the stream of ionized gases flowing outward from the Sun.

Breccia: A type of rock formed by sharp, angular fragments embedded in fine-grained material such as clay or sand. Among the "Moon rocks" returned by the Apollo astronauts, breccias were the most common.

Burn: As a noun, the term used to refer to the firing of a rocket engine, including any burn used during a spaceflight to set the spacecraft on a trajectory toward a planet or other target.

Caldera: A very large crater formed by the collapse of the central part of a volcano.

Canopus: The brightest star in the sky after Sirius, visible south of 37 degrees latitude. Canopus is often the target of a spacecraft's star tracker, which uses it as a reference point in steering a course toward the spacecraft's destination.

Capsule communicator (CapCom): A ground-based astronaut who acts as a communications liaison between ground and flight crews during manned missions.

Cassegrain telescope: The most common type of reflecting telescope named for its inventor, Guillaume Cassegrain (1672). The telescope contains two mirrors: a concave mirror near its base, which reflects light from the sky onto a convex mirror above it; the convex mirror, in turn, reflects the light back down through a hole in the middle of the concave mirror to the focal point. The Hubble Space Telescope is of Cassegrain design.

C-band: A radio frequency range of 3.9 to 6.2 gigahertz. (See also Hertz.)

CCD. See Charge coupled device.

Celestial mechanics: The branch of physics concerned with those laws that govern the motion (especially the orbits) of celestial bodies, both artificial and natural.

Celestial sphere: An imaginary sphere surrounding an observer at a fixed point in space (the sphere's center), with a radius extending to infinity, a celestial equator (a "belt" cutting the sphere into two even halves), and celestial poles (north and south). By reference to these points on the celestial sphere, the observer can describe the position of an object in space.

Celsius scale: A temperature scale, named for its inventor, Anders Celsius (1701-1744), which sets the freezing point of water at zero and the boiling point at 100. Also referred to as the centigrade scale, its increments correspond directly to Kelvins. To convert Kelvins to degrees Celsius, subtract 273.15. (See also Kelvin)

Centrifuge: A device for whirling objects or human beings at high speeds around a vertical axis, exerting centrifugal force to test spacecraft hardware or train astronauts to withstand the forces of launch.

Cepheid variable: A star that has passed its main sequence phase (the greater part of its lifetime) and has entered a transitional phase in its evolution, during which the star expands and contracts, at the same time pulsating in brightness. By measuring the star's period of pulsation and extrapolating from that its absolute magnitude, then comparing the absolute magnitude to the apparent magnitude, astronomers find the distance of the star and nearby celestial objects.

Chandrasekhar limit: The maximum possible mass for a white dwarf star, calculated by Bubramanyan Chandrasekhar in 1931 as approximately 1.4 solar masses (later modified upwards for rapidly rotating white dwarf stars). When the mass exceeds the Chandrasekhar limit, gravity compresses it into a neutron star.

Charge: A property of matter defined by the excess or deficiency of electrons in comparison to protons. Negative charge results from excess electrons; positive charge, from a deficiency of electrons.

Charge coupled device (CCD): A highly sensitive device, commonly one centimeter square, that is sensitive to electromagnetic radiation and contains electrodes and conductor channels, overlying an oxide-covered silicon chip, for collecting, storing, and later transferring the data to create images of celestial objects and other astronomical phenomena.

Chromosphere: The lower layer of the solar atmosphere (between the photosphere and the solar corona), several thousands of kilometers thick, that is composed mainly of oxygen, helium, and calcium and is visible only when the photosphere is obscured, as during a solar eclipse. The term also applies to corresponding regions of other stars.

Circular orbit: An orbit in which the path described is a circle.

Closing rate: The speed of approach of two spacecraft preparing for rendezvous.

Coma: See Comet.

Comet: A luminous celestial object orbiting the Sun, consisting of a nucleus of water ice and other ices mixes with solid matter, and, as the comet approaches the Sun, a growing coma and tail. The coma is a collection of gases and dust particles that evaporate from the nucleus and form glowing ball around it; the tail forms as these materials are swept away from the nucleus. Comets appear periodically, depending on the parameters of their solar orbits, and vary in size from a few kilometers to thousands.

Command and service module: The potion of the Apollo spacecraft consisting of the command module, in which the astronauts traveled to the Moon, and the service module, which contained the lunar module and main rocket engine.

Communications satellite: See Telecommunications satellite.

Comsat: An abbreviation for "communications satellite." (See also Telecommunications satellite.)

Conjunction: The alignment of two planets or other celestial bodies so that their longitudes on the celestial sphere are the same. Inferior conjunction occurs between two bodies whose orbits are closer to the sun than Earth's; superior conjunction occurs between bodies whose orbits are farther from the Sun than Earth's.

Constellation: A collection of stars that form a pattern as seen from Earth. The stars in these groupings are often quite distant from one another, their main common characteristic being the illusory picture they form (such as the Big Dipper, or Ursa Major) against the backdrop of the night sky. Constellations are useful in that they provide points of reference for astronomers and other stargazers.

Convection: A process that results from the movement of unevenly heated matter, gases or liquids, whereby hotter matter moves toward and into cooler matter. The resultant circular motion and transfer of heat energy is convection.

Core: The central portion of any celestial body, especially the terrestrial planets but also stars. Also, the central part of a launch vehicle, to which may be added strap-on boosters.

Core sample: A sample of rock and soil taken from Earth, the Moon, or another terrestrial planet by pressing a cylinder down into the planet's surface.

Corona: The outermost portion of the Sun's atmosphere, extending like a halo outward from the Sun's photosphere. The corona consists of extremely hot ionized gases that eventually escape as solar wind. The term is also used to refer to the corresponding region of any star's atmosphere.

Chronograph: A device for viewing the solar (or another star's) corona, consisting of a solar telescope outfitted with an occulting mechanism to obscure the photosphere, as during a solar eclipse, so that the corona is more easily perceived.

Cosmic dust: Tiny solid particles found throughout the universe, thought to have originated from the primordial universe, the disintegration of comets, the condensation of stellar gases, and other sources. Also know as interstellar dust.

Cosmic radiation: Atomic particles that are the most energetic known, consisting mainly of protons, along with electrons, positrons, neutrinos, gamma-ray photons, and other atomic nuclei. These particles emanate from a number of sources, both within and beyond the Milky Way, and they bombard atoms in Earth's atmosphere to produce showers of secondary particles such as pions, muons, electrons, and nucleons. If a primary cosmic particle is sufficiently energetic when it hits an atmospheric atom, the secondary particles can reach Earth's surface, and do, passing through matter.

Cosmic ray detector: A device for sensing, measuring, and analyzing the composition of cosmic radiation in an attempt to discover its sources and distribution.

Cosmodrome: Any launch site in the former Soviet Union, one of the best known being Tyuratam/Baikonur in Siberia.

Cosmology: The study of the origins and structure of the universe.

Cosmonaut: An astronaut in the former Soviet Union.

Countdown: The tracking of the time immediately preceding a space launch, during which all conditions, both outside and inside the spacecraft, are closely monitored to ensure proper functioning of all systems. Usually expressed as "T" (time) "minus" so many minutes and seconds before engine ignition. (See also Ground-elapsed time, Mission-elapsed time.)

Crater: A depression in the surface of a planet or moon caused by the force of a meteorite's fall. Also, the depression that forms at the mouth of a volcano.

Cruise missile: A bomb-bearing missile that flies at low altitude by means of an on-board guidance system that senses terrain and identifies its target.

Crust: The outermost layer, or shell, of a planet or moon, such as Earth.

Cryogenic fuels: Liquid rocket propellants that operate at extremely low temperatures, such as liquid oxygen.

Cyclotron radiation: The radiation produced by charged particles as they spiral around magnetic lines of force at extremely high speeds.

Data acquisition: The detection, gathering, and storage of data by scientific instruments.

DBS. See Direct broadcast satellite.

De-orbit: To execute maneuvers, such as firing of retrograde rockets to reduce orbital speed and leave an orbit, usually in preparation for reentry into Earth's atmosphere or a change in course.

Deep space: Regions of space beyond the Earth-Moon system.

Deep space probe: A device launched beyond the Earth-Moon system that is designed to investigate other parts of the solar system or beyond. Sometimes called an interplanetary space probe in reference to spacecraft investigating the planets and the space between them.

Density: The ratio of an object's mass to its volume.

Descent stage: One of two stages of the Apollo lunar module; it carried fuel for landing as well as some scientific equipment. (See also Ascent stage.)

Dewar: A container, similar to a vacuum bottle, with inner and outer walls between which is evacuated space to prevent transfer of heat. Used to store cryogenic fuels at very low temperatures.

Dielectric: Used to characterize any device, substance, or state (insulation materials, a vacuum) that is a nonconductor of electricity.

Digital imaging: See Imaging.

Direct broadcast satellite (DBS): A telecommunications satellite designed to broadcast television signals directly into private residences equipped with dish antennae, particularly useful where television reception is difficult.

Dirty snowball theory: The model of comets, accepted by most astrophysicists, that considers a comet's nucleus to be composed of a small sphere of ice and rock.

Diurnal: Occurring daily. The diurnal motion of a planet or other celestial body is its daily path across the sky as seen from a fixed point such as Earth, which depends on the position of the observer.

Dock: To link one spacecraft with another or others while in space, first achieved in 1965 by Geminis IV-A and VII.

Doppler effect: First described in 1842 by Christiaan Doppler, the principle that describes the effect, from the perspective of an observer, of an object's movement on the electromagnetic or sound energy that it emits. The wavelength or frequency of this energy appears to increase (shorten) if the energy source is approaching the observer, and the rate (speed) of approach will determine the rate of increase. The opposite is true as the energy source moves away from the observer. Hence, the sound from an ambulance streaking past a motorist seems to rise sharply as it approaches, then fall as it rushes away: The motorist perceives increasingly "compressed" (shortened) wavelengths of sound as the ambulance approaches, then increasingly "stretched" (elongated) wavelengths as the ambulance speeds down the road. Electromagnetic (light) energy behaves similarly. Red and infrared rays have longer wavelengths, while the wavelengths of blue and ultraviolet rays are shorter. A source of light that is moving away from an observer will appear to get "redder" (called the redshift). If the light source is approaching, it will get "bluer" (blueshift). These phenomena are measured by observing the spectral lines of the energy source over time. The Doppler effect is at the base of much of our understanding of the universe, providing strong support for the theory of the expanding universe.

Downlink: Transmissions to Earth from a spacecraft; often used in reference to telecommunications satellites.

Drogue chute: A small parachute designed to pull a larger parachute from stowage or to decrease the velocity of a free-falling spacecraft during reentry.

Early-warning satellite: A satellite designed to detect launches of ballistic missiles and tests of nuclear weapons.

Earth day: Twenty-four hours, or the time required for Earth to complete one rotation on its axis. Scientists measure the planets' periods of rotation in Earth days.

Earth-orbital probe: An unmanned spacecraft carrying instruments for obtaining information about the near-Earth environment.

Earth resources satellite: A satellite designed to detect and store data on Earth resources and their conditions, such as mineral deposits, forests, crops and crop diseases, and pollution. The Landsat series of satellites is a preeminent example.

Eccentricity: The degree to which an ellipse (or orbital path) departs from circularity. Eccentricity is characterized as "high" when the ellipse is very elongated.

Eclipse: The obscuring of one celestial body by another. In a lunar eclipse, Earth's shadow obscures the Moon when Earth is situated directly between the Sun and Moon. In a solar eclipse, the Moon is situated between the Sun and Earth in such a way that part or all of the Sun's light is blocked; the total blockage of sunlight (with the exception of the Sun's corona) is called a total eclipse of the Sun.

Ecliptic plane: The plane in which Earth orbits the Sun. From Earth, the ecliptic plane is perceived as the Sun's yearly path through the sky.

Ejecta: Material thrown out from a volcano.

Electromagnetic radiation: Radiation, or a series of waves of energy, consisting of electric and magnetic waves, or particles (photons), vibrating perpendicularly to each other and traveling at the speed of light. Electromagnetic radiation varies in wavelength and frequency as well as source (which may be thermal or nonthermal) and whether or not it is polarized. (See also Electromagnetic spectrum.)

Electromagnetic spectrum: The continuum of all possible electromagnetic wavelengths, from the longest, radio waves (longer than 0.3 meter), to the shortest, gamma rays (shorter than 0.01 nanometer). The shorter the wavelength, the higher the frequency and the greater the energy. Within the electromagnetic spectrum is a range of wavelengths that can be detected by the human eye, visible light. Its wavelengths correspond to colors: Red light emits the longest-wavelength visible radiation; violet light, the shortest-wavelength radiation. None of these types of electromagnetic radiation is dis-

crete; each blends into the surrounding forms. Detection of nonvisible radiation by special instruments (used in such branches of astronomy as infrared astronomy and X-ray astronomy) reveals much about the behavior of celestial bodies and the origins of the universe.

Electron: An atomic particle that carries a negative charge and a mass about one eighteen-thousandth of a proton. One or more electrons whirl around the nuclei of all atoms and can also exist independently.

Electronic intelligence satellite (ELINT): A satellite designed to identify sources of radar and radio emissions, primarily used in military applications. (See also Ferret.)

Electrophoresis: A process for separating cells using a weak electric charge, more easily accomplished in space than on Earth.

Elementary particles: The smallest units of matter or radiation, characterized by electrical charge, mass, and angular momentum. Among elementary particles are electrons, neutrons, protons, neutrinos, the various mesons, and their corresponding antiparticles (which form antimatter). Photons, the smallest units of electromagnetic radiation, are also considered as elementary particles.

ELINT. See Electronic intelligence satellite.

Ellipse: An oval-shaped geometric curve formed by a point moving so that the sum of the distances between two points around which it moves is always the same. The planets trace out ellipses in their orbits around the Sun.

ELV: See Expendable launch vehicle.

Elliptical orbit: An orbit that departs from circularity, as most orbits do. A highly elliptical orbit is one whose apoapsis is much greater than its periapsis, resulting in an orbit that traces out an elongated ellipse.

Emission spectrum: A spectrum showing the array of wavelengths emitted by a thermal source of electromagnetic radiation, such as a star. Atoms in this source subjected to thermal (heat) energy will emit energy as their electrons move from one energy level to another. The wavelengths that characterize these energy jumps correspond to specific elements (such as hydrogen), and appear as a characteristic line on the emission spectrum. The relationships of these lines to one another tell astronomers much about the composition, density, temperature, and other conditions in the energy source.

EMU. See Spacesuit.

Equatorial orbit: An orbit that follows the equator of the body orbited.

Escape velocity: The speed at which an object must travel to escape the gravitational attraction of a celestial body. In order for a spacecraft to leave Earth orbit, for example, its engines must exert enough in-orbit thrust to achieve escape velocity.

EVA. See Extravehicular activity.

Event horizon: The boundary beyond which an observer cannot see. Also, the boundary beyond which nothing can escape from a black hole, where escape velocity equals the speed of light and thus nothing, not even light, can escape. Therefore, the event horizon is theoretically the spherical delineation of a black hole. (See also Escape velocity.)

Exobiology: The study of the conditions for and potential existence of life-forms beyond Earth.

Exosphere: The outermost layer of Earth's atmosphere.

Expendable launch vehicle (ELV): A launch vehicle not intended for reuse. The Atlas, Delta, Titan, and Saturn launch vehicles fall into this category.

External tank: The large tank that stores liquid fuel for the three space shuttle main engines; it separates from the shuttle after all the fuel has been used (about 8.5 minutes after lift-off) and burns up in the atmosphere.

Extraterrestrial life. See Exobiology.

Extravehicular activity (EVA): Popularly known as a "space walk," any human maneuver taking place partially or fully outside the portion of a spacecraft that houses the astronauts.

Extravehicular mobility unit (EMU): See Spacesuit.

F region. See Thermosphere.

False color image: An image resembling a photograph, created from data collected by instruments (such as an infrared sensor) aboard a spacecraft and deliberately assigned unnatural colors in order to make nonvisible radiation visible or to highlight distinctions. (See also Imaging.)

Ferret: An electronic intelligence satellite designed to detect hostile electromagnetic radiation. (See also Electronic intelligence satellite.)

Fission (atomic): The breaking of an atomic nucleus into two parts, resulting in a great release of energy.

Flight path: The trajectory of an airborne or spaceborne object relative to a fixed point such as Earth.

Fluid mechanics: The study of the behavior of fluids (gases and liquids) under various conditions, including that of microgravity in spaceflight. Understanding fluid mechanics in space is important to the technology of spaceflight and may have applications on Earth as well.

Flyby: A close approach to a planet or other celestial object, usually made by a probe for the purpose of gathering data; the maneuver does not include orbit or landing. Also used to refer to a mission that undertakes a flyby.

Fluorescence: The property of emitting visible light absorbed from an external source.

Focal ratio (f-number): The ratio of (1) the distance between the center of a lens or mirror and its point of focus (focal-length) and (2) the aperture, or diameter, of the mirror lens. The focal ratio of a telescope determines its power of magnification.

Footprint: An area on Earth's surface where a spacecraft is expected to land. Also, an area served by a telecommunications satellite.

Frauenhofer lines: Prominent absorption lines in the Sun's spectrum, first observed by Joseph von Frauenhofer in 1814, indicating the presence of certain elements in the Sun's corona. Also used to refer to such absorption lines in other stars' spectra.

Free-flyer: Any spacecraft capable of solitary flight, and not attached to another for electrical power.

Free return trajectory: An orbital flight path that allows a disabled spacecraft to reenter Earth's atmosphere without assistance.

Frequency: The number of times an event recurs within a specific period of time. Frequency characterizes both sound and electromagnetic radiation and is defined as the ratio of its wavelength to its speed. Frequency is measured in hertz or multiples of hertz.

Fuel cell: A device that joins chemicals to produce electric energy, different from a battery in that the chemicals are joined in a controlled fashion depending on electrical load. A by-product of fuel cells that use oxygen and hydrogen is potable water. Used on Apollo and space shuttle missions.

Fusion: A thermonuclear reaction in which the nuclei of light elements are joined to form heavier atomic nuclei, releasing energy. The process can be controlled to produce power. It is also the process whereby the stars formed the elements with atomic numbers up to iron.

Gabbro: A dark-colored, igneous, crystalline rock.

Gain: The increase in power of a transmitted signal as it is picked up by an antenna.

Galaxy: A collection of stars, other celestial bodies, interstellar gas and dust, and radiation rotating or clustered around a central hub, classified by Edwin Hubble in 1925 as one of four shapes: spiral, elliptical, lenticular, or irregular. Galaxies are thousands of light years across and contain billions of stars, and there are thousands of galaxies in the universe.

Gamma radiation: Electromagnetic radiation with wavelengths less than 0.01 nanometer, the most energetic known in the universe outside cosmic radiation. The ability of gamma rays to penetrate the interstellar matter and radiation of the universe makes them especially valuable to astronomers.

Gamma-ray astronomy: The branch of astronomy that investigates gamma radiation and its sources by detectors sent aloft in satellites such as Orbiting Solar Observatory 3, SAS2, and COS-B. Among the sources of gamma rays are pulsars and neutron stars.

Gegenschein: A patch of faint light about twenty degrees across, visible from the night side of Earth at a point opposite the Sun in the ecliptic plane, and possibly caused by the reflection of sunlight from a dust "tail" swept away by the solar wind. (See also Zodiacal light.)

Geiger counter: A device that detects high-energy radiation (including cosmic rays) by means of a tube containing gas and an electric current. The radiation causes the gas to ionize, which is transmitted to the current and detected as a sound or a needle jump.

Geocentric orbit: An orbit with Earth as the object orbited.

Geodesy: The science concerned with the size and shape of Earth and its gravitational field.

Geostationary orbit: A type of geosynchronous orbit that is circular and lies in Earth's equatorial plane, at an altitude of approximately 36,000 kilometers. As a result, a satellite in geostationary orbit appears to hover over a fixed point on Earth's surface. (See also Geosynchronous orbit.)

Geosynchronous orbit: A geocentric orbit with a period of 23 hours, 56 minutes, 4.1 seconds, equal to Earth's rotational period. Such an orbit is also geostationary if it lies in Earth's equatorial plane and is circular. If inclined to the equator, a geosynchronous orbit will appear to trace out a figure eight daily; the size of the figure eight will depend on the angle of inclination. These orbits are used for satellites whose purpose

it is to gather data on a particular area of Earth's surface or to transmit signals from one point to another. Communications satellites are geosynchronous.

GET. See Ground-elapsed time.

Gigahertz. See Hertz

Glide path: The path of descent of an aircraft under no power.

Globular clusters: Spherically shaped congregations of thousands, sometimes millions, of stars, which occur throughout the universe, although more often near elliptical galaxies than spiral galaxies such as the Milky Way. It is believed that globular clusters contain the oldest stars, and because they also contain a variety of stars of different sizes, all occurring at relatively the same distance from Earth, it is possible to learn much from them about the history of stars and the size of the Galaxy.

Gravitation: The force of attraction that exists between two bodies, such as Earth and the Moon. In 1687, Sir Isaac Newton described this force as proportional to the distance between them squared. Although gravitation is the weakest of the naturally occurring forces (the others being electromagnetic and nuclear in nature), it has the broadest range and is responsible for much celestial movement, including orbital dynamics.

Gravitational constant: The universal constant defined as the force of attraction between two bodies of 1 kilogram mass separated by 1 meter.

Gravitational field: A force field of attraction exerted around a mass, such as a celestial body.

Gravity assist: A technique, first used with the Mariner 10 probe to Mercury, whereby a spacecraft uses the gravitational and orbital energy of a planet to gain energy to achieve a trajectory toward a second destination or to return to Earth.

Great Red Spot: A vast, oval-shaped cloud system occurring at 22 degrees south latitude in Jupiter's atmosphere, rotating counterclockwise. Its name comes from an unknown substance that the convection of the phenomenon pulls to the surface; the substance absorbs violet and ultraviolet radiation and consequently delivers a red hue to the Spot. The Great Red Spot has been observed for more than three centuries.

Greenhouse effect: The heating of a planet's surface and lower atmosphere as a result of trapped infrared radiation. Such radiation becomes trapped when there is an excess of carbon dioxide in the atmosphere, which absorbs and remits infrared radiation rather than allowing it to escape. As a result, the atmosphere acts like a greenhouse, heating the planet. The effect on Earth is exacerbated by the burning of fossil fuels, which releases carbon dioxide into the atmosphere.

Ground station: A location on Earth where radio equipment is housed for receiving and sending signals to and from satellites, probes, and other spacecraft.

Ground test: To test, on Earth, craft, devices, and instruments designed for operation in space.

Ground-elapsed time (GET): The time that has elapsed since lift-off

Gyroscope: A device that uses a rapidly spinning rotor to assist in stabilization and navigation.

Hard landing: A crash landing. Early probes to the Moon, for example, were designed to take pictures of the lunar surface during free fall before impacting the surface.

Hatch: A tightly sealed door to the outside or to another module of a spacecraft.

Heatshield: A layer of material that protects a space vehicle from overheating, especially upon reentry into Earth's atmosphere. (See also Ablative heatshield.)

Heliocentric orbit: An orbit with the Sun at its center.

Heliopause: The border between the solar system and the surrounding universe, where the solar wind gives way to interstellar matter and winds.

Hertz: An SI unit of frequency, equaling one cycle per second. Multiples include kilohertz (103 hertz), megahertz (106 hertz), and gigahertz (109 hertz).

High-Earth orbit: Any Earth orbit at a relatively great distance from Earth, such as the geosynchronous orbits of telecommunications satellites.

High-gain antenna: A single-axis, strongly directional antenna that is able to receive or transmit signals at great distances.

Horizon: The line formed where land meets sky, from the perspective of an observer. In astronomy, the horizon also means the circle on the circumference of the celestial sphere that is formed by the intersection of the observer's horizontal plane with the sphere. The particle horizon is the theoretical horizon on the celestial sphere at a distance beyond which particles cannot yet have traveled.

Hubble's law: The principle, articulated in 1929 by Edwin Hubble, that the galaxies are moving away from one another at speeds proportional to their distance: that is, uniformly across time. Hubble deduced this principle from observations of the redshifts in galactic spectra. Along with the discovery of the cosmic microwave background radiation by Arno Penzias and Robert Wilson in 1965, Hubble's law forms the basis for the "big bang" theory of the expanding universe. (See also Big bang theory, Doppler effect, Redshift.)

Hypergolic fuel: Rocket fuel that ignites spontaneously when its components are mixed with an oxidizer. As a result, such a fuel can be stored in place for ignition.

ICBM. See Intercontinental ballistic missile.

IGY. See International Geophysical Year.

Imaging: The process of creating a likeness of an object by electronic means.

Impact basin: A large depression in the surface of a planet or a moon, created by the force of meteorite impact.

Inclination: See Orbital inclination.

Inertial Upper Stage (IUS): A rocket engine used on the space shuttle, which boosts satellite payloads from the shuttle's payload bay into higher Earth orbit.

Infrared astronomy: The branch of astronomy that examines the infrared emissions of stars and other celestial phenomena. Studying the infrared emissions tells astronomers much about the composition and dynamics of their sources. Because infrared rays cannot readily penetrate most of Earth's atmosphere, infrared astronomy has burgeoned with space age technology such as the Infrared Astronomical Satellite and the Kuiper Airborne Observatory.

Infrared radiation: Electromagnetic radiation of wavelengths from 1 to 1,000 micrometers, wavelengths that occur next to visible light at the red end of the electromagnetic spectrum.

Infrared scanner: An imaging instrument that is sensitive to the infrared spectrum or to heat.

Infrared spectrometer: A spectrometer that takes spectra of infrared radiation emitted by celestial bodies.

Intercontinental ballistic missile (ICBM): A ballistic missile with a range of more than 10,000 kilometers, designed to carry a warhead.

Interferometry: A data acquisition technique that uses more than one signal receiver (such as a series of radio telescopes). Signals are combined to form one highly detailed image. (See also Very long-baseline interferometer.)

Intergalactic medium: Matter that exists between galaxies. Although space between galaxies is generally transparent and apparently empty, intergalactic matter must exist, because the combined mass of the galaxies in a galaxy cluster (of which there are many) is much less than that required to exert the gravitational force that forms the cluster. Hence, not only must intergalactic matter exist, but there must be enough of it to account for the missing mass. Postulations include invisible masses, such as black holes and other forms of dead stars, as well as the intergalactic gases.

Intermediate range ballistic missile (IRBM): A ballistic missile with a range of approximately 2,800 kilometers.

International Geophysical Year (IGY): The eighteen-month period from July, 1957, to December, 1958, during which many countries cooperated in the study of Earth and the Sun's effect on it. During this time, the space age can be said to have begun with the launch of Sputnik 1 on October 4, 1957.

Interplanetary space probe. See Deep space probe.

Interstellar dust. See Cosmic dust.

Interstellar wind. See Solar wind, Stellar wind.

Ion: An atom that is not electrically balanced but rather has either more electrons than protons or more protons than electrons. Such atoms are unstable and thus in search of the particles (mission electrons or protons) that will return them to electric balance.

Ionization: The process whereby atoms are made into ions, by removal or addition of electrons or protons. Such a process often occurs as a result of excitation of atoms into an energy state whereby they lose electrons.

Ionosphere: The ionized layer of gases in Earth's atmosphere, occurring between the thermosphere (below) and the exosphere(above), between about 50 and 500 kilometers above the planet's surface. Within the ionosphere, ionized gases are maintained by the Sun's ultraviolet radiation. The resulting free electrons reflect long radio waves, making long-distance radio communication possible. Other planets are known to have ionospheres, including Jupiter, Mars, and Venus.

IRBM. See Intermediate range ballistic missile.

IUS. See Inertial Upper Stage.

K-band: A radio frequency range of about 11 to 15 gigahertz. (See also Hertz.)

Kelvin: A unit of temperature on the Kelvin temperature scale, which begins at absolute zero (-273.15° Celsius). One unit Kelvin is equal to one degree Celsius. The Kelvin scale is particularly suited to scientific (especially astronomical) measurement. (See also Absolute temperature scale.)

Kepler's law of motion: Three laws of motion discovered by Johannes Kepler and published by him in 1618-1619: (1) Each planet moves in an ellipse around the Sun, with the Sun at one of the two foci of that ellipse. (2) A line from the sun to the planet sweeps out equal areas in equal times. (3) The square of the period of a planet's orbit is proportional to the cube of its mean distance from the Sun. (See also Angular momentum.)

Kick motor: A rocket motor on a spacecraft designed to boost it, or a payload, from parking orbit into a higher orbit or a different trajectory. Also known as an apogee motor.

Kilogram: A metric unit of weight, the equivalent of 1,000 grams or approximately 2.25 pounds.

Kilometer: A metric unit of distance, the equivalent of 1,000 meters or approximately 0.62 mile.

Lander: A spacecraft or module designed to make a soft landing on the surface of a planet: it carries scientific instruments to measure surface conditions.

Laser: Originally an acronym for "light amplification by stimulated emission of radiation." A beam of infrared, visible, ultraviolet, or shorter-wavelength radiation produced by using electromagnetic radiation to excite the electrons in a suitable material to a higher energy level in their cycles around their atomic nuclei. These electrons are then stimulated in such a fashion that they jump back down to their normal energy levels. When they do, they emit a stream of "coherent" radiation: photons with the same wavelength and direction as the originating radiation. This results in a narrow, intense beam of light (or nonvisible radiation), which bounces off a reflector and directly back to the propagating material, where the process is repeated and thus the laser is maintained. Laser technology has a vast range of applications in telecommunications, medicine, and astronomical measurements.

Laser-ranging: A technique whereby scientists at two different Earth stations can determine, very precisely, their distance from each other by bouncing a laser beam off a satellite retroreflector. The time it takes to receive an "echo" allows each scientist to calculate his distance from the satellite; knowing both distances allows calculation of the distance between the two points on Earth. Over time, these measurements are repeated; changes in the distance between the two Earth locations are noted, providing much information about crustal movements and the likelihood of earthquakes.

Laser reflector: An instrument off which a scientist can "bounce," or reflect, a laser beam in order to measure (usually great) distances.

Latitude: The angular distance from a specified horizontal plane of reference on Earth, the distance north or south of the equatorial plane; in the solar system, the angular distance of a celestial body from the ecliptic plane. (See also Longitude.)

Launch: To boost a body, such as a spacecraft, from Earth into space or from one orbit into another orbit or a trajectory. Also, the act of doing so.

Launch escape system (LES): A mechanism used during the Apollo Moon missions, consisting of a solid-fueled rocket booster set atop the command module at launch. After lift-off, the LES could fire, to remove the module to safety in the event of an emergency. The mechanism proved successful at both low and high altitudes.

Launch site: A location housing a facility designed to handle preparations for launch as well as the launch itself.

Launch vehicle: See Expendable launch vehicle.

Launch window: A period of time (usually days or hours) during which conditions, such as weather and planetary alignment, are in sync for meeting the goals of a particular space mission and launch is therefore possible.

Launchpad: The physical platform from which a spacecraft is launched.

LES. See Launch escape system.

Life-support system: The combined mechanisms that maintain an environment capable of sustaining life, including devices that control air pressure, oxygen, temperature, and the like.

Lift-off: The point at which a launch vehicle or spacecraft separates from the ground or from another spacecraft on its way into space.

Light-year: The distance that light travels in one year, or approximately 9.5×10^{12} kilometers.

Limb: The outer edge of the visible disk of the Sun, Moon, planets, or other celestial body.

Liquid-fueled rocket booster: A rocket booster that uses a liquid, or cryogenic, propellant. Expendable launch vehicles from Redstone through Titan have used liquid-fueled rockets.

Lithosphere: Earth's crust and top layer of the underlying mantle, about 80 kilometers thick. The term can be used in reference to the solid part of any planet.

Long-range intercontinental ballistic missile (LRICBM): An intercontinental ballistic missile with a range greater than approximately 4,630 kilometers.

Longitude: The angular distance from a specified vertical plane of reference; on Earth, the angular distance east or west of the plane that dissects Earth through the poles at the meridian.

Look angle: Angular limits of vision.

Low-Earth orbit: Generally, any orbit at an altitude of about 300 kilometers or less. Such an orbit has a period (time required to complete on orbit around Earth) of 90 minutes or less.

LRICBM. See Long-range intercontinental ballistic missile.

LRVs. See Lunar Rover Vehicles.

Luminosity: The brightness of a celestial object.

Lunar day: The time it takes the Moon to complete one rotation on its axis, or approximately 27.33 days.

Lunar module: The portion of the Apollo spacecraft, housed in the service module during the trip to the Moon, which made a controlled descent to the Moon's surface and later lifted off to rejoin the command module. (See also Command and service module.)

Lunar Rover Vehicles (LRVs): The Moon vehicles used on Apollo missions 15, 16, and 17 to transport the astronauts several kilometers over the lunar surface. Battery-powered with four wheels and a television camera, these lunar "cars" enabled the astronauts to transmit their observations to Earth.

Mach: The ratio of the speed of a moving object to the speed of sound in the surrounding medium. At Mach 1, the speed of an aircraft equals the speed of sound.

Magellanic Clouds: The two nearest galaxies outside the Milky Way, visible from the Southern Hemisphere as the Large Cloud and the Small Cloud, respectively 160,000 and 185,000 light-years away. The Magellanic Clouds have been instrumental in establishing an extragalactic distance scale.

Magnetic field: Any force field of attraction created by the mass of a body or the combined masses of multiple bodies. Magnetic fields are responsible for much of the shape of the universe, from the orbits of planets in the solar system to the shapes of galaxies and clusters of galaxies. (See also Gravitation.)

Magnetometer: An instrument that detects disturbances in a magnetic field, within which the body's magnetic lines of force control the movement of ionized particles.

Magnetotail: A "tail" of nearly parallel lines of magnetic force extending from Earth in the direction away from the Sun.

Magnitude: The brightness of a celestial body expressed numerically. (See also Absolute magnitude, Apparent magnitude.)

Main sequence star: A star, such as the Sun, which produces energy mainly by a hydrogen-to-helium fusion reaction. Most stars spend the greater part of their lifetimes in this state.

Manned Maneuvering Unit (MMU). See Spacesuit.

Man-rating: Approval for use during a manned mission. A device that is man-rated is deemed safe for use by or around humans.

Mantle: The section of Earth between the lithosphere and the central core.

Mare (pl. maria): A large flat area on the Moon or Mars, so named (after the Latin for "sea") because these areas appear dark, thus sealike, to the Earth observer.

Maritime satellite: A satellite designed for telecommunications by and for shipping industries. These satellites occupy geostationary orbits over oceans to transmit ship-to-shore communications and data.

Mascon: One of several concentrations of mass located beneath lunar maria, which causes a distortion in the orbit of a spacecraft around the Moon.

Maser: An acronym for "microwave amplification by stimulated emission of radiation." A device similar to a laser in which energy is generated as in a laser, but at microwave levels. A maser can exist in nature as a celestial object. Artificial masers are used to amplify weak radio signals. (See also Laser.)

Mass: The amount of matter contained within a body, which determines the amount of gravitation force it exerts. Mass is measured in such units as kilograms and pounds; it differs from weight, however, which is the force exerted on a mass by gravity.

Mass spectrometer: An instrument that identifies the chemi-

cal composition of a substance by separating ions by mass and charge.

Materials processing: The manufacture of crystals and other materials in the microgravity environment of a spacecraft, whereby uniform crystal growth and other processes that are difficult on Earth can be accomplished to improve space technology and for industrial applications.

Matter: A substance that has mass and occupies space, which along with energy is responsible for all observable phenomena.

Maunder minimum: Named for E.W. Maunder, who in 1890 discovered a period in the 300-year history of sunspot observations when few sunspots were recorded. Confirmed independently in 1976 by evidence from tree rings, the Maunder minimum covers the years 1645 to 1715, a period also known as the Northern Hemisphere's "Little Ice Age."

Megahertz. See Hertz.

Mesosphere: The layer of Earth's atmosphere occurring above the stratosphere and below the thermosphere, from about 40 kilometers to 85 kilometers above sea level. This is the coldest layer of the atmosphere.

MET (Mission-elapsed time). See Ground-elapsed time.

Meteor: A streak of light in Earth's upper atmosphere caused by the burning of a meteoroid.

Meteorite: A meteoroid that does not burn completely and reaches Earth.

Meteoroid: A particle of interplanetary dust greater than 0.1 millimeter which enters Earth's atmosphere and burns as a result of friction, creating a "shooting star."

Meteorological satellite: A satellite that collects data on weather systems for forecast and other analysis.

Meter: The metric unit of length, equivalent to approximately 39.37 inches, or a little more than 1 yard.

Metric system: The decimal system of weights and measures, which forms part of the *Système International d'Unites*. (See also SI units.)

Metric ton: A metric unit of weight equivalent to about one short ton (2,205 pounds). Just as thrust is often measured in pounds in the United States, metric tons are often used as units of thrust in Russia and other countries. (See also Newton, Pound.)

Microgravity: Nearly zero gravity. Microgravity exists in a space vehicle because of the minute gravitational forces exerted by objects on one another. The microgravity environment is of great importance as an ideal environment for certain types of materials processing.

Micrometeorite: A micrometeroid that has reached Earth's surface.

Micrometeoroid: A meteoroid with a diameter of less than 0.1 millimeter. Because of their size, micrometeoroids rarely burn up but reach Earth's surface instead, as spherules or as cosmic dust particles.

Micropaleontology: The study of microscopic fossils, of potential importance in exobiology as well as life sciences on Earth.

Microwaves: A form of electromagnetic radiation with wavelengths ranging between 1 millimeter and 30 centimeters, located between infrared and long-wave radio on the electromagnetic spectrum.

Milky Way: The galaxy in which our solar system is located, of the spiral variety, containing about 1011 stars and about 100,000 light-years across. (See also Galaxy).

Mission-elapsed time (MET): See Ground-elapsed time.

Mission specialist: An astronaut who has overall responsibility for a mission payload.

MLR. See Monodisperse latex reactor.

MMU. See Spacesuit.

Molecule: The smallest unit of a substance, formed by a characteristic complex of atoms joined together. The smallest unit of the substance water, for example, is a molecule formed by two hydrogen atoms and one oxygen atom.

Monodisperse latex reactor (MLR): A device designed to develop monodisperse, or identically sized, beadlike rubber particles for use in medical and industrial research.

Moon: Any natural satellite orbiting a planet, especially Earth's Moon.

MSS. See Multispectral scanner.

Multispectral scanner (MSS): A type of radiometer, used on such satellites as Landsat, which produces detailed false-color images of a planet's surface.

Nanometer: One thousand millionth of a meter; a unit used to express electromagnetic wavelengths.

Navigation satellite: A satellite which provides positional information for any moving object on land, sea, or in the air, including inner space.

Navsat: An abbreviation for "navigation satellite."

Near-Earth space: Roughly defined as the space environment from the outer reaches of Earth's atmosphere to the path of the Moon's orbit, the area beyond which is known as deep space.

Nebula: A celestial body composed of aggregated gas and dust, which may be either luminous, reflecting or emitting light under the influence of nearby stars (an emission nebula), or dark, obscuring the light of distant stars and appearing as a silhouette.

Neutral gas analyzer: An instrument that determines the chemical composition of the atmosphere.

Neutrino: An elementary particle of enormous penetrating power as a result of its lack of electric charge and its nearly total lack of mass. Traveling directly out from the cores of stars as a by-product of nuclear reactions, neutrinos have enormous potential as a source of information on the stars and other astrophysical phenomena.

Neutron: An uncharged elementary particle found in atomic nuclei; its mass is approximately equal to that of a proton.

Neutron stars: The smallest stars known, with diameters of about 20 kilometers and densities matching that of the Sun, consisting of a thin iron shell enclosing a liquid sea of neutrons. The properties of neutron stars are beyond scientists' complete understanding, but they are thought to originate from main sequence stars much larger than the sun, which become supernovae. Rapidly spinning neutron stars are observable as pulsars.

New astronomy: A term used collectively to refer to the areas of astronomy (such as gamma-ray astronomy, infrared astronomy, and X-ray astronomy) investigating electromagnetic emissions by celestial phenomena. The application of space technology and electronics to the development of instruments capable of detecting such data has greatly increased astronomers' understanding of the universe.

Newton: An SI unit of force used to measure thrust.

Northern lights. See Aurora.

Nose cone: The conically shaped front end of a launch stack, missile, or other spacecraft, built for aerodynamic efficiency and as a protective shield.

Nova: A star which emits a sudden radiation of light and quickly (over months or years) returns to its former brightness. (See also Supernova.)

Nuclear energy: Energy that is released as a result of interactions between elementary particles and atomic nuclei.

Nuclear reactor: A device, usually located at a nuclear power station, designed to contain nuclear fission reactions during the production of nuclear energy.

Nucleus: The central part of an atom, around which electrons rotate. An atomic nucleus can consist of one proton (in a hydrogen atom) or many protons and neutrons (as in an atom of uranium).

Oblate: Flatten at the poles.

Occultation: The obscuring of one celestial body by another, such as occurs during a solar eclipse.

OMS. See Orbital maneuvering system.

Oort Cloud: A theoretical cloud of millions of comets orbiting the Sun between 30,000 and 100,000 astronomical units from the Sun, postulated by Dutch astronomer Jan Oort in 1950.

Opposition: The alignment of Sun, Earth, and a superior planet (one whose orbit is farther from the Sun than Earth's) in a straight line; that is, the superior planet appears in the sky at 180 degrees celestial longitude from the Sun. In this position, the planet is closest to Earth and therefore most easily observed by ground-based instruments.

Orbit: The path traced out by one celestial or artificial body as it moves around another that exerts greater gravitational force. The distinguishing characteristics of an orbit are called its orbital parameters and include apoapsis, periapsis, inclination to the ecliptic of the body orbited, eccentricity, and period. All orbits trace out an ellipse, of which the body orbited forms at least one of two foci. (See also Apoapsis, Circular orbit, Eccentricity, Ellipse, Elliptical orbit, Equatorial orbit, Orbital inclination, Parabolic orbit, Parking orbit, Periapsis, Period, Polar orbit, Prograde orbit, Retrograde orbit, Synchronous orbit, Transfer orbit.)

Orbital inclination: The angle formed between the orbital plane of a satellite and the equatorial plane of the object orbited.

Orbital maneuvering system (OMS): A system of rocket engines, located on the space shuttle's aft fuselage, which pro-

vide small amounts of thrust for fine maneuvers in orbit.

Orbital transfer vehicle (OTV): A liquid-fueled thrusting mechanism designed to boost satellites into orbit from the Space Station. The OTV will be housed at the Space Station for regular use.

Orbiter: A spacecraft intended to orbit, rather than land on, a planet or other celestial body, often used to relay signals from a lander to Earth. (See also Lander.)

O-ring: A ringed rubber gasket, one-quarter inch in width, which acts as a sealant between the bottom and next to bottom segments of a shuttle solid-fueled rocket booster. Failure of an O-ring or the insulating putty which surrounds it can allow combustible gases to escape through the joint from within the rocket, as apparently occurred to cause the *Challenger* accident of January, 1986.

OTV. See Orbital transfer vehicle.

Outer Space: All space beyond Earth's atmosphere. (See also Deep space.)

Outgassing: The process whereby gases are emitted from solids into a vacuum, referring mainly to the exudation of gases from terrestrial bodies into space, a remnant of the way these bodies were formed.

Ozone layer: The thin layer of Earth's atmosphere, located between 12 and 50 kilometers above Earth's surface (in the stratosphere), in which ozone (O^3) is found in its greatest concentrations. This layer, which absorbs most of the ultraviolet radiation entering the atmosphere, forms a protective blanket around the planet, shielding it from excess radiation.

PAM. See Payload Assist Module.

Panspermia: A theory proposed by chemist Svante Arrhenius in 1906, and later modified by Sir Fred Hoyle, which holds that organic molecules (hence the beginnings of life) were transported to Earth by means of comets. The organic material found in Halley's comet supports the theory, which is further supported by the argument that each process on the road to a life-form is so improbable as to require a greater combination of conditions than those possible on Earth alone. Nevertheless, the theory is far from accepted by most scientists.

Parabolic orbit: An orbit that describes a parabola around the object orbited and hence escapes from the gravitational field of that object. A comet's orbit "around" the Sun is parabolic.

Parallax: The apparent displacement of a celestial object as seen from two different points on Earth. Knowing this angle allows astronomers to calculate the object's distance from Earth.

Parking orbit: An interim orbit around a celestial body between launch and injection into another orbit or into a trajectory toward another destination.

Parsec: A unit for measuring astronomical distances equivalent to 3.26 light-years.

Particle. See Elementary particles.

Passive experiment: An experiment package which requires only exposure to the space environment to perform its investigations.

Passive relay satellite: An early telecommunications satellite, such as Echo, which relayed radio signals from one point to another by bouncing them off its surface.

Payload: Any experiment package, satellite, or other special cargo carried into space by a spacecraft.

Payload Assist Module (PAM): A solid-fueled rocket engine designed to boost satellites into geostationary orbit from the space shuttle's payload bay.

Payload bay: The portion of the space shuttle that carries satellites into space, approximately 18 by 4 meters. Hinged doors at the top of the fuselage open to expose the satellite, which can then be boosted into orbit from the bay by a Payload Assist Module.

Payload specialist: An astronaut who assists a mission specialist in conducting an experiment aboard the space shuttle.

Periapsis: The point in one object's orbit around another at which the orbiting object is closest to the object being orbited.

Pericynthion: The point in an object's orbit around the Moon at which it is closest to the Moon.

Perigee: The point in an object's orbit around Earth at which it is closest to Earth.

Perihelion: The point in a solar orbit at which the orbiting object is closest to the Sun.

Perilune: Pericynthion of an artificial satellite.

Period: The time span between repetitions of a cyclic event. An orbital period is the time required for a satellite or moon to make one complete orbit around a planet, a moon, the Sun, or another celestial body.

Photometer: An instrument that measures the brightness of a light source.

Photon: The smallest theoretical quantity of radiation, visualized as both wavelength energy and an elementary particle.

Photomultiplier: An instrument for increasing the apparent brightness or strength of a source of light by means of secondary excitation of electrons; effectively, a light (or other radiation) amplifier.

Photopolarimeter: An instrument for producing an image of a celestial body (or other light source) by means of polarized light.

Photoreconnaissance: The gathering of information, especially on enemy installations, by means of photography from the air or from space.

Photosphere: The region of the Sun that separates its exterior (the chromosphere and corona) from its interior, forming the boundary between the transparent and opaque gases. The photosphere appears as the bright central disk form Earth, and it is the source of most of the Sun's light.

Photovoltaic cell: A solid state energy device that converts sunlight into electricity.

Pitch, roll, and yaw: Movements that a spacecraft undergoes as a result of launch or other stresses. Pitch is up-down movement; roll is longitudinal rotation; yaw is side-to-side movement.

Pixel: A small unit arranged with others in a two-dimensional array which contains a discrete portion of an image (as on a television screen) or an electrical charge (as on a charge coupled device). Together these pixels form an image or other meaningful information.

Planets: A planet is a nonluminous natural celestial body that orbits the Sun (or another star) and is not categorized as an asteroid or comet. There are nine known planets in the solar system: Mercury, Venus, Earth, Mars. Jupiter, Saturn, Uranus, Neptune, and Pluto.

Plasma: Ionized gas, consisting of roughly equal numbers of free electrons and positive ions. Plasma forms the atmospheres of stars, interstellar and intergalactic matter, nebulae—in fact, most of the matter in the universe. The extremely high excitation of its constituent particles has earned for it the label "fourth state of matter," after solid, liquid, and gas.

Plasma sheath: The definite outer boundary of Earth's ionosphere, identified by Orbiting Geophysical Observatory 1.

Plate tectonics: The study of the continental drift, seafloor spreading, and other dynamics of Earth's lithosphere, which is divided into seven major sections, or "plates."

Pogo effect: The up-and-down vibrating motion that occurs during the launch of a spacecraft.

Polar orbit: An orbit in which a satellite passes over a planet's or moon poles.

Polarimeter: An instrument for measuring the degree to which electromagnetic radiation is polarized.

Pole: One of two points on the surface of a planet where it is intersected by its axis of rotation. In a magnetic field, one of two or more points of concentration of the lines of magnetic force.

Posigrade: Moving in the direction of travel.

Pound: A unit used to measure thrust in the United States and some other English-speaking nations.

Precession: A type of motion that occurs in a rotating body in response to torque: A planet or other rotating body orbiting around a gravitational force, such as the Sun, slowly turns in the direction of its rotation so that, over a long period, each of the planet's poles describes a circle. The fact that precession is exhibited by many planets and moon means that adjustments must be made in the locations to which astronomers look to observe stars and other celestial phenomena. Earth's period of precession is approximately 25,800 years.

Probe. See Deep space probe.

Prograde orbit: An orbit that moves in the same direction as the rotation of the body orbited.

Propellant: See Cryogenic fuels, Hypergolic fuel, Solid propellant.

Propulsion system: The combined mechanisms that propel a space vehicle, including engines and fuel systems.

Proton: An elementary particle of mater that carries positive charge and forms the nucleus of the hydrogen atom, as well as the nuclei of other chemical elements in combination with protons and neutrons.

Pulsar: A rapidly spinning neutron star that emits a narrow beam of electromagnetic radiation in the form of visible and radio waves (single pulsars), as well as X rays and gamma rays (pulsars occurring in binary star systems). The regular emission of radio waves can be detected on Earth by radio tele-

scopes, and the pulsar's distance from Earth can be detected by the difference in the arrival time of radio waves at different wavelengths.

Quarks: Subparticles hypothesized to form the known elementary particles (electrons, protons, neutrons, and their antiparticles), characterized by electric charge, "flavor," and "color." The forces required to break elementary particles into their component quarks is so great that quarks do not exist as free particles in nature, although it is thought that neutron stars may consist of a "quark soup" within a solid iron shell.

Quasar: An acronym for "quasi-stellar" or "quasi-stellar object." An object continuously releasing a tremendous amount of energy, equivalent to the output of between one million and 100 trillion suns, which includes virtually all kinds of electromagnetic radiation (gamma rays, X rays, ultraviolet, optical, and infrared radiation, microwaves and radio waves) from a very small volume of space about the size of the solar system. As far as is known, all objects satisfying these criteria are located in the nuclei of galaxies, although it is thought that they may be galaxies themselves, with black holes at their centers. Discovered in 1963, the first quasar caused much excitement among astronomers, and these phenomena continue to be among the most fascinating and mysterious in the universe.

Radar: An acronym for "radio detection and ranging." A means of locating and determining the distance of objects by bouncing radio waves off them and measuring the time required to receive the echo.

Radiation. See Electromagnetic radiation.

Radio astronomy: The branch of astronomy that examines the radio emissions of celestial objects. Because radio radiation, along with visible radiation, can penetrate Earth's atmosphere, radio receivers have provided much of the data detectable by ground-based, as well as space-based, instrument. Radio emissions also form a significant portion of certain celestial phenomena, such as radio galaxies, quasar, and pulsars.

Radio telescope: A radio receiving aerial, either a dish or a dipole, connected to recording devices, whereby distant radio emissions can be detected.

Radioisotope thermoelectric generator (RTG): A device for creating power from a radioactive substance, used on spacecraft to supplement the power generated by the solar-energy-collecting solar panels. The RTG is especially important when access to sunlight is weak or nonexistent.

Radiometer: An instrument, used by meteorological and

other satellites, that measures Earth's infrared and reflected solar radiance and uses small, selected wavelengths (ultraviolet to microwave) to measure temperature, ozone, and water vapor in the atmosphere. Radiometers are sensitive to one or more wavelength bands in the visible and invisible ranges. If visible wavelengths are used, the satellite can detect cloud vistas from reflected sunlight, resulting in a slightly blurred version of those images photographed directly by astronauts. Using the invisible range, satellites can capture terrestrial radiation, producing images from Earth's radiant energy.

RBV. See Return beam vidicon.

Real time: Referring to the transmission of signals or other data at the same time that they are used.

Reconnaissance satellite: A satellite that gathers in information about enemy military installations

Red giant: Stars with surface temperatures less than 4,700 Kelvins and between 10 and 100 times the diameter of the Sun.

Red Planet: Mars, so named because of its color seen through Earth-based telescopes.

Redshift: The apparent lengthening of electromagnetic wavelengths issuing from a celestial object or other source as a result of the object's movement away from the observer. As a result, the spectral lines in the spectra of such an object will shift toward the red end of the electromagnetic spectrum. (See also Doppler effect.)

Reentry: The return of a spacecraft into Earth's atmosphere.

Reflecting telescope: An optical telescope that uses a mirror or mirrors to capture, magnify, and focus light from the object observed. These telescopes, such as the 200-inch reflecting telescope on Palomar Mountain in Southern California, are widely used for Earth-based optical astronomy. (See also Cassegrain telescope, Refracting telescope.)

Refracting telescope: An optical telescope that uses a lens to magnify and focus light from the object observed. Refracting telescopes were used by the earliest astronomers. When reflecting telescopes were perfected in the twentieth century, refracting telescopes became less important in astronomy, although they are still widely used for guided and amateur observations. (See also Reflecting telescope.)

Regolith: A thick layer of broken rock that overlies the surface of a moon or planet, caused by the impact of a meteoroid.

Relativity: The physical law, first proposed by Albert Einstein, that states that measurements of time and space are dependent upon the frame of reference in which they are measured. The general theory of relativity applies this law to gravity and mass; the special theory of relativity applies it to the propagation of electric and magnetic phenomena in space and time.

Remote manipulator system (RMS): The space shuttle's 15-meter-long robot arm. Operated from within the shuttle by an astronaut, the arm duplicates the operator's hand and wrist movements, allowing payloads to be moved and repairs to be made without extravehicular activity. Because the RMS was manufactured by Canadian aerospace company, it is sometimes called the "Canada arm."

Remote sensing: Acquiring data at a distance by electronic or mechanical means.

Rendezvous: The planned meeting of two spacecraft in orbit and often their maneuvering into proximity of each other in preparation for docking.

Resolution: The degree to which a photographic or other imaging system, or the image produced, clearly distinguishes objects of a certain size. In a photograph with a resolution of 200 meters, for example, the smallest distinguishable objects are 200 meters across.

Restart: Reignition of a rocket engine after it has been inactive during orbit.

Retrofire: The firing of a rocket to slow down or change the orbit of a spacecraft.

Retrograde orbit: An orbit that moves opposite to the rotational direction of the body orbited.

Retroreflector: A device, carried by a satellite, used to reflect laser beams directed at the satellite from Earth. (See also Laser ranging.)

Retro-rocket: A rocket that exerts thrust to slow down or change the orbit of a spacecraft.

Return beam vidicon (RBV): A camera, used by Earth resources satellites, that takes very high-resolution photographs of the planet's surface from space.

Revolution: One complete orbit of a planet around the Sun or of a natural or artificial satellite around another celestial body.

Rille: A long, narrow valley on the Moon.

RMS. See Remote manipulator system.

Robot arm: A mechanical arm extending from many space-craft which can be remotely controlled to manipulate instruments and repair equipment. (See also Remote manipulator system.)

Robotics: The development, construction, and use of computerized machines to replace humans in a variety of tasks requiring precise "hand-eye" coordination.

Roche limit: Named for Edouard Roche, who discovered it in 1848, the minimum distance from a planet at which a natural satellite can form by accretion: roughly 2.44 times the planet's radius. Within this limit an existing satellite will be torn apart by gravitational stresses. Saturn's rings, which lie within the planet's Roche limit, may be the remnants of a former moon.

Rocket booster: A propulsion engine used to launch a spacecraft into orbit from Earth or into a different orbit or trajectory form space.

Roll. See Pitch, roll, and yaw.

Rollout: The termination of a flight, occurring after touchdown on the landing site and before brakes are set, during which an aircraft or space shuttle rolls to decrease speed.

RTG. See Radioisotope thermoelectric generator.

Sample return mission: A mission designated to collect soil and rock samples from another body and return them to Earth.

Satellite: Any body that orbits another of larger mass, usually a planet. Satellites include moons, the small bodies that form planetary rings, and man-made satellites. (See also Artificial satellites.)

Scanning radiometer: An instrument used on meteorological satellites to measure radiation emitted by the atmosphere, especially in the infrared region, building up a picture of atmospheric conditions and temperature that can be used in forecasting.

Scarp: A broken slope or a line of cliffs caused by a fault line or by erosion.

Schwarzschild radius: The radius of a collapsing mass, such as a degenerating star, at which it becomes a black hole—that is, at which its gravitational force will not allow light to escape. The length of this radius depends on the body's mass, and the formula for calculating it was established by Karl Schwarzschild in 1916.

Scientific satellite: A broad term for any satellite dedicated primarily, if not solely, to collecting scientific data, especially data on astrophysical phenomena. Although most satellites can be described as "scientific" in some sense, the main purpose of a scientific satellite is to broaden our knowledge of the universe, rather then serve a practical (e.g.,telecommunications), military (e.g., reconnaissance), or commercial purpose.

Seismic activity: Any movement in the outer layer of a planet or moon.

Seismometer: A sensitive electronic instrument that measures movements in the outer layer of a planet or moon. The graphs produced by this instrument can be interpreted—on the Richter scale, generally—to determine the magnitude and intensity of seismic activity.

Selenography: The study of the lunar surface features; the counterpart of geography on Earth.

Selenology: The study of the Moon, analogous to geology on Earth.

Service propulsion system: A large rocket engine which propelled the Apollo command and service module, slowing it as it neared the Moon, and later sending it back on a trajectory toward Earth.

Shuttle imaging radar (SIR): A high-resolution radar system, used aboard the space shuttle, which operates at frequencies high enough to penetrate not only Earth's atmosphere but at times a few feet beneath its surface as well.

Shroud: A heat-resistant covering used to protect a spacecraft, payload, or missile, especially during launch.

SI units: The collective units of measurment used in the *Système International d'Unites*, the system of measurement most widely accepted by scientists. Its fundamental, or base units are seven: the meter (the base unit of length), kilogram (mass), second (time), ampere (electric current), Kelvin (temperature), mole (amount of substance), and candela (luminosity). From these seven base units, other units are derived, which are multiples, fractions, or powers of the base units such as the kilometer (1 meter x 10_3) and the square meter (the unit of area). Further derived units are derived from combinations of the base units and have their own names: hertz (the unit of frequency, which is cycles per second), newton (force or thrust, kilogram-meters per second squared), pascal (pressure,newtons per square meter), joule (energy, the kilogram-meter), watt (power, joules per second), coulomb (quantity of electricity, the ampere-second), volt (electric potential, watts per ampere), farad (capacitance, or the ability to store energy, coulombs per volt), and ohm (electrical resistance, volts per ampere). In the United States, the base SI units are coming into increasing use. Some measures, however, remain more familiarly rendered by English units of measure, even in scientific use: It is common, for example, to refer to rocket thrust in pounds or even metric tons rather than newtons, and atmospheric pressure is often measured in pounds per square inch (or bars and millibars in the centimeter-gram-second system).

SIR. See Shuttle imaging radar.

Soft landing: A controlled landing on a planet's or moon's surface, designed to minimize damage to the spacecraft and its instrument payload.

Solar array: An assembly of solar cells, as on a solar panel extending from a satellite.

Solar cell: A photovoltaic device that converts solar energy directly into electricity for use in powering a spacecraft.

Solar constant: The amount of solar energy received by a square meter per second on Earth (or one astronomical unit from the Sun), approximately 1,370 watts per square meter per second.

Solar cycle: A period of approximately eleven years during which the number of sunspots visible near the Sun's equator increases to a maximum and then decreases. Other solar activity follows the solar cycle. (See also Sunspots.)

Solar flare: A large arc of charged particles and electromagnetic radiation ejected from the Sun's surface (in the low corona and upper chromosphere) and lasting from a few minutes to several hours. Solar flare activity affects radio transmission on Earth and can produce auroras in Earth's atmosphere.

Solar mass: A unit equivalent to the mass of the Sun, or $1,989 \times 10^{33}$ grams. Masses of other stars are sometimes given in solar masses.

Solar system: The Sun, the planets, asteroids, comets, and other matter that orbit the Sun, and the satellites of those bodies, along with interplanetary space, radiation, and gasses. (See also Planets.)

Solar wind: The hot ionized gases, or plasma, that escape the Sun's gravitational field and flow in spirals outward at about 200 to 900 kilometers per second. It consists primarily of free protons, electrons, and alpha particles escaping from the Sun's corona. (See also Stellar wind.)

Solid propellant: Rocket propellant in solid form: cast, extruded, granular, powder, or other.

Solid-fueled rocket booster (SRB): One of two rocket boosters, fueled by solid propellants, which assists the main engines of the space shuttle during the first two minutes of ascent. The SRBs then separate from the external tank; their descent is slowed by drogue chutes issuing from nose caps, the SRBs splash down, and they are recovered for later reuse. (See also External tank.)

Sonar: An acronym for "sound navigation ranging." A system for bouncing sonic and supersonic waves off a submerged object in order to determine its distance.

Sounding rocket: A suborbital rocket carrying scientific instruments which take measurements of Earth's atmosphere. Sounding rockets were used before satellites came into prominence.

Sounding sensor: A sonarlike device that probes the atmosphere to detect data about temperature, moisture, and other conditions.

Space adaptation syndrome. See Spacesickness.

Space age: The age of space exploration, whose beginning is generally dated from October 4, 1957, the day on which the first artificial satellite, Sputnik 1, was launched into Earth orbit.

Space capsule: A small, manned or unmanned, spacecraft, such as those used on the Mercury and Gemini missions, which is pressurized and otherwise environmentally controlled.

Space center: A complex that houses a variety of facilities for development of space technology and preparations for or monitoring of space missions, such as launch facilities.

Space medicine: The study of human health in microgravity and other conditions surrounding space travel, including systems for maintaining health.

Space race: A term applied primarily to the early year of space exploration (1957 through the 1960's), during the Cold War between the United States and the Soviet Union. Space "firsts" were a preoccupation of the space programs of both countries, becoming a matter of public concern and national pride, culminating in the race to place the first man on the Moon—achieved in 1969 by the United States. The achievements of both nations in space have been impressive; it is arguably impossible to quantify them in any meaningful fashion.

Space shuttle main engine (SSME): A reusable, liquid-fueled engine which generates 70,000 horsepower. Three SSMEs are clustered in the tail of the U.S. space shuttle, supplying a total thrust of about 5 kilonewtons, somewhat less than the F-1 engines of a Saturn 5.

Space station: A large, Earth-orbiting structure designed to provide an environment in which atmospheric and gravitational conditions on Earth can be duplicated for long-term habitation. Space stations will serve a variety of purposes: as way stations on the way to Mars and other destinations, as materials processing plants, as sites for scientific and biological studies. Skylab, Spacelab, and Salyut were early versions of the Mir and international space stations, which in turn are prototypes of the larger space stations described above.

Spacecraft: Any self-contained, manned or unmanned, space vehicle; more specifically, a deep space probe.

Spacenaut: Any space traveler; specifically, one who is neither an American nor a Russian.

Spacesickness: Any health problem experienced as a result of space travel, which varies with individual and spacecraft environment. Specifically, however, spacesickness is the nausea and other symptoms associated with adjustment to microgravity. Also know as space adaptation syndrome.

Spacesuit: The pressurized garment worn by an astronaut during extravehicular activity, and sometimes within a spacecraft. The spacesuit worn by space shuttle astronauts outside the spacecraft, called an extravehicular mobility unit, incorporates a portable life-support system, a manned maneuvering unit, a displays and controls module, a liquid cooling and ventilation garment, a urine collection device, and a delivery system for drinking water.

Spacewalk: See Extravehicular activity.

Spectrograph: A type of spectrometer that splits light into its component wavelengths and records the separated wavelengths photographically or by means of a charge coupled device.

Spectrometer: An instrument that splits electromagnetic radiation into its component wavelengths for viewing or electronic recording of the emitting body's spectrum.

Spectroscope: A device that splits electromagnetic radiation into its component wavelengths, which can then be "read" by a spectrometer or a spectrograph.

Spectroscopy: The creation and interpretation of spectra using a variety of instruments, including spectrographs and a variety of spectrometers. By analyzing the spectra of celestial bodies, scientists are able to discover much about their composition and the chemical reactions taking place in them.

Spectrum: An image that represents the distribution and intensity of electromagnetic radiation from a body. This image can be photographic or a "map" of lines showing all or selected wavelengths.

Spin axis: The line around which a body rotates.

Spin stabilization: The method whereby an artificial satellite is made to spin at a constant rate about a symmetry axis, relying on gyroscopic effects to keep that axis relatively fixed in space.

Spiral galaxy: A galaxy consisting of a bulge of gas and stars at the center, around which "arms" of stars and other celestial bodies, matter, and radiation rotate in a spiral fashion. Spiral galaxies, of which the Milky Was is one, are by far the most common in the universe.

Splashdown: The free-fall landing of a space capsule or other spacecraft in one of Earth's oceans.

SRB. See Solid-fueled rocket booster.

SSME. See Space shuttle main engine.

Stage: A self-contained section of a space vehicle, used for propulsion of a spacecraft into orbit and separated from the launch stack after use.

Standby: A piece of equipment available to replace its counterpart on short notice.

Star: A large, nearly spherical mass of extremely hot gas bound together by gravity.

Star tracker: An electronic device programmed to detect and lock onto a celestial body, such as the star Canopus, to provide a spacecraft with a fixed point of reference for purposes of navigation.

Steady state theory: A model of the universe, proposed by Hermann Bondi, Thomas Gold, and Fred Hoyle, which posits that the density of matter in the universe remains constant in an expanding universe, being created at the same rate as that at which old stars die. The theory is less widely subscribed to than the big bang theory, because it does not explain the presence of the cosmic microwave background radiation. (See also Big bang theory.)

Stellar wind: The ionized gases, or plasma, that flow out from stars at high speeds, composed mainly of free protons and electrons. (See also Solar wind.)

Stratosphere: The layer of Earth's atmosphere between the troposphere and the mesosphere, extending from about 15 to 50 kilometers above Earth's surface, roughly coinciding with the ozone layer, which absorbs the Sun's ultraviolet radiation and heats the stratosphere from a low of about -60 Celsius at the bottom to about 0 Celsius at its top. There is no meteorological activity or vertical air movement in this region of the atmosphere.

Suborbital flight: A spaceflight comprising less than one orbit of Earth or not intended to reach orbit.

Subsatellite: A satellite carried into orbit by another satellite. Also, a satellite of a moon.

Sunspots: Dark spots that appear on the Sun's surface in cycles, increasing and decreasing with the eleven-year solar cycle. These dark spots are about 500 Kelvins cooler than a surrounding lighter area of the spot, which in turn is another 500 Kelvins cooler than the surrounding photosphere. These cooler regions result from magnetic fields.

Supergiant: The brightest stars in the universe, which also include the largest: Red supergiants are 1,000 times the size of the Sun.

Supernova: A nova resulting from the explosion of any star whose mass is 1.4 times the Sun's. (See also Nova.)

Surveillance: Ongoing monitoring of enemy territory or military installations by aircraft or satellites carrying any of a variety of sensors.

Swing-by: The close approach of a spacecraft as it passes a planet on a tour of the solar system. (See also Flyby.)

Synchronous orbit: Any orbit whose period equals the rotational period of the object orbited.

Système International d'Unites. See SI units.

Tectonics: The branch of geology that examines folding and faulting in Earth's crust. (See also Plate tectonics.)

Telecommunications satellite: An artificial satellite dedicated to the receiving and transmission of radio, television, and other communications signals. Such satellites are usually placed in geostationary orbits so that they remain over a fixed point on Earth.

Telemetry: Real-time transmission of data from a distance via radio signals.

Teleoperations: Manipulation of an orbiter, booster, or instruments in space from Earth via remote control.

Terminator: The line, on a planet or moon, between dark and light that forms the boundary between day and night.

Terrestrial planets: In addition to Earth, Mercury, Venus, and Mars, so called because they resemble Earth in certain fundamental features such as density and composition.

Test flight: Experimental operation of an aircraft or spacecraft to determine whether it functions as designed and to identify systems in need of adjustment.

Test range: A site dedicated to the testing of aircraft or spacecraft.

Thermal mapping: Gathering data from which to construct maps by means of instruments capable of sensing heat-producing electromagnetic radiation.

Thermal tiles: Heat-resistant tiles glued to the underside of the space shuttle to protect it from overheating upon reentry into Earth's atmosphere. They are favored over ablative material because they are lighter.

Thermosphere: The highest layer of Earth's atmosphere except for the exosphere, beginning at 85 kilometers above sea level. The oxygen and nitrogen that compose the atmosphere at this level are extremely rarefied, and are heated by the Sun's ultraviolet radiation to the point of ionization; hence the ionosphere (which lies between 50 and 500 kilometers above sea level) roughly coincides with the thermosphere. This is also the region in which auroras and meteors occur.

Three-axis stabilization: Stabilization of a satellite against pitch, roll, and yaw. (See also Pitch, roll, and yaw.)

Thrust: The force required to propel a vehicle, especially that force exerted by a rocket engine to launch a space vehicle into orbit or a deep space trajectory. Measured in newtons (SI), metric tons, or pounds.

Time dilation: The phenomenon, predicted by Albert Einstein's special theory of relativity, whereby time appears to slow down in a system moving near the speed of light from the vantage point of an observer outside that system.

Topside observation: Electronic scanning of Earth's (or another) atmosphere from above. Used mainly in reference to meteorological satellites.

Tracking network: A network of tracking stations at different points on the globe which send and receive radio signals to and from spacecraft, via large dish antennae, allowing continuous communications with spacecraft. Examples are the Spaceflight Tracking and Data Network and the Deep Space Network.

Trajectory: The path traced out by a ballistic missile or by a spacecraft launched from Earth or from orbit toward the Moon or other destination.

Transducer: Any device that transforms one type of energy into another, such as a solar cell (sunlight into electrical power), a thermocouple (thermal energy into electric signal), or a Geiger counter (radioactive into sound energy).

Trans-Earth injection: A boost from lunar (or other planetary) orbit which places a spacecraft on a trajectory toward Earth.

Transfer orbit: The orbit into which a spacecraft is boosted from Earth orbit on its way to orbit around another celestial body. Since the spacecraft does not complete a full revolution of the transfer orbit, but only part of the ellipse, the path it follows describes a trajectory which intersects the final orbit.

Transit: The passage of one celestial body across the face of another or across the observer's meridian.

Translunar injection: The process whereby a spacecraft in orbit around Earth is boosted into a trajectory that heads it toward the Moon.

Transponder: A device that receives radio signals and automatically responds to them using the same frequency.

Triangulation: A means of determining the position of an object by calculation from known quantities: The distance between two fixed points and the angles formed between the line described by those points and the line between a third point. Triangulation is the oldest method of determining distances, both on Earth and in space.

Troposphere: The layer of Earth's atmosphere that lies closest to Earth's surface, extending upward to about 8 kilometers. The troposphere is the densest region of the atmosphere and the region in which all meteorological phenomena occur.

Ultraviolet astronomy: The branch of astronomy that examines the ultraviolet emissions of celestial phenomena. Ultraviolet astronomy has developed with the advent of the space age; since ultraviolet rays are unable to penetrate Earth's atmosphere, instruments carried aloft by satellites such as the International Ultraviolet Explorer have enabled scientists to learn much about celestial bodies, since many of the elements of which they are composed are most evident in the spectroscopic measurements taken of them.

Ultraviolet radiation: Electromagnetic radiation which is emitted between the wavelengths of 900 and 3,000 angstroms (between 400 and 2 nanometers), which is the band lying between visible violet light and X-radiation on the electromagnetic spectrum.

Ultraviolet spectrometer: An instrument that measures electromagnetic wavelengths in the ultraviolet range, used on satellites such as the International Ultraviolet Explorer and the Extreme Ultraviolet Explorer.

Uplink: Signals sent up to a satellite from an Earth station.

Vacuum: An area in which absolutely nothing exists. Although a true vacuum never occurs in nature, the behavior of bodies within a vacuum is of concern to physicists studying interplanetary and deep space, in which near-vacuum conditions exist.

Van Allen radiation belts: The two layers of Earth's magnetosphere, discovered by James Van Allen in the late 1950's, in which ionized particles spiral back and forth between Earth's magnetic poles. These zones are of importance for the potential hazards they pose to electronic instruments aboard spacecraft.

Variable star: A star whose brightness varies over time, as a result of several intrinsic or extrinsic factors. There are thousands of such stars, and they can be categorized into seven classes based on the causes of their variation. (See also Cepheid variable.)

Vernier rocket: A small thruster rocket used in space to make fine corrections to a spacecraft's orientation or trajectory.

Very long-baseline interferometry: A technique used by radio astronomers to increase the sharpness (resolution) of received signals by using several or many radio telescopes at widely spaced locations. Unlike ordinary radio interferometry, in which two or only a few radio telescopes are used, this more advanced technique allows highly accurate mapping of celestial bodies that emit radio signals. (See also Interferometry.)

Vidicon: A video camera, or a device that converts light into electronic signals that can be transmitted or recorded as pictures.

Volcanism: The dynamic process in which molten material from the interior of a planet is transferred to the planet's solid surface, issuing forth explosively from cracks or other openings.

Wavelength: A characteristic that in part defines sonic waves and electromagnetic radiation: the length (measured in angstroms or nanometers in the electromagnetic range) between successive crests in a photon's wave pattern as it moves up and down in a direction of propagation.

Weather satellite. See Meteorological satellite.

Weightlessness. See Zero gravity.

White dwarf: A dying star (one that is collapsing in on itself) with a mass 1.4 times that of the Sun or less, and with a radius approximately that of Earth. Such stars are destined to end their lives as cold, dark spheres, having expended all of their energy but not initially massive enough to end life as neutron stars or black holes.

Wind tunnel: A large tubular structure through which air is forced to flow at high speeds for the purpose of testing the behavior of aircraft and other structures that travel through the atmosphere.

Window. See Launch window.

X-band: The range of radio frequencies between 5.2 and 10.9 gigahertz.

X-radiation: Electromagnetic radiation with wavelengths between 0.1 and 10 nanometers, the range lying between gamma and ultraviolet radiation on the electromagnetic spectrum.

X-ray astronomy: The branch of astronomy that examines the X-ray emissions of celestial phenomena. This radiation cannot be studied from the ground, since Earth's atmosphere absorbs most X-radiation, and therefore has blossomed with the advent of spacecraft that can carry X-ray telescopes and other detectors into space. The examination of celestial X-ray sources has led, among other things, to the discovery of neutron stars and black holes as members of binary star systems. The Einstein Observatory launched by NASA in 1978 revealed that nearly all stars, not a few, emit X-radiation. The importance of X-ray astronomy was therefore established.

Yaw. See Pitch, roll and yaw.

Zero gravity: The condition of absolute weightlessness, which occurs in free-fall and is approached in deep space, far from massive bodies. Because all masses exert gravitational force on one another, the condition of zero gravity does not occur in nature. (See also Microgravity.)

Zodiacal light: The glow seen in the west after sunset and in the east before dawn, caused by sunlight reflecting off microscopic dust particles.

CHRONOLOGY OF THE U.S. SPACE PROGRAM

1608

Air and Space Telescopes

Telescopes, first invented in 1608, have undergone enormous diversification in the twentieth century.

1928-1971

Ancestors of the Space Shuttle

Ancestors of the space shuttle helped engineers learn how to build a reusable vehicle that can withstand the rigors of outer space and reentry into the atmosphere and gave pilots experience in flying at supersonic and hypersonic speeds.

1945

Space Centers and Launch Sites

The first testing centers away from populated areas became a necessity with the advent of the German V-2 rocket in World War II.

1950

Cape Canaveral and the Kennedy Space Center

Cape Canaveral Air Force Station becomes the launch site for early missile tests and eventually becomes one of three principal facilities for launching satellites and manned spacecraft.

1955

Private Industry and Space Exploration

Since the mid-1950's, utilization of launch vehicles, satellites, and space shuttles by private companies has grown to multibillion-dollar proportions, creating challenges and opportunities.

The Vanguard Program (1955-1959)

Remembered more for its failed attempts than it three successful launches of U.S. satellites, the program generated important developments in rocket propulsion, satellite design, and satellite telemetry and tracking.

Geodetic Satellites (1955-1988)
Satellites to map and measure the Earth benefitted by adaptions of electronic innovations linked to extensive and complex land support networks.

1957

Funding Procedures of U.S. Space Programs
With the onset of the space race, U.S. began seriously funding the space program, peaking in the 1960's and declining thereafter.

Delta Launch Vehicles
This workhorse of the U.S. space program has played a role in space exploration since the 1950's.

Saturn Launch Vehicles (1957-1975)
The Saturn launch vehicle made possible the placement in orbit of very large payloads.

Atlas Launch Vehicles
Originally developed to launch and carry thermonuclear warheads a distance of 8,200 kilometers, it eventually boosted the first man into space and served as a booster for many other space missions.

Titan Launch Vehicles
Beginning its life as an intercontinental ballistic missile in 1957, the Titan series of rockets provided the U.S. with a means of launching satellites into orbit.

1958

The Development of Project Mercury (1958-1959)
The technical developments of the Mercury spacecraft culminate in the *Big Joe* launch.

U.S. and Russian Cooperation in Space
Such cooperative accomplishments as Apollo-Soyuz Test Project and the Space Shuttle/Mir missions represent the best of cooperation between the U.S. and Russia in space exploration.

Explorers 1-7 (1958-1961)
The first successful U.S. program to launch a man-made satellite into orbit, with five of the first seven Explorers achieving Earth orbit.

Launch Vehicles
Powerful launch vehicles and booster rockets are crucial to a space program. Without powerful rocketry, it is impossible to leave Earth's atmosphere.

The National Aeronautics and Space Administration
NASA was formed in 1958 to unite all U.S. space exploration activities.

Ames Research Center
The Ames Research Center becomes part of NASA.

Langley Research Center
The oldest aeronautics research and testing center, Langley became part of NASA and now conducts programs on advanced aerodynamics and the futue of manned and unmanned space travel.

Lewis Research Center
Lewis Research Center joined NASA in 1958 and performs basic and applied research to develop technology in aircraft propulsion, space propulsion, space power, microgravity science and satellite communications.

Vandenberg Air Force Base
Vandenberg became the first operational ICBM facility in the U.S. in 1958: The launch site for more than five hundred orbital and one thousand nonorbital launches of rockets and ballistic missiles, the launch complex for the Manned Orbiting Laboratory, a West Coast launch center for the space shuttle, and the Western Commercial Space Center since 1992.

Project Mercury (1958-1963)
The first U.S. manned orbital space program yielded important data on human adaptability to space travel.

Pioneer Missions 1-5 (1958-1960)
All five Pioneers made important discoveries about the radiation belts around Earth even though only 5 was an unqualified success, achieving heliocentric orbit between the paths of Earth and Venus.

The Space Task Group (1958-1961)
The first U.S. civilian agency for manned spaceflight was responsible for the Mercury, Gemini, and Apollo projects and the ancestor of the Manned Spacecraft Center, now the Johnson Space Center.

The Deep Space Network
The deep space communications complexes that provide the Earth-based radio communications link to all NASA unmanned interplanetary spacecraft.

The Jet Propulsion Laboratory
JPL came under NASA jurisdiction in 1958 and its emphasis was changed to lunar and planetary exploration.

Military Telecommunications Satellites
Communications satellites provide reliable, worldwide communications between troops and decision makers.

1959

Spy Satellites
Spy satellites were to be operational by 1959 to provide the U.S. sophisticated information in a more effective and less dangerous way than other reconnaissance methods.

Astronauts and the U.S. Astronaut Program
NASA recruits the first seven astronauts and launches the astronaut training program.

Goddard Space Flight Center
Founded in 1959, Goddard Space Flight Center is called the intellectual brain trust of NASA.

Astronomy Explorers
The Astronomy Explorers collected data on solar radiation, gamma rays, meteoroids, X rays, and radio waves.

1960

Early-Warning Satellites
A product of the Cold War, these satellites are designed to detect firings of intercontinental ballistic missiles and transmit warnings to Earth.

Meterological Satellites
Since 1960, meterological satellites have helped predict the weather and revolutionized meterological science.

Military Meteorological Satellites
Since 1960, military meteorological satellites have provided cloud-cover photographs and other weather data for orbit to help schedule reconnaissance satellite launches and military operations.

TIROS Meteorological Satellites (1960-1986)
These satellites provided high-altitude views that have increased meteorologists' capability to forecast weather.

Navigation Satellites
These military satellites, made accessible to civilians, allow ships, aircraft, and land vehicles to pinpoint their positions on Earth.

Marshall Space Flight Center
Ground was broken for the new NASA facility which was built around the core of German scientists and engineers under the direction of Wernher von Braun.

Private and Commercial Telecommunications Satellites
Since 1960, communications satellites have been designed and built by private corporations to serve the needs of their customers.

Passive Relay Telecommunications Satellites (1960-1969)
The U.S. Echo satellite project included the first passive relay communications satellite launched into space and the first cooperative space venture between the U.S. and the Soviet Union.

Electronic Intelligence Satellites
ELINT satellites receive various electronic signals, providing a major portion of the intelligence upon which the United States and Soviet Union relied.

Global Positioning System
This network of satellites was designed originally for military application but has evolved into a navigation system that allows users to determine their position anywhere in the world.

Ionosphere Explorers
These Explorer satellites carried out the first comprehensive direct measurements of Earth's ionosphere.

1961

The Global Atmospheric Research Program (1961-1982)
A program to increase scientists' understanding of the atmosphere through space- and land-based observations.

The Ranger Program (1961-1965)
The Ranger Program was the first part of NASA's three-stage plan leading to the manned exploration of the Moon.

Air Density Explorers (1961-1970)

Data from the four Explorers determined the effect upon Earth's atmosphere of solar heating over a sunspot cycle, provided the first evidence of the winter helium and summer atomic oxygen bulges, and showed that the atmospheric constituents above 300 kilometers exist in layers and maximize where maximum temperature exists.

Mercury-Redstone 3

Alan Shepard became the first American to reach space on Mercury-Redstone 3, a fifteen-minute suborbital mission.

The Development of Spacesuits

Russian cosmonauts and American astronauts first wore full pressured spacesuits in space in 1961.

Micrometeoroid Explorers (1961-1972)

These Explorers performed direct measurements of the micrometeroid environment in near-Earth space.

Mercury-Redstone 4

Gus Grissom duplicated the successful suborbital flight of Alan Shepard.

Johnson Space Center

In 1961, Johnson Space Center became an independent entity, and a suburb of Houston was selected as the site for this manned spacecraft headquarter.

The Gemini Program (1961-1966)

This program placed the first men into Earth orbit and taught man how to track, maneuver, and control orbiting spacecraft; dock with other orbiting vehicles; and re-enter Earth's atmosphere and land at specified locations.

Amateur Radio Satellites

The first amateur radio satellite was launched in 1961. It was the first privately owned, nongovernmental satellite launched. Today, amateur radio satellites allow residents of different countries to communicate with one another, bring space exploration into the classroom, and assist in emergency relief projects.

1962

Mercury-Atlas 6

John Glenn proved that man could work in a zero-gravity environment and that the Mercury spacecraft was sound in this first U.S. manned Earth-obiting flight.

The Orbiting Solar Observatories (1962-1978)

These observatories were designed to study the structure of the Sun and its outward flow of high-energy particles.

Mercury-Atlas 7

This was the first U.S. space mission devoted to manned scientific research in space.

Mariner 1 and 2 (1962-1963)

The first interplanetary spacecraft directed to study Venus, Mariner 1 suffered a launch failure, and Mariner 2's flyby revolutionized knowledge of the conditions on Venus.

Mercury-Atlas 8

Wally Schirra demonstrated that man could work in a zero-gravity environment for an extended time on the Mercury-Atlas 8 mission.

1963

Atmosphere Explorers (1963-1981)

These satellites gathered invaluable data on Earth's atmosphere and ionosphere.

Mercury-Atlas 9

Gordon Cooper demonstrated that man could work in a zero-gravity environment for more than one day on the Mercury-Atlas 9 mission.

Nuclear Detection Satellites

Since the 1963 Nuclear Test-Ban treaty, satellites have been used to detect secret nuclear explosions in space and within Earth's atmosphere.

The Interplanetary Monitoring Platform Satellites (1963-1973)

The ten spacecraft in this part of the Explorer program measured cosmic radiation levels, magnetic field intensities, and solar wind properties in the near-Earth and interplanetary environment.

The Manned Orbiting Laboratory (1963-1969)

The first space station project that evolved beyond the study stage.

1964

The Apollo Program (1964-1972)

Apollo was America's bid for international leadership in space exploration; this leadership was demonstrated by landing men on the Moon and returning them safely to Earth.

Nimbus Meteorological Satellites (1964-1986)

By remote sensing from orbit, the Nimbus satellites were used to develop new techniques for observing Earth, especially its atmosphere and oceans.

The Orbiting Geophysical Observatories (1964-1972)

These six spacecraft returned significant data on various geophysical phenomena.

Mariner 3 and 4 (1964-1965)

Mariner 4 was the first U.S. unmanned probe to Mars, providing data that greatly increased man's understanding of Earth's neighboring planet.

1965

Gemini 3

The first U.S. two-man orbital spaceflight completed three Earth orbits with a total flight time of 4 hours and 53 minutes.

The Intelsat Communications Satellites

Intelsat was created to develop and operate a global satellite communications system that would guarantee access to international satellite communications for all member nations.

Gemini IV

The second manned capsule in the Gemini series and the first U.S. mission to include an extravehicular activity.

The Manned Maneuvering Unit

The strap-on device which allows astronauts to move independently in space has developed from the early model.

The Skylab Program (1965-1974)

This first American space station saw three crews spend a total of 171 days in space.

Gemini V

This flight demonstrated the human ability to survive eight days in space, rudimentary rendezvous techniques, and fuel-cell electrical power generation systems.

Gemini VII and VI-A

Gemini IV-A tested the ability for rendezvous in space, whereas Gemini VII's flight of fourteen days demonstrated that man could endure a long-duration flight.

Solar Explorers

The information obtained increased understanding of the effects of solar activity on manned space activity and radio communications.

Pioneer Missions 6-E

Pioneers 6-E were the first spacecraft specifically prepared to obtain synoptic information on the effects in planetary space of solar activity.

1966

Environmental Science Services Administration Satellites (1966-1976)

The first true weather satellite system created a base for meteorological satellites used by scientists on Earth.

Gemini VIII

The first manned capsule to rendezvous and dock with the Agena, a rocket that could be restarted either from the Gemini spacecraft or ground control.

The Orbiting Astronomical Observatories (1966-1972)

These observatories provided astronomers with an opportunity to conduct observations at specific wavelengths above Earth's atmosphere.

The Surveyor Program (1966-1968)

The Surveyor Program developed the technology for soft-landing on the Moon.

Gemini IX-A and X

Three-day flights designed to demonstrate rendezvous and docking techniques and to evaluate extravehicular activity.

Gemini XI and XII

The final two missions of the Gemini program shared the primary objectives of performing rendezvous and docking with the Agena target vehicles and conducting extravehicular activity.

Applications Technology Satellites (1966-1979)

Part of a multifaceted satellite program designed to demonstrate the promise of artificial satellites for direct broadcast communications and meteorological monitoring of Earth's surface.

Biosatellites (1966-1970)

A series of spacecraft investigating the effects of spaceflight on basic life processes.

Tethered Satellite System

Since Gemini XI and XII, scientists have experimented with tethered satellites systems to provide electrical power to the space shuttle or to recharge failing satellite batteries.

The Lunar Orbiter (1966-1968)

One of three unmanned spacecraft programs designed to help scientists select safe landing sites on the Moon for the Apollo program.

1967

Apollo 1-6 (1967-1968)

The early Apollo missions were unmanned test flights that furthered spaceflight technology for the eventual safe transport of astronauts to the Moon.

Mariner 5 (1967-1968)

The primary scientific mission of Mariner 5 was to gather data on the atmosphere and the ionosphere of Venus and to study the interaction of the solar plasma with Venus's environment.

Ocean Surveillance Satellites

Since 1967, these satellites have been used to locate ships, identify them, and determine their speed and course.

1968

Radio Astronomy Explorers (1968-1975)

RAE-A and RAE-B were orbited to detect and measure extraterrestrial radio noise.

Apollo 7

The first manned flight of the Apollo spacecraft, the vehicle that would be used to carry humans to the Moon.

Apollo 8

The first manned mission to the Moon.

1969

Mariner 6 and 7 (1969-1970)

The second and third unmanned missions to Mars provided valuable photographic information on the surface characteristics of the planet.

Apollo 9

An Earth-orbital flight to test the command and service module and lunar module for both docking and rendezvous.

Apollo 10

Apollo 10 successfully orbited the Moon and provided the final testing of all systems needed for an actual landing.

Apollo 11

The first mission to land humans on the Moon and return them to Earth.

Apollo 12

The second successful lunar landing.

1970

ITOS/NOAA Meteorological Satellites (1970-1979)

A series of meteorological satellites developed by the U.S. to maintain constant surveillance of weather conditions around the world. NOAA is the federal agency that operated the ITOS satellites during the 1970's.

Apollo 13

The third scheduled lunar-landing, but an explosion on board caused the landing to be aborted and required the crew to use the lunar module as a lifeboat.

1971

Apollo 14

The third successful landing of an American scientific team on the lunar surface.

Materials Processing in Space

The lack of gravitational effects in space promises great technological advances for processing materials.

Mariner 8 and 9 (1971-1972)

As the first artificial satellite of another planet, Mariner 9 did the work planned for both Mariners.

Apollo 15

The fourth landing of an American scientific team on the lunar surface.

Apollo 15's Lunar Rover

The Lunar Rover greatly expanded the area the astronauts could explore.

1972

The Spaceflight Tracking and Data Network

Beginning in 1972, this network of ground communications and tracking stations provided data relay, data processing, communications, and command support to the U.S. space shuttle program and other orbital and suborbital spaceflights.

The Space Shuttle

The Space Transportation System was established to develop an economic and reusable system that could transport humans, satellites, and equipment to and from Earth orbit on a regular basis.

Pioneer 10

Pioneer 10 was the first spacecraft to provide close-up reconnaissance of Jupiter and to sample directly its magnetic and particle environment.

Apollo 16

The second-to-last of the Apollo lunar flights, 16 allowed the astronauts to explore more difficult terrain than previous crews.

Landsat 1, 2, and 3 (1972-1982)

These satellites collected data about Earth's agriculture, forests, flatlands, minerals, waters, and environment.

Apollo 17

The last and perhaps most ambitious of the Apollo flights confirmed findings of earlier flights and returned the largest number of lunar rocks for study.

1973

Pioneer 11

Pioneer 11 collected critical data on the outer solar system.

Skylab 2

The crew established a record for the longest Earth-orbiting flight with their twenty-eight-day mission in space.

Skylab 3

Skylab 3's crew exceeded the objectives of their mission in fifty-nine days in the orbital workshop.

The Spacelab Program

Spacelab is a major space shuttle payload designed to provide scientists with facilities approximating those of a terrestrial laboratory.

Mariner 10 (1973-1975)

Mariner 10 collected vital data on the inner solar system, including detailed photographs of Venus and Mercury.

Skylab 4 (1973-1974)

In the last flight of the first U. S. space station, the three astronauts spent eighty-four days in Earth orbit, making it the longest manned U.S. spaceflight up to that time.

1974

SMS and GEOS Meterological Satellites

Beginning in 1974, these satellites provide continuous coverage of weather conditions on Earth.

1975

The Apollo-Soyuz Test Project

The first U.S.-Soviet cooperative manned spaceflight.

The Viking Program (1975-1982)

The Viking program, using a pair of heat-sterilized landers, acquired the first data from the surface of Mars.

Viking 1 and 2 (1975-1982)

The Viking mission to Mars was the first long-duration intensive exploration of the surface of another planet.

1976

Maritime Telecommunications Satellites

A network of satellites designed to upgrade the communications capabilities of commercial and military maritime vessels was first proposed in 1972; satellites were launched between 1976 and 1979, when the International Maritime Satellite

Organization was formed.

The Get-Away Special Experiments

A program to allow a wide variety of users to have their space experiment packages launched inside small canisters aboard space shuttle missions at relatively low cost.

1977

The Voyager Program

The Voyager probes executed the first Grand Tour in planetary exploration by encountering Jupiter, Saturn, Uranus, and Neptune.

The High-Energy Astronomical Observatories

These observatories provided a detailed survey of the celestial sphere, studying X-ray sources and detecting gamma ray and cosmic radiation.

Voyager 2: Jupiter (1977-1979)

The Voyager 2 flyby of Jupiter provided vital information about the Jovian system, in spite of a number of technical problems and equipment failures.

Voyager 2: Neptune

In 1986, Voyager 2 completed the mission of the originally proposed Grand Tour of the outer solar system, flying by Jupiter, Saturn, Uranus, and Neptune, on a trajectory that will ultimately take it beyond the solar system.

The Space Shuttle Approach and Landing Test Flights (1977-1985)

Enterprise, the first space shuttle orbiter, tested the approach and landing techniques of the Space Transportation System.

International Sun-Earth Explorers (1977-1985)

Three ISEEs performed observations and experiments in deep space and Earth's magnetosphere. ISEE 3 was also sent around the Moon to make the first spacecraft contact with a comet.

1978

The International Ultraviolet Explorer

Conceived as an orbiting observatory designed to observe the ultraviolet portion of the spectrum, IUE proved one of the most productive and oldest funtioning satellites in the history of the space program.

Air Traffic Control Satellites

Beginning with the earliest satellite program, the U.S. has conducted experiments using artificial satellites to aid in air traffic control operations. These efforts led to Navstar, a network of navigational satellites, and to Nusat, the first satellite designed exclusively for air traffic control.

Dynamics Explorers

The first satellites placed in Earth orbit to provide data on the energy and momentum of charged particles in Earth's upper atmosphere.

The Heat Capacity Mapping Mission

HCMM was the first satellite designed to measure thermal inertia.

Pioneer Venus 1

The first Venus mission to map the planet's surface and investigate its atmosphere and ionosphere.

Seasat

This satellite performed a variety of experiments, including photographic and remote-sensing procedures, above the world's oceans.

Pioneer Venus 2

Called the multiprobe, this cluster of five spacecraft was designed to penetrate Venus's cloud cover and gather information about its atmosphere at separate locations.

1979

The Stratospheric Aerosol and Gas Experiment

The SAGE instrument was designed to measure the concentration of some of the constituents of Earth's atmosphere.

Voyager 1: Jupiter

Voyager 1 collected detailed information on the planet Jupiter, its rings, satellites, and surrounding environment, including detailed photographs of the four Galilean satellites.

1980

The Solar Maximum Mission

A mission designed to study the Sun during the 1980 peak of the eleven-year solar cycle.

Voyager 1: Saturn

On its second planetary flyby, Voyager 1 encountered Saturn and sent back to Earth information on the planet's rings, satellites, and atmosphere.

1981

Space Shuttle Living Conditions

Every effort has been made to provide safe and comfortable living conditions in a microgravity environment for space shuttle astronauts since STS-1 in 1981.

Space Shuttle Mission STS-1

The objectives of the first launch of the space shuttle were to demonstrate a safe launch to orbit, test basic systems in space, achieve reentry, and land safely.

Voyager 2: Saturn

The Voyager 2 flyby of Saturn produced high-resolution images of Saturn's ring system and of the satellites Iapetus, Hyperion, Enceladus, and Tethys.

Space Shuttle Mission STS-2

The second flight of the shuttle *Columbia* carried the first scientific experiments and tested the remote manipulator system for the first time.

1982

Space Shuttle Flights, 1982

Two additional test flights determined the flight worthiness of the space shuttle and led to a flight carrying the first commercial payload.

Landsat 4 and 5

These satellites collect data about the Earth's agriculture, forests, flatlands, minerals, waters, and environment.

The High-Energy Astronomical Observatories

These observatories provided a detailed survey of the celestial sphere, studying X-ray sources and detecting gamma and cosmic radiation.

Search and Rescue Satellites

Since 1982, satellites have been used to detect emergency beacons of downed aircraft, capsized boats, and individuals involved in exploration.

1983

The Infrared Astronomical Satellite

From its vantage point in Earth orbit, this satellite enabled scientists to study in detail, and without atmospheric interference, the heat emitted by astronomical sources.

The Strategic Defense Initiative

This project using space- and Earth-based technology to counter nuclear missile attacks was essentially cancelled following the collapse of the Soviet Union and the election of Bill Clinton to the presidency.

Space Shuttle Mission STS-6

The first voyage of *Challenger* deployed the first Tracking and Data Relay Satellite and permitted two astronauts to work in space thanks to the extravehicular mobility unit.

Tracking and Data-Relay Communications Satellites

Both the U.S. and Russia have developed and begun implementation of separate communications satellite systems, designed to improve tracking and data relay capabilities from low-Earth orbiting spacecraft, scheduled to be operational in the late 1990's.

Space Shuttle Flights, 1983

This year of firsts saw the introduction of the second orbiter, the first woman astronaut, the first African-American astronaut, the first non-American astronaut, the first dual spacewalk, and the first fully equipped, spaceborne science laboratory.

Shuttle Amateur Radio Experiment

Since 1983, amateur radio operators have communicated with space shuttle astronauts orbiting the Earth.

Spacehab

Since 1983, this commercial mini-laboratory fits into the cargo bay of the space shuttle and has been leased to private corporate interests.

1984

Space Shuttle Flights, 1984

In 1984, the space shuttle program achieved a number of firsts, overcame nagging problems, and displayed the ability of humans to play an important role in spaceflight with the capture and return of two failed satellites.

Space Shuttle Mission STS 41-B

The tenth mission of the space shuttle was the first to return to Earth where it was launched, at Kennedy Space Center, and demonstrated the ability of astronauts to maneuver in space.

Space Shuttle Mission STS 41-C

STS 41-C launched into low Earth orbit the Long-Duration Exposure Facility to study materials degradation and saw the crew repair the Solar Maximum Mission satellite.

Space Shuttle Mission STS 51-A

During STS 51-A the crew captured and returned to Earth two satellites that otherwise would have been useless forever.

Active Magnetospheric Particle Tracer Explorers (1984-1985)

First spacecraft to perform active experiments on the Sun's effects on Earth's magnetosphere and radiation belts.

Design and Use of the International Space Station

This planned Earth-orbiting facility is designed to house experimental payloads, distribute resource utilities, and support permanent human habitation for conducting research in a microgravity environment.

Development of the International Space Station

An international space station is planned to be completed by June, 2002.

Living and Working in the International Space Station

A crew of six persons is planned to live and work both in and out of the Habitation Module on the International Space Station.

Modules and Nodes of the Internationl Space Station

The International Space Station will include a pressurized habitation module; three laboratory modules, one built by the United States, one by Europe, and one by Japan; and propulsion modules from Russia

U.S. Contributions to the International Space Station

The International Space Station Program will be the largest international scientific program in history, drawing on the resources and experience of thirteen nations, led by the United States.

The National Commission on Space (1984-1986)

The group that prepared a report on the long-term space goals, including a "Declaration for Space."

1985

Space Shuttle Flights, January-June, 1985

STS 51-C was the first mission dedicated to a classified Department of Defense payload; STS 51-D carried the first non-pilot, nonscientist astronaut; STS 51-B carried the Spacelab 3 science payload into orbit; STS 51-G released four satellites, retrieved one, and completed important technological and scientific investigations.

Space Shuttle Flights, July-December, 1985

During Space Shuttle Missions STS 51-F, 51-I, 51-J, 61-A, and 61-B many experiments were conducted, satellites were launched, in-flight maintenance was completed on a malfunctioning satellite, and astronauts practiced assembling large structures in space.

Space Shuttle Mission STS 51-I

The twentieth flight in the space shuttle program deployed three communications satellites, and crew members captured and repaired a malfunctioning satellite.

The United States Space Command

Beginning in 1985, the U.S. Space Command began to provide the needed operational focus across all armed services, consolidating control of space assets and activities in support of nonspace missions.

Voyager 2: Uranus (1985-1986)

Voyager 2 was the first spacecraft to collect and return data from the planet Uranus.

1986

Space Shuttle Mission STS 61-C

Astronauts aboard *Columbia* launched a communications satellite, tested a new payload carrier system, and photographed Halley's comet.

Space Shuttle Mission STS 51-L

STS 51-L exploded 73 seconds after launch, killing its crew and completing destroying the space shuttle *Challenger* and its satellite cargo.

National Aerospace Plane

The National Aerospace Plane is a program to design and build a single-stage-to-orbit vehicle that will land and take off on conventional runways.

1988

Space Shuttle Mission STS-26

The first shuttle flight after the *Challenger* accident, STS-26 launched a vital communications satellite into geosynchronous orbit and returned the U.S. manned space program to an active flight status.

1989

Space Shuttle Flights, 1989

Space shuttle crews orbited the Earth for more than four weeks during 1989 as NASA successfully launched and landed five space shuttle missions.

Galileo: Jupiter

Galileo has provided more extensive sampling of Jupiter's outer atmosphere, and far closer encounters of its principal satellites, than previous missions.

Magellan: Venus (1989-1994)

Magellan mapped 99 percent of Venus's surface using powerful radar imaging instruments.

1990

Space Shuttle Flights, 1990

NASA flew six space shuttle flights in 1990, highlighted by STS-31, the flight that deployed the Hubble Space Telescope into orbit.

The Hubble Space Telescope

The largest optical astronomical observatory to orbit the Earth.

Mission to Planet Earth

This long-term NASA program studies how the global environment is changing

Ulysses: Solar-Polar Mission

This joint mission between NASA and the European Space Agency is the first spacecraft to explore interplanetary space out of the plane of ecliptic.

1991

Space Shuttle Flight, 1991

Three missions were flown during which satellites were deployed; astronauts conducted experiments to determine the physiological effect of spaceflight and demonstrated techniques for constructing the space station.

Compton Gamma Ray Observatory

The most massive robotic civilian spacecraft the U.S. had ever built provided dramatic new insights into some of the highest energy phenomena in the universe.

1992

Space Shuttle Flights, 1992

The eight missions of 1992 saw the longest shuttle mission and the most spacewalks to date; the *Endeavour* replaced the *Challenger*.

Atmospheric Laboratory for Applications and Science (1992-1994)

A package of instruments to record data relating to the upper atmosphere, especially ozone depletion, and to irradiance from the Sun and outer space.

Space Shuttle Mission STS-49

The newest space shuttle orbiter, *Endeavour*, saw its maiden voyage.

Pioneer Venus 1

Pioneer Venus 1 obtained important information on Venus's topography and atmosphere.

Mars Observer (1992-1993)

Mars Observer was to have photographed the Martian surface and to have carried out scientific experiments on the Martian atmosphere, but it was unsuccessful.

1993

Space Shuttle Flights, 1993

The seven space shuttle missions of 1993 included science missions, a mission to retrieve a satellite left in space a year earlier, missions to deploy new satellites, and the mission to service and repair the Hubble Space Telescope.

Space Shuttle Mission STS-61

This important mission repaired the Hubble Space Telescope, demonstrating the value of manned spaceflight.

1994

Space Shuttle Flights, 1994

Seven space shuttle missions all were dedicated to science.

Mars Surveyor Program

A ten-year effort to explore the planet Mars.

Clementine Mission to the Moon

A new generation of spacecraft, designed to employ new technologies within a low-cost budget, accomplishes the first complete mapping of the Moon.

1995

Space Shuttle Flights, 1995

Of the year's seven missions, three involved a rendezvous with the Russian space station Mir.

Space Shuttle Mission STS-63

STS-63 began a phase of cooperation between NASA and the Russian Space Agency, in which techniques for joint operations between the space shuttle and Mir demonstrated the feasibility of an International Space Station.

Space Shuttle Mission STS-71/Mir Primary Expedition 18

STS-71 conducted the first U.S. docking with Mir, bringing supplies and a new crew from Earth, returning materials and the old crew, and gaining experience in joint operations in space.

1997-2008

Cassini: Saturn (1997-2008)

This deep space probe is scheduled to investigate Saturn, its rings, and its moons.

Pluto Flyby Missions (2000's)

The Pluto flyby missions will study the most distant, previously unprobed, and anomalous planet-moon pair.

WORLD WIDE WEB PAGES

Research tools have greatly improved since 1989 with the popularization of the Internet and the World Wide Web. With the aid of a computer and modem, vast stores of documents and photographs are available to the general public. There is insufficient room to list all of them, but an attempt has been made to list those which will probably still exist when the next edition of this book is undertaken. Although every attempt has been made to ensure accuracy, we can not be responsible for changes in the home pages listed. Please note when entering these Universal Resource Locators (URLs) that proper capitalization and punctuation are required.

Air Force Space Command
http://www.dtic.dla.mil/airforcelink/pa/factsheets/Air_Force_Space_Command.html

AIR&SPACE Magazine Home Page
http://www.airspacemag.com/

Ames Imaging Library System
http://ails.arc.nasa.gov/

Ames News Release And Picture Archives
http://ccf.arc.nasa.gov/dx/archive.html

Ames Research Center
http://www.arc.nasa.gov/

Ames Research Center Public Affairs
http://ccf.arc.nasa.gov/dx/

Army Space Command
http://www.ssdc.army.mil/SSDC/index/ARSPACEIndex.html

ASI - Italy
http://hp.mt.asi.it/

Association of Space Explorers
http://www.airspacemag.com/ASE/ASE_Home.html

Astronaut Candidates
http://sauron.msfc.nasa.gov/astronaut-candidates/

Astronauts
http://www.conveyor.com/space/astronaut.html

Aviation Week & Space Technology
http://www.mcgraw-hill.com/aviation/avweek.htm

Boeing Space Systems
http://www.boeing.com/space.systems.html

Canadian Space Agency
http://www.cisti.nrc.ca/programs/indcan/csa.html

Cassini: Voyage to Saturn
http://www.jpl.nasa.gov/cassini/

CERT - France
http://www.cert.fr/

COMSAT Corporation
http://www.comsat.com/

DLR - Germany
http://www.dlr.de/

Dryden Flight Research Center
http://www.dfrc.nasa.gov/dryden.html

Dryden Flight Research Center Photos Services
http://www.dfrc.nasa.gov/PhotoServer/

Edwards AFB Home Page
http://wwwcpo.elan.af.mil/

European Space Agency
http://www.esrin.esa.it/

Global Positioning System (GPS)
http://www.auslig.gov.au/geodesy/gps.htm

Goddard Lithographs
http://pao.gsfc.nasa.gov/gsfc/service/gallery/lithos/lithos.htm

Goddard Space Flight Center
http://pao.gsfc.nasa.gov/gsfc.html

Hughes Space and Communications
http://www.hughespace.com/

INTELSAT
http://www.intelsat.int/

International Space Station - MSFC
http://liftoff.msfc.nasa.gov/station/welcome.html

International Space Station Bulletin Board
http://issa-www.jsc.nasa.gov/

International Space Station Reference Guide
http://issa-www.jsc.nasa.gov/ss/techdata/ISSAR/ISSAReferenceGuide.html

International Space Station Techdata Book Home Page
http://issa-www.jsc.nasa.gov/ss/techdata/techdata.html

Jet Propulsion Laboratory
http://www.jpl.nasa.gov/

Jet Propulsion Laboratory – Welcome to the Planets – Photos
http://pds.jpl.nasa.gov/planets/

Jet Propulsion Laboratory Public Image Archive
http://www.jpl.nasa.gov/archive/images.html

Johnson Space Center
http://www.jsc.nasa.gov/

Kennedy Space Center
http://www.ksc.nasa.gov/ksc.html

Langley Research Center
http://www.larc.nasa.gov/larc.html

Langley Research Center Photo Archive
http://lisar.larc.nasa.gov/LISAR/BROWSE/browser.html

Lewis Research Center
http://www.lerc.nasa.gov/LeRC_homepage.html

Lewis Research Center Photo Gallery
http://www.lerc.nasa.gov/WWW/PAO/html/paogalry.htm

LIFE Space Covers
http://pathfinder.com/@@AwsjwYAzfkwgM@N/Life/space/covers/coversspace.html

Lockheed Martin Corporation
http://www.lockheed.com/

Lockheed Martin Corporation Photo Archive
http://www.lockheed.com/photo/

Lockheed Missiles & Space Public Relations Image Archive
http://www.lmsc.lockheed.com/newsbureau/photos.html

Marshall Space Flight Center
http://www.msfc.nasa.gov/

Marshall Space Flight Center Historical Image Gallery
http://history.msfc.nasa.gov/history/images.html

McDonnell Douglas Aerospace
http://pat.mdc.com/

Mission to Planet Earth
http://www.hq.nasa.gov/office/mtpe/

NASA Astronaut Biographies
http://www.jsc.nasa.gov/Bios/index.html

NASA Biographies
http://www.gsfc.nasa.gov/hqpao/bios.html

NASA Expendable Launch Vehicle Images
http://www.ksc.nasa.gov/elv/vidpicp.htm

NASA Headquarters
http://www.hq.nasa.gov

NASA Headquarters Newsroom
http://www.gsfc.nasa.gov/hqpao/newsroom.html

NASA Headquarters Photo Gallery
http://www.nasa.gov/hqpao/library.html

NASA Headquarters Photo Library
http://www.hq.nasa.gov/office/pao/Library/photo.html

NASA History
http://www.ksc.nasa.gov/history/history.html

NASA Mission Patches
http://www.hq.nasa.gov/office/pao/History/mission_patches.html

NASA Office of Space Flight
http://www.osf.hq.nasa.gov/

NASA Pocket Statistics
http://www.hq.nasa.gov/office/pao/History/pocket_statistics.html

NASA Public Affairs
http://www.gsfc.nasa.gov/hqpao/hqpao_home.html

NASA Shuttle Web
http://shuttle.nasa.gov

NASA Shuttle-Mir Web
http://shuttle-mir.nasa.gov/

NASA Spacelink
http://spacelink.msfc.nasa.gov/

NASA Spacelink: Space Station
http://spacelink.msfc.nasa.gov/NASA.Projects/Human.Space.Flight/Space.Station/

NASA Telerobotics Photo Archive
http://ranier.oact.hq.nasa.gov/telerobotics_page/photos.html

NASA Thesaurus
http://www.sti.nasa.gov/nasa-thesaurus.html

NASA's Catalog of Planetary GIF images
http://delcano.mit.edu/http/amesinfo.html

NASA's Space Science Photo Gallery
http://nssdc.gsfc.nasa.gov/photo_gallery/PhotoGallery.html

NASDA – Japan
http://hdsn.eoc.nasda.go.jp/

National Air & Space Museum Homepage
http://ceps.nasm.edu:/NASMpage.html

National Technical Information Service (NTIS)
http://www.fedworld.gov/ntis/ntishome.html

Nine Planets – Photos
http://seds.lpl.arizona.edu/nineplanets/nineplanets/nineplanets.html

NOAA Home Page
http://www.noaa.gov/

Pioneer Image Page
http://pyroeis.arc.nasa.gov/pioneer/PNimgs/PNimgs.html

Planetary Photojournal: NASA's Image Access Home Page
http://www-pdsimage.jpl.nasa.gov/PIA/PIA.html

Planetary Society Home Page
http://wea.mankato.mn.us/tps/index.html

Planetary Society Image Archive
http://wea.mankato.mn.us/tps/archive.html

Rocketdyne Home Page
http://www.rdyne.rockwell.com/indext.html

Rocketdyne Sights and Sounds
http://www.rdyne.rockwell.com/sightsns/sightsns.html

Rockwell Home Page
http://www.rockwell.com/

Russian (FSU) Space Missions and Vehicles
http://solar.rtd.utk.edu/~jgreen/rusguide.html

Russian Space Science Internet (RSSI) Home Page
http://www.rssi.ru/

SAREX Home Page
http://www.nasa.gov/sarex/sarex_mainpage.html

SOHO Artwork
http://sohowww.nascom.nasa.gov/artwork/

Space Calendar (JPL)
http://newproducts.jpl.nasa.gov/calendar/calendar.html

Space Image Libraries
http://www.okstate.edu/aesp/image.html

Space Shuttle Launches
http://www.ksc.nasa.gov/shuttle/missions/missions.html

Space Station Pictures
http://liftoff.msfc.nasa.gov/station/about/pictures.html

Space Telescope Institute
http://www.stsci.edu/EPA/Pictures.html

Space Telescope Science Institute/Hubble Space Telescope Public Release Images
http://www.stsci.edu/EPA/Pictures.html

SPACEHAB Home Page
http://hvsun.mdc.com:/SPACEHAB/SPACEHAB.html

Spacenet
http://www.space.ru/

Stennis Space Center
http://www.ssc.nasa.gov/

TRW Inc.
http://www.trw.com/

U.S. Army Space and Strategic Defense Command
http://www.ssdc.army.mil/

U.S. Space Camp
http://www.spacecamp.com/

Vandenberg AFB History
http://mercury.vafb.af.mil/history/history.htm

Wallops Flight Facility, Sounding Rocket Pictures
http://www.wff.nasa.gov/~web/sndroc_pics.html

Wallops Island Flight Facility
http://www.wff.nasa.gov/

White Sands Missile Range
http://sd-www.jsc.nasa.gov/wsmr.html

USA IN SPACE

ALPHABETICAL INDEX

CATEGORY INDEX

ORGANIZATION

PROGRAM

SATELLITE

PERSONAGE INDEX

LIST OF ILLUSTRATIONS